Structural Virology

RSC Biomolecular Sciences

Editorial Board:
Professor Stephen Neidle (Chairman), *The School of Pharmacy, University of London, UK*
Dr Marius Clore, *National Institutes of Health, USA*
Professor Roderick E Hubbard, *University of York and Vernalis, Cambridge, UK*
Professor David M J Lilley FRS, *University of Dundee, UK*

Titles in the Series:
 1: Biophysical and Structural Aspects of Bioenergetics
 2: Exploiting Chemical Diversity for Drug Discovery
 3: Structure-based Drug Discovery: An Overview
 4: Structural Biology of Membrane Proteins
 5: Protein–Carbohydrate Interactions in Infectious Disease
 6: Sequence-specific DNA Binding Agents
 7: Quadruplex Nucleic Acids
 8: Computational and Structural Approaches to Drug Discovery: Ligand–Protein Interactions
 9: Metabolomics, Metabonomics and Metabolite Profiling
10: Ribozymes and RNA Catalysis
11: Protein-Nucleic Acid Interactions: Structural Biology
12: Therapeutic Oligonucleotides
13: Protein Folding, Misfolding and Aggregation: Classical Themes and Novel Approaches
14: Nucleic Acid-Metal Ion Interactions
15: Oxidative Folding of Peptides and Proteins
16: RNA Polymerases as Molecular Motors
17: Quantum Tunnelling in Enzyme-Catalysed Reactions
18: Natural Product Chemistry for Drug Discovery
19: RNA Helicases
20: Molecular Simulations and Biomembranes: from Biophysics to Function
21: Structural Virology

How to obtain future titles on publication:
A standing order plan is available for this series. A standing order will bring delivery of each new volume immediately on publication.

For further information please contact:
Book Sales Department, Royal Society of Chemistry,
Thomas Graham House, Science Park, Milton Road, Cambridge,
CB4 0WF, UK
Telephone: +44 (0)1223 420066, Fax: +44 (0)1223 420247, Email: books@rsc.org
Visit our website at http://www.rsc.org/Shop/Books/

Structural Virology

Edited by

Mavis Agbandje-McKenna and Robert McKenna
Department of Biochemistry and Molecular Biology, University of Florida, FL, USA

RSCPublishing

RSC Biomolecular Sciences No. 21

ISBN: 978-0-85404-171-8
ISSN: 1757-7152

A catalogue record for this book is available from the British Library

Published by The Royal Society of Chemistry,
Thomas Graham House, Science Park, Milton Road,
Cambridge CB4 0WF, UK

Registered Charity Number 207890

For further information see our web site at www.rsc.org

Printed and bound in Great Britain by Henry Ling Limited, at the Dorset Press, Dorchester, DT1 1

Preface

Viruses can be grouped among the simplest biological systems that have the ability to evolve and adapt to exist in different environments. That is, they have the ability to 'jump' from one host to another, some carrying the necessary molecular machinery to transfer and modify their genetic information from one generation to the next, while others hijack the host machinery to effect the necessary modifications. Because of this innate ability, it would not be unreasonable to state that viruses have most likely infected every life form that has ever existed on our planet, from the simplest single-cell organisms to plants, animals, and humans.

To achieve such biodiversity, viruses have evolved different and efficient strategies for host recognition, internalization, cellular trafficking, genome replication, capsid assembly, genome packaging, release of progeny (for re-infection) and host immune surveillance evasion, to optimize their life cycle in their unique niche. This has resulted in viruses of different shapes and sizes, from simple single-protein spherical or helical assemblages, to multiple complex systems, assembled from hundreds of proteins without/with (enveloped) the incorporation of host lipids. Invariably the viral coat protein(s) (referred to throughout this monograph interchangeably as either CPs or VPs) form some sort of integral protective shell (a viral capsid) around the infectious genomic nucleic acid, which can be single-stranded (ss) DNA, ssRNA, double-stranded (ds) DNA or dsRNA, packaged as single or multiple, linear or circular molecule(s). The packaged viral genome encodes all the required structural CPs/VPs and auxiliary non-structural proteins that are required in combination with host proteins for host infection. The enveloped viruses incorporate their host's lipids as either an internal and/or external envelope during their assembly. For a number of viruses, CP/VP recognition and encapsidation of the genomic nucleic acid is a prerequisite for infectious capsid formation, whereas for others the genome is packaged into preformed capsids via interactions with viral or host encoded proteins. In addition to genome encapsidation and protection during cellular entry and trafficking, the CP/VP can also dictate many other

---

RSC Biomolecular Sciences No. 21
Structural Virology
Edited by Mavis Agbandje-McKenna and Robert McKenna
© Royal Society of Chemistry 2011
Published by the Royal Society of Chemistry, www.rsc.org

viral functions, including host receptor/vector recognition, transmission and the genomic transduction efficiency during infection.

For spherical viruses, the CP/VP organization in the capsid architecture takes on the form of an icosahedron (a platonic solid with point group symmetry 5.3.2), a regular polyhedron which is assembled from 20 equilateral triangles. This symmetrical shell is a consequence of it consisting of identical (or almost identical) gene products, consistent with the argument that there is insufficient volume inside a virus to accommodate a more complicated protein coding strategy. The exact twofold, threefold and fivefold symmetry of the icosahedron permits the (quasi) equivalent symmetry required to construct structures with 60 or multiples [denoted by a T (triangulation) number] of 60 subunits. This monograph will discuss viruses assembled from the simplest of icosahedral capsids, with T = 1 triangulation (assembled from 60 CP/VP subunits), to those with more complicated VP shells assembles and lipid membrane envelopes.

Viruses have been responsible for more human deaths, either through direct infection (such as influenza virus) or infection of crops, than any other known human disease-causing agent. In addition, their ability to package efficiently and deliver genomic material to different living organisms and tissues also makes them attractive vehicles for the delivery of therapeutic genetic material in situations where defective genes lead to disease phenotypes. Thus viruses are the subject of intense scientific study in many different disciplines, including structure biology, in efforts to (i) understand the basic biological processes governing viral infection and (ii) develop treatment strategies, including vaccines, anti-virals and gene delivery vectors.

The use of structure approaches in virology has given insight into the structural basis of assembly, nucleic acid packaging, particle dynamics and interactions with cellular molecules and allowed the elucidation of mechanistic pathways at the atomic and molecular level. Biological processes, such as the life cycle of a virus infection, are governed by numerous intricate macromolecular interactions. The role of the structural virologist is thus to visualize these interactions in three dimensions (3D), to provide a full understanding of these interactions as 'seeing is believing'. These structural characterizations of viruses then provide crucial platforms for the development of treatment and therapeutic strategies (Section 3 of this monograph).

The range of biophysical methods used in structural virology is vast, ranging from hydrodynamic to scattering techniques (Section 1 of this monograph), and have played a fundamental role in our understanding of viral infection in recent years. The method undertaken for a particular study is often dependent on the resolution and type of information desired and also the size and complexity of the macromolecule under investigation, the amount of material available, its solubility in aqueous environments (Chapter 1) and the type of interactions being visualized. For example, for the imaging of whole viruses during infection, confocal microscopy (Chapter 2) and cryo-electron tomography (cryo-ET) (Chapter 4) are applied, which permit studies at molecular resolution. And while both nuclear magnetic resonance (NMR) spectroscopy (Chapter 8) and X-ray crystallography (Chapters 6 and 7) can give atomic resolution detail on protein backbone and side-chain placement, NMR also provides dynamic (ensemble) information and crystallography provides a 'snapshot' and is often considered static. Solution approaches, such as limited proteolysis combined with mass spectrometry and small-angle scattering approaches (Chapter 3), also provide dynamic information. In cryo-ET and cryo-electron microscopy (cryo-EM) (Chapter 5), macromolecules are frozen in their native state, allowing for discrete selection of dynamic states to be visualized, albeit at lower resolution. Generally, NMR spectroscopy is utilized for small protein molecules that are flexible, X-ray crystallography for medium-sized proteins and complexes that are compact, whereas very large macromolecular assemblages or membranous protein structures are determined by cryo-EM. The largest issue separating cryo-EM and cryo-ET from crystallography, in addition to size and the limitations of crystal formation, is resolution. Cryo-EM has generally been considered a low-resolution technique, giving reconstructions around 15–30 Å, but with advances in sample handling, instrumentation, image processing and model building, near-atomic resolution structures are now being achieved. For cryo-ET the resolution achieveable is still low.

In reality, hybrid approaches, combining NMR, X-ray crystallography and cryo-EM, cryo-ET and solution data, are often adopted, which provides a powerful means of filling gaps which can arise in the structural characterization of large macromolecules. For example, in studies where large viruses cannot be crystallized, subcomponents can be crystallized to obtain high-resolution information, which can then be used to interpret the structure at lower resolution obtained by cryo-EM or cryo-ET. Or atomic structures obtained from homologous viral proteins/virus capsids can be used for 3D homology model building. These approaches permit the pseudo-atomic visualization of interaction interfaces between protein–protein subunits, protein–nucleic acids and protein–lipid in virus capsids and also the visualization of virus capsid–host interactions.

Combined with biochemical, biophysical and molecular biology analysis, structural studies indicate a high degree of fidelity in the steps that result in the assembly of mature infectious virus capsids (Chapter 10). They also show that the fundmental principles governing successful viral capsid assembly, efficient polymerization of CP subunits utilizing specific interface interactions that spontaneously terminate, often employ structural polymorphisms to facilitate the required interactions. Structural virology approaches have also been platforms for the elegant description of the virus infection process, from initial receptor attachment to the interaction of the capsid with host antibodies (Section 2), and provided the targets for therapeutic intervention and improved viral capsid vectors for gene delivery (Section 3).

This monograph is designed to provide a basic introduction to the use of structural virology and its applications in virus research towards functional annotation and is not intended to provide a detailed discussion of approaches utilized.

Mavis Agbandje-McKenna
Robert McKenna

Contents

RSC Biomolecular Sciences No. 21
Structural Virology
Edited by Mavis Agbandje-McKenna and Robert McKenna
© Royal Society of Chemistry 2011
Published by the Royal Society of Chemistry, www.rsc.org

Section 1

CHAPTER 1

Production and Purification of Viruses for Structural Studies

BRITTNEY L. GURDA AND MAVIS AGBANDJE-MCKENNA

Department of Biochemistry and Molecular Biology, Center for Structural Biology, The McKnight Brain Institute, University of Florida, Gainesville, FL 32610, USA

1 Introduction

Advances in protein production and purification techniques over the past two decades have allowed the structural study of numerous proteins and macromolecular assemblages that would have otherwise been intractable to the necessary approaches (detailed in the following chapters). This chapter focuses on the production and purification of intact viral capsids (particles) with/without genome for structure determination. The production and purification of viral proteins for structure determination by X-ray crystallography and NMR spectroscopy are the subjects of Chapters 7 and 8, respectively. Crystallization is often considered a method of purification and a function of purity, often of a protein or virus capsid, and, as such, sample preparation for structure determination by X-ray crystallography places high demands on sample quality. Screening trials to identify the optimal crystallization conditions also require large quantities of sample compared with the majority of other structure determination approaches discussed in the subsequent chapters of this monograph. Virus samples produced for such analyses also have to be both stable and soluble in their storage buffer since degradation and aggregation are detrimental to the crystallization process. Hence this chapter will focus

RSC Biomolecular Sciences No. 21
Structural Virology
Edited by Mavis Agbandje-McKenna and Robert McKenna
© Royal Society of Chemistry 2011
Published by the Royal Society of Chemistry, www.rsc.org

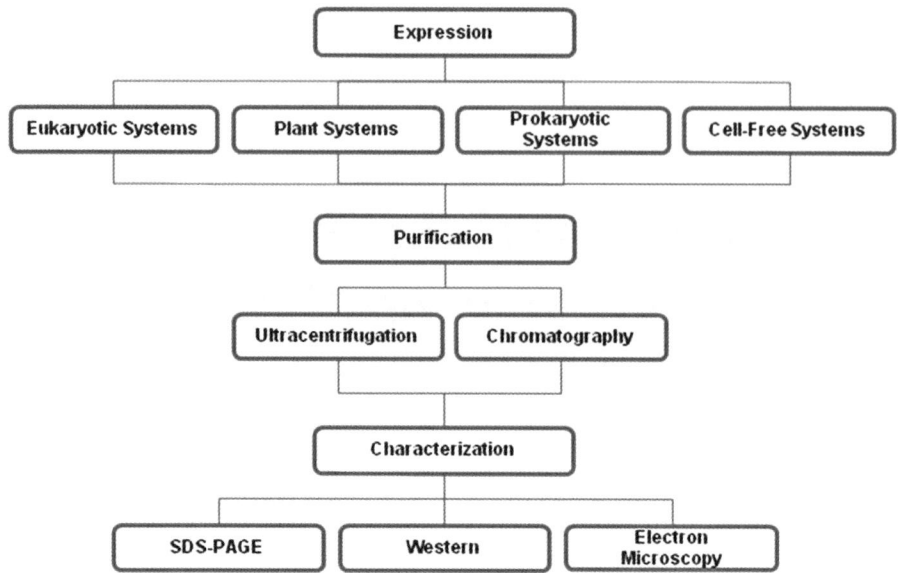

Figure 1   The steps involved in the expression, purification and characterization of virus capsids prior to structural analysis.

on methodologies to produce and purify virus capsids (Figure 1) in quantities suitable for structure determination by X-ray crystallography, with the premise that such a sample would also be suitable for structural or biophysical analysis using other methodologies.

2   Expression Systems

Most viruses are considered hazardous material in their wild-type (wt) infectious form (for information on safe handling and containment of infectious microorganisms and hazardous biological materials, see http://www.cdc.gov/ biosafety) and are therefore often studied in a recombinant form. Significant effort has been extended into the development of heterologous expression systems to produce recombinant viral proteins which will assemble into viral capsids. The system selected for use is often dependent on the properties of the viral genes and the environmental requirements of the final product. However, the most important factor to consider is the capacity of the host cells to translate the RNA transcript, to ensure proper folding of the gene product and to sustain the protein(s) expressed in an intact and functional state.[1] Protein expression systems contain at least four general components: (1) the genetic elements necessary for transcription/translation and selection; (2) in vector-based systems, a suitable replicon: plasmid, virus genes, *etc.*; (3) a host strain containing the appropriate genetic traits needed to function with the specific expression signals and selection scheme; and (4) the culturing conditions for the transformed cells or organisms.[2]

Eukaryotic Systems

Mammalian Cells

Since most viruses currently studied are of human or animal origin, mammalian tissue culture is an ideal source to generate viral capsids for structural studies which are generally aimed at functional annotation. In this system, proper folding is achieved and modifications such as complex glycosylation, phosphorylation, acylation, acetylation and γ-carboxylation are obtained. However, yields can be low, depending on gene product(s), ranging from 0.1 to 100 mg L^{-1} of culture volume. For some of the structural approaches discussed in Section 1 of this monograph, low yields may not be a problem since small amounts of sample are adequate. However, low yields can become problematic in crystallization, especially with a virus that does not have an established crystallization condition. In such a situation, numerous preparation steps may be required to obtain the quantities needed to screen crystallization conditions efficiently. Supplies and reagents can then become expensive, depending on individual cell line requirements. In addition, considerable time and resources can be spent on the construction of a suitable expression system and equally on optimization for suitable yields. In such situations, it is always advisable to seek the expertise of an established molecular biologist before designing new constructs.

Established cell lines and protocols exists for many different tissue systems and, although most of these cell lines are derived from human or mouse tissues, other mammalian cell culture lines are available, such as monkey, raccoon, horse, pig and rabbit. The American Type Culture Collection (ATCC) has over 3400 cell lines from 80 different species, including over 950 cancer cell lines (http://www.atcc.org/). Other cell suppliers include the Health Protection Agency Culture Collections (HPACC; http://www.hpacultures.org.uk/), the German Research Center for Biological Material (DSMZ; http://www.dsmz.de/) and the Riken BioResource Center Cell Bank (Riken; http://www.brc.riken.jp). It is strongly recommended that investigators purchase cell lines from recognized centers such as these listed above to ensure pure, authentic and quality controlled cell lines. The decision to use cells directly from an organism, *i.e.* primary cells or an immortalized cell line, should be based upon requirements of the virus system and available current protocols. As discussed below, there are three main approaches for virus production in mammalian cell lines: (i) infection of permissive cell lines with wt virus, (ii) transfection of cells with plasmid constructs containing viral genome sequences and (iii) viral vector systems which expression heterologous viral genes.

Although the majority of viruses currently studied are obtained from recombinant expression systems (see below), direct infection of cell lines with wt virus can be used to generate suitable quantities of sample for structural studies under certain conditions and for well-characterized viral systems.

For example, the human rhinovirus 3 (HRV3) virion particles used for determining its structure were purified from virus-infected HeLa cells (immortalized human cancer cells). The atomic structure of HRV3 was initially determined to 3 Å,[3] and later refined to 2.15 Å.[4] It was reported that 10–12 L of HeLa cells (at $6-8\times10^5$ cells mL^{-1}) were used to generate the amount of virus necessary to carry out crystallization and structure determination. Echovirus-1, also of the *Picornaviridae* family, was also successfully produced in HeLa cells for its structure determination to ~ 3.55 Å resolution.[5]

In the use of plasmid constructs, one or more plasmids usually containing capsid proteins alone and, if needed, replication factors, are used to transfect cells, which results in the assembly of virus-like particles (VLPs). Often, another plasmid is added when a packaged gene is desired, *e.g.* reporter gene, or if genome is needed to produce stable virions. Recovered virus can either be purified for structural studies or, if infectious, used to infect permissive cells for continual propagation of virions. As an example, molecular clones containing the capsid sequence of canine parvovirus was used for the transfection of Norden Laboratories feline kidney cells (NLFK)[6] to produce particles for X-ray crystallographic structural studies to 3.2 Å resolution.[7] For the crystallographic structure determination of the immunosuppressive strain of minute virus of mice (MVMi), infectious virions were harvested from plasmid transfected cell lines and subsequently propagated in a permissive cell line to produce virus for crystallization.[8]

The development of heterologous surrogate expression systems for virus capsid production has enabled researchers to overcome the lack of efficient expression in homologous systems for several viruses of interest. As an example, for hepatitis C virus (HCV), a herpes simplex virus-1 (HSV-1)-based amplicon vector system that expresses HCV capsid proteins and the two envelope proteins, E1 and E2, under the HSV-1 IE4 promoter was developed.[9] This system has several advantages; (i) the ability to infect a wide range of cells, without the limitation of transfection efficiency, including primary cells in a quiescent state, (ii) the simplicity of cloning desired genes into amplicons, (iii) the high capacity of incorporation of exogenous sequences in the vector genome and the transfer of high copy numbers of the exogenous gene and (iv) the potential for using amplicons in vaccine design and development.[10] A mini-review has covered HSV amplicons from genomes to engineering.[11] Norovirus is another example of a non-cultivable virus that remained refractory to structural studies due to the lack of a reverse genetics system and a permissive cell line until recent advances. A novel expression strategy, which combined the use of a two baculovirus trans-activation system to deliver viral cDNA and an inducible DNA polymerase (pol) II promoter, led to the ability to grow this virus in several cell lines, including HepG2, BHK-21, COS-7 and HEK293T cells.[12,13]

Yeast Cells

Among the microbial eukaryotic host systems, yeasts can combine the advantages of unicellular organisms (*e.g.* ease of genetic manipulation and growth) with the capabilities of a protein processing typical of eukaryotic organisms (*e.g.*, protein folding, assembly and posttranslational modifications).[14] The majority of recombinant proteins produced in yeast have been expressed using *Saccharomyces cerevisiae*. More commonly referred to as baker's or budding yeast, *S. cerevisiae* was the first eukaryote to have its entire genome sequenced[15] and is still today considered a model organism. A scientific database has been established for *S. cerevisiae* and is available at http:// www.yeastgenome.org/. With its biochemistry, basic genetics and cellular biology already well established, this simple eukaryote has become a major tool in answering questions of fundamental biological importance and is a central player in post-genomics research.

Appealing aspects of the yeast expression system are its rapid cell growth (with a doubling time of $\sim$90 min), simple growth media, secretion of recombinant proteins to the medium and glycosylation capability. N-linked glycosylation is minimal with high mannose, but O-linked modifications appear similar to mammalian cells. Phosphorylation, acetylation and acylation are also present. Protein yields are comparable with the baculovirus system (see below) at $\sim$10–200 mg L^{-1} depending on recombinant gene properties. Issues in large-scale protein production involving *S. cerevisiae* appear to be hyperglycosylation and retention in the periplasmic space.[16,17] This ultimately leads to a loss of final protein due to retention and degradation. The search for alternative hosts has led to the use of 'non-conventional' yeasts in expression protocols. The most established examples include *Hansenula polymorpha*, *Pichia pastoris*, *Kluyveromyces lactis*, *Yarrowia lipolytica*, *Pichia methanolica*, *Pichia stipitis*, *Zygosaccharomyces rouxii*, *Zygosaccharomyces bailaii*, *Candida boidinii* and *Schwanniomyces (Debaryomyces) occidentalis*.[14] These systems are broken down even further into two categories: methyltrophic, *e.g. P. pastoris*, and non-methyltrophic, *e.g. S. cerevisiae*. These categories are based on the fermentation processes involved and generally dictate the promoter that should be used in the experimental design. The choice of yeast host is one of the most important determinants of the success of the entire project, and many reviews debating the subject can be found in the current literature. Generally, the expression of foreign proteins in yeasts consists of (i) cloning of a foreign protein-coding DNA sequence within an expression cassette containing a yeast promoter and transcriptional termination sequences and (ii) transformation and stable maintenance of this DNA in the fusion host.[14] The transformation process is highly dependent on the yeast strain and detailed studies should be conducted in order to achieve high-efficiency transformation.

This system is extensively used for studying biological processes in higher eukaryotes and also allows replication of eukaryotic viruses. The first eukaryotic virus for which replication and genome encapsidatation was conducted in *S. cerevisiae* was brome mosaic virus (BMV), a positive strand RNA [(+)RNA] virus that infects plants.[18,19] The BMV VLPs were subsequently purified for structure-to-function studies using cryo-electron microscopy (cryo-EM) studies.[19] Other (+)RNA viruses that have been successfully replicated in *S. cerevisiae* include the plant viruses tomato bushy stunt virus and carnation

Italian ringspot virus and animal viruses Flock House virus (FHV) and Nodamura virus.[20] Human papillomavirus-16 (HPV-16) VLPs have also been successfully expressed in the yeast system[21] in addition to the bovine papillomavirus-1 (BPV-1).[22,23] The yeast virus L-A was isolated and purified from *S. cerevisiae* and the structure was solved to 3.4 Å resolution.[24]

Insect Cells

Originally isolated from the alfalfa looper (*Autographa californica*) insect, *Autographa californica* multiple nucleopolyhedrovirus (AcMNPV) is the most widely used and best characterized baculovirus for recombinant gene expression (a recent review on baculovirus molecular biology is available[25]). The rather large genome ($\sim 134\,kbp$[26]) can stably accommodate an insertion of ~ 38kb,[27] making expression of large genes possible. This virus is also known to infect several other insect species including *Spodoptera frugiperda*. The most commonly used insect host cell lines, Sf9 and Sf21AE, are derived from *S. frugiperda* pupal ovarian tissue[28] and the BTI-Tn-5B1-4 line, also known as 'High 5 cells', derived from *Trichoplusia ni* egg cell homogenates.[29] The wt nucleopolyhedrovirus (NPV) produces small inclusion bodies composed of a polyhedron protein which allows for the encapsulation of many virions into a crystalline protein matrix. This protein is expressed in the very late phase of gene expression and is controlled by a very strong promoter, the polydron promoter (a review on baculovirus late expression factors is available[30]). The baculovirus expression vector system (BEVS)[31,32] takes advantage of this very strong polyhedron promoter to drive foreign protein expression. It has also been shown that the non-structural p10 protein is expressed at similar levels in the same very late phase of expression. Both proteins have been shown to be non-essential in the production of baculovirus particles,[33,34] making the replacement of their open reading frame (ORF) ideal for use in foreign gene expression.

The coupling of the very strong polyhedron promoter with a foreign gene-coding region results is the production of high levels of recombinant protein (~ 5–$200\,mg\,L^{-1}$) in a relatively short amount of time using the BEVS. Since the baculovirus genome is generally considered too large to insert the foreign gene of choice by direct ligation, transfer vectors are used. There are many different vectors available for gene insertion, which are variants of a basic design (a review appeared recently[35]). These offer single gene, multiple genes and fusion gene expression. Multiple copies of the promoter can also be engineered into BEVS for the expression of multiple recombinant proteins concurrently in infected cells,[36,37] which permits the assembly of structures that are made up of heterologous proteins, such as viruses.

Advances in experimental design such as a wide variety of transfer vectors, simplified recombinant virus isolation and quantification methods, advances in cell culture technology and commercial availability of reagents have led to the increased use of BEVS for recombinant viral capsid protein production. Belyaev and Roy[37] were able to construct a multiple gene transfer vector which co-expressed the four major structural proteins of bluetongue virus (BTV) to produce VLPs. These samples permitted structure-to-function correlations for BTV and advanced BTV research in efforts to characterize its assembly properties and in vaccine development.[38,39] VLPs have also been successfully expressed for many other viruses, including Norwalk virus,[40] HPV and BPV,[41] rabbit hemorrhagic disease virus,[42] the adeno-associated viruses,[43] avian influenza virus[44] and MVM[45] and FHV[46] which were useful in crystallization studies, producing crystals which diffracted X-rays to beyond $\sim 3.3\,\text{Å}$. One of the most appealing factors in this system is the presence of post-translational modifications. N- (simple; no sialic acid) and O-linked glycosylation, acylation, acetylation, disulfide bond formation and certain phosphorylation processes are all carried out in a manner similar to that found in mammalian cells. Recently, recombinant baculovirus vectors that contain mammalian expression cassettes for gene delivery and expression in mammalian cells have been developed (detailed information can be found elsewhere[47,48]). These versatile constructs have been termed 'BacMams' to avoid confusion with the original baculovirus that drives gene expression in insect cell lines.

Plant Systems

Advances in plant molecular biology and genomics have opened up the possibility of modifying their genomes. A large number of plant viruses studied today are propagated in the host plant. Host plant species can be easily grown, under proper conditions, and readily inoculated with the infecting virus. The disadvantage of this approach appears to be time and resources. Depending on the plant species and desired size for infection, it may take several weeks to obtain optimal conditions. Extensive space and supplies may also be required to generate adequate amounts of infected plants from which virus can be purified.

There is, to date, no general protocol for gene transfer into plants. Each cell type, tissue and plant species requires careful characterization to ensure optimal transfer to attain the highest efficiencies and reproducibility in terms of gene expression.[49] The genetic information of plants is distributed among three cellular compartments: the nucleus, the mitochondria and the plastids. The plastid is a circular double-stranded DNA (dsDNA) molecule which can account for 10–20% of the total cellular plant DNA content. Development of reliable methods in plastid genome transformation made feasible the targeted manipulation of the endogenous genetic information of plastids and, in addition, the possibility of introducing novel information to be expressed from engineered chloroplast genomes.[50]

Inoculation of plant host species through natural transmission routes generally requires an insect vector and is often not feasible for the average researcher. More common practice involves genetic manipulation of plastids from *Agrobacterium tumefaciens* and microparticle bombardment or biolistics.

A. tumefaciens is a rod-shaped Gram-negative ubiquitous soil bacterium that has become a useful tool due to a set of genes (T-DNA) located on the tumor-inducing (Ti) plasmid. These genes are capable of transferring and integrating into a foreign host and have become an ideal vehicle for gene transfer in plant research.[51,52] The use of *A. tumefaciens* inoculation has successfully been used to express HPV-16 L1 VLPs for use in edible vaccine production[53] and the plant geminivirus maize streak virus for structural studies by cryo-EM.[54] Biolistics involves the high-speed transfer of naked DNA that has been adsorbed on small metal particles, combining biology and ballistics, into plants. This method is very effective especially when the plant strain being utilized is resistant to *A. tumefaciens*.

Regardless of the method used to introduce exogenous DNA into plants, the gene of interest must be cloned into an expression cassette whose minimal requirements are a promoter and a terminator of transcription functional in the plant system.[55] Another requirement is the early selection of transformed cells from non-transformed tissue, achieved by the inclusion of a selection marker. A fertile plant can then be grown from these transformed cells. A major disadvantage is that homologous recombination is not efficient in plants and can lead to random insertion of genes and instability. In this case, several independently inoculated plants should be compared for expression levels.

An alternative system to whole plants is the use of plant cell systems. These can be cultivated like mammalian and bacterial cells and also offer eukaryotic post-translational modifications. Virus-based expression systems are also applicable. This is synonymous to using phage viruses in bacteria and provides an alternative to stable genetic transformation in plants. These vectors can be full (DNA-containing) virus vectors or deconstructed vectors, which generally lack several infectious aspects and places the gene of choice between viral DNA elements. Loss of gene insertion can occur with full viral vectors, especially if it is a large insertion, due to systemic movement in the plant cell. The use of deconstructed vectors has produced decent yields of VLP for several mammalian viruses using a tobacco mosaic virus system from Icon Genetics (http://www.icongenetics.com).[55] Protein production can reach relatively high levels in 3–14 days, depending on the system. For example, Norwalk virus coat protein levels reached ~ 20–$30\,\mu g\,g^{-1}$ of dry fruit in transgenic tomato plants.

Prokaryotic Systems

Escherichia coli

The use of *Escherichia coli* in the laboratory setting is a definite hallmark in biotechnology and almost marks the birth of this field. It is generally the preferred prokaryotic expression system due to (i) rapid ($\sim 30\,min$) and high-level expression (50–$500\,mg\,L^{-1}$) as a result of the speed of cell growth to high density, (ii) low complexity and low cost of growth media and (iii) the ability to target proteins to the desired subcellular localization.[56] However, the system has many disadvantages when used for the production of large eukaryotic proteins, as follows. (i) The cytoplasm is a reducing environment that strongly disfavors the formation of stable disulfide bonds. This can be detrimental to the formation of important assembly interactions, especially in a large macro-molecule such as the viral capsid, which will affect stability and proper folding. The creation of certain strains of *E. coli* with mutations in thioredoxin reduc-tase (encoded by *trxB*) and glutathione reductase (*gor*) aid the expression of proteins whose solubility depend upon an oxidative environment.[57–59] (ii) Wt *E. coli* lacks the ability to phosphorylate tyrosine residues, although strains have been engineered which allow this modification.[60] (iii) The overproduction of heterologous proteins in *E. coli* often results in misfolding and segregation into insoluble inclusion bodies. A number of techniques are cited in the lit-erature, which discuss the conversion of inactive protein, expressed in an insoluble fraction, into a soluble and active form.[61] (iv) Of great concern to virologists is the lack of post-translational modification in bacteria, such as glycosylation, acetylation and amidation. This can alter many of the functional properties of viral proteins and also structural features.

The selection of the promoter is also critical in bacterial expression system design. It is generally controlled by a regulatory gene or inducer, that is either inherent in, or supplied to, the host. The most widely used promoters are the lactose (*lac*)[62] and trytophan (*trp*)[63] promoters. Stronger, more tightly regu-lated promoters, *trc* and *tac*, have been created from the *lac* and *trp* promoters, but have incomplete repression in the induced state. This is not an issue when the gene product is not toxic to the cell. Another factor to consider is sub-cellular localization. Recombinant proteins may be directed to one of three compartments: cytoplasm, periplasm or the extracellular medium. Proteins found in the cytoplasm may require extra purification steps from inclusion bodies and refolding. This can hinder proper folding and ultimately cause issues in assembly when dealing with viral proteins. Periplasmic targeting offers advantages in proper folding due to the oxidative environment. However, proteins must change conformation to be shuttled across the cytoplasmic membrane and can be degraded due to incompatibility with the membrane. Since *E. coli* does not secrete many proteins into the extracellular fluid, there is less proteolytic activity and thus less degradation. Secretion also makes pur-ification easier as less undesired proteins are present. The disadvantage is the low yield due to successful passage across both the inner and outer membranes.

Although the *E. coli* system offers many advantages in cost and quantity produced, the expression of complete viruses is not common due to the com-plexity of the interactions often required for the assembly of viruses. However, viruses that will self-assemble into VLPs with a monomeric unit of their capsid protein can be successfully expressed. For example, Chen *et al.*[64] successfully expressed small VLPs of HPV-16 from one of two virally encoded capsid proteins, L1, in an *E. coli* system. The VLPs were successful crystallized and the structure of the HPV16 L1 capsid was determined to $\sim 3.5\,\text{Å}$ resolution. Bac-teriophages naturally use *E. coli* as a host and have been used to produce many wt viruses which have been successfully used for structure determination studies by many different approaches, including X-ray crystallography and cryo-EM.

A classic example is ΦX174, for which virions produced in *E. coli* were used to grow crystals which diffracted X-rays to ~2.7 Å[65] and were used for its structure determination.[66]

Cell-free Systems

This system utilizes purified components from cell homogenates that are necessary for protein synthesis, allowing the study of many biological processes free from the complex reactions that occur in a living cell. Cell-free synthesis of infectious virus has been useful in studying the mechanism of viral replication and assembly, and also screening anti-viral drugs. The methodology has been used for the production of poliovirus[67] and encephalomyocarditis virus.[68] Yields have not yet been optimized, however, for the amounts required for structural studies by X-ray crystallography, but enough virions can be produced for cryo-EM applications for a well-characterized viral system.

Tissue Samples

Isolation of infectious virions from patient samples, such as blood, feces or urine, is generally used in initial attempts to determine the presence of virion particles. This technique was used for discovering the newly described human bocavirus, which was isolated from nasopharyngeal aspirates,[69] and capsids were subsequently visualized by negative stain electron microscopy.[70] This method does not generally produce a large enough amount of sample for structural studies. In addition, safety concerns associated with the handling of wt infectious virus isolated from patient samples can restrict their use to facilities with established containment appropriate for the biosafety level of the virus system of interest. In general, once described and genetically characterized, viruses isolated from tissue samples are expressed in cells using one of the methods described above for further molecular and structural studies.

3 Purification

Purification is an essential process for generating virus capsids for structural characterization and is an integral requirement for successful crystallization for structure determination by X-ray crystallography. Prior to the use of the capsids for structural or other biophysical study, the sample should be checked by SDS-PAGE developed by Coomassie blue or silver staining to ascertain purity, by Western blot against an antibody to verify that the capsid viral proteins are present and by negative stain electron microscopy for integrity. The steps involved in a purification protocol are contingent on the nature of the virus under study and the medium from which it is being purified. For example, enveloped viruses may require extra factors, such as detergents for solubilization, which may not be necessary for non-enveloped viruses, or mild solvent conditions may have to be employed when purifying unstable complexes.

Samples that are not secreted from cells or are in inclusion bodies will require more steps than those that have been secreted into the cell media. Other factors that must also be taken into account after expression includes solubility, *i.e.* whether refolding is required, and capsid stability, *i.e.* whether pelleting will damage the integrity of the assembled capsid. The two most common approaches used for virus capsid purification are ultracentrifugation and chromatography, which are described below.

Ultracentrifugation

Ultracentrifugation is the usual technique of choice for the purification of virus capsids, particularly because this approach can utilize their defined size and shape to aid separation from other cellular material. The rate of sedimentation depends upon the size, density and morphology of the capsid, in addition to the nature of the density medium and the force that is applied during centrifugation. Cushions, using sucrose, dextran or Ficoll (GE Healthcare), can be incorporated in virus purification after the initial centrifugation to allow for the collection of morphologically intact capsids, without causing mechanical stress, for further purification. Density gradients, *i.e.* a variation in density over an area, are often used after the cushion step and are a cornerstone in virus purification. The two types of density gradients routinely used are rate-zonal and isopycnic centrifugation.

In rate-zonal approaches, the sample is layered over a gradient that allows for the separation of particles into bands or zones, based on the particle sedimentation rate. Step gradients generally result in better separation, but linear gradients can also be used. Gradient makers can be used to create linear gradients or steps poured can be allowed to sit vertically overnight (4 °C) to aid diffusion between the boundaries. The rate at which separation occurs is dependent on the particle size, shape, density, force applied and the profile of the gradient medium. In this method, capsids continue to migrate into the gradient, hence an idea of the virus capsid sedimentation velocity is required to ensure that the sample is not pelleted. Sucrose is generally used for virus purification by sedimentation velocity, but established protocols are also available for Ficoll, iodixanol (OptiPrep; Axis-Shield) and dextran detergents may be added, especially if the virus has a tendency to associate with membranes, to separate cellular debris from virus particles. To achieve the best separation, gradients should not be overloaded with sample and while the amount is dependent on the virus being studied, it is suggested that for large swing-out rotors, which hold volumes up to 30 mL, loaded samples should not exceed 5 mg, whereas for smaller capacity rotors, which may hold ∼5 mL, 1 mg or less should be loaded.[71]

In contrast to rate-zonal gradients, isopycnic gradients separate capsids based on their calculated densities. Samples can be layered on the gradient or mixed directly with the gradient medium since the gradient will equilibrate upon centrifugation. Particles migrate to their density in the gradient and do not migrate further. Cesium chloride (CsCl) is the medium of choice for this application, mainly due to its ability to form dense solutions of up to ~ 1.91 $g\,cm^{-3}$ that extend beyond the range of density values for most non-enveloped and enveloped viruses. Suggestions for the concentrations of CsCl to be used range from $1.32\,g\,cm^{-3}$ (32% w/v) for virus containing 5% RNA to $\sim 1.7\,g$ cm^{-3} (55% w/v) for DNA-containing virus.[72] A milder method for labile or non-enveloped viruses uses a 'positive density/negative viscosity' approach. This involves the layering of potassium tartrate such that it generates increasing density from top to bottom, or the use of glycerol, which yields decreasing viscosity from top to bottom. CsCl can still be used for these viruses, but potassium tartrate or glycerol provides a gentler medium.[72]

Chromatography

Chromatographic techniques, which utilize separation on a column, may be implemented as intermediate or final steps in virus capsid purification. There are several properties of viruses which can be exploited to aid in their purification using chromatographic methods, such as size, charge, hydrophobicity and ligand specificity. For excellent guidance in selecting the proper media and more in-depth methods on column chromatography and protein purification, the reader is directed to GE Healthcare Handbooks (www.gelifesciences.com).

In size-exclusion chromatography (SEC) or gel filtration, samples are run over a solid-phase column composed of beads with pores of a defined size, thus samples are separated based solely on size. There are two distinct approaches that can be used, (i) group separation, where high or low molecular weight species can be distinguished, or (ii) high-resolution fractionation, in which isolation of individual species occurs based on molecular weight. The first step towards a successful separation is the selection of the appropriate media. The Sephadex G series (GE Healthcare) are useful for group separation applications, whereas for high-resolution fractionation Sephacryl, Superose and Superdex (GE Healthcare) are used. The high-resolution matrices are applicable for the separation of samples in the 1–8000 kDa molecular weight range, which can separate from peptides to large proteins or large complexes. The choice of medium to use for a particular application is dictated by the range of sizes to be separated and a predetermined selectivity curve, which is available from the supplier.

Ion-exchange column chromatography (IEX) separates samples based on differences in the net surface charge and is capable of distinguishing molecules that have minor differences in their charge properties. The functional groups that are bound to the matrix determine the charge of the IEX media. The columns can be either cation exchangers, binding to net-positive surfaces, or anion exchangers, binding to net-negative surfaces. Separation by this technique thus relies on the condition of the sample under certain pH and ionic strength conditions.

Hydrophobic interaction chromatography is an excellent step which can be incorporated between other methods to allow for buffer exchange and concentration of the final product. This method uses the inherent hydrophobic properties of a molecule for binding and its subsequent reversal for elution. Samples can also be separated based on their differing degrees of hydrophobicity. Reversed-phase chromatography is essentially based on the same properties, but in this approach the surface medium is more hydrophobic. This leads to stronger binding interactions, which then requires more stringent elution techniques, for example, organic solvents.

Lastly, ligand specificity or affinity purification involves the use of columns with bound ligands that interact specifically with the capsid of interest, based on biological function or chemical composition. This can result in high purity due to its specificity. There are many ligands available for different applications, including antibodies, enzymes and cell-surface receptor molecules. For virus purification, this can be very useful for the isolation of properly folded capsid components and properly assembled capsids due to the high selectivity involved in the interactions with the column medium.

4 Example Virus Capsid Production and Purification – Adeno-associated Virus Serotype 1

The adeno-associated virus (AAV) is a small, non-enveloped, single-stranded DNA virus (ssDNA) that belongs to the family *Parvoviridae*. Several different serotypes are under development as viral vectors for gene delivery applications due to their simplicity, non-pathogenicity and ability to package and deliver non-genomic DNA to non-host cells and tissues. In an effort to improve the efficacy of gene delivery by these promising AAV vectors, there is a need to understand their basic biology, particularly in terms of the capsid structure and its role in dictating the functions of the virus during the infectious process. These include the interactions required for (i) cellular receptor recognition and entry, (ii) trafficking to the nucleus for genome replication, (iii) genome packaging following assembly and (iv) with host cell antibodies, which can lead to neutralization. Hence methods have been developed for large-scale capsid production and purification to facilitate these studies.

The AAVs package a 4.7 kb ssDNA viral genome with two open reading frames (ORFs), *rep* and *cap*. The *rep* ORF codes for four overlapping proteins required for replication and DNA packaging. The *cap* ORF encodes three capsid viral proteins (VPs) from two alternately spliced mRNAs. One of these mRNAs contains the entire *cap* ORF and encodes VP1. The other mRNA encodes for VP2, from an alternative start codon (ACG), and VP3, from a conventional downstream ATG. AAV capsids are assembled as a T = 1 icosahedral particle (~ 260 Å in diameter) from a total of 60 copies of VP1, VP2 and VP3, in a predicted ratio of 1:1:8/10.[73] VP3 is a 61 kDa protein that constitutes 90% of the capsid's protein content. The less abundant capsid proteins, VP1 (87 kDa) and VP2 (73 kDa), share the same C-terminal amino acid (aa) sequence with VP3 but have additional N-terminal sequences. AAV capsids can be assembled from heterologous systems from expressed VPs in the absence of genome. An example is given below for the production of VLPs assembled from VP1, VP2 and VP3 of AAV serotype 1 without packaged genome using the BEVS and their purification using ultracentrifugation and chromatographic approaches for structural studies.

VLP Expression Using the BEVS

A recombinant baculovirus encoding the AAV1 *ORF* was constructed using the Bac-to-Bac system (Gibco BRL). The AAV2 capsid ORF in pFBDVPm11[43] was replaced by the respective ORF encoding AAV1 capsid proteins derived from pAAV2/1.[74] Similar mutations were introduced into 50 non-coding and coding sequences to permit the expression of the AAV1 capsid proteins in the insect-cell background,[43] and the resulting construct expressed all three AAV capsid proteins, VP1, VP2 and VP3 (Figure 2). DH10Bac-competent cells containing the baculovirus genome were transformed with pFastBac transfer

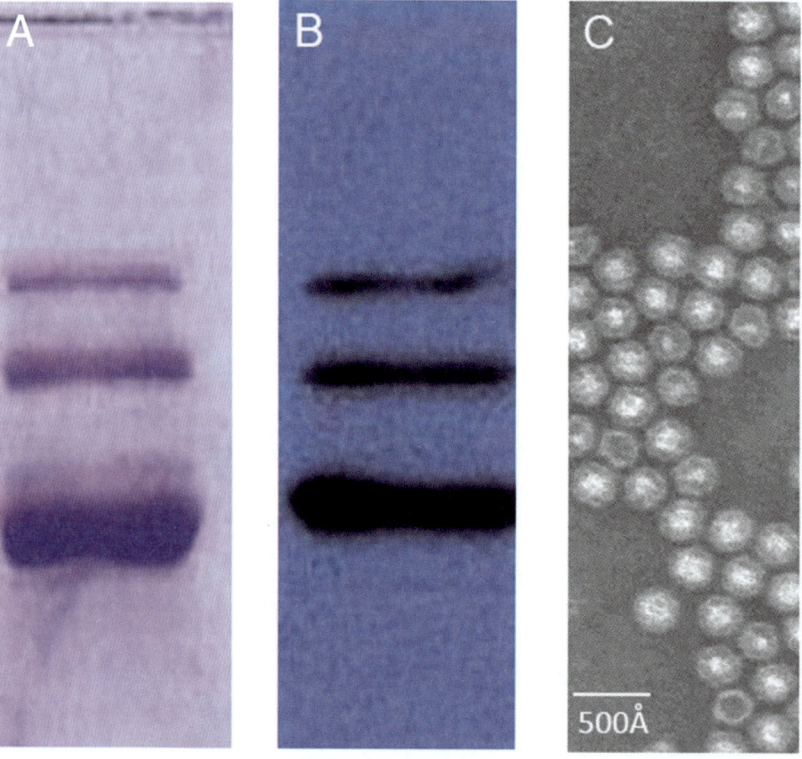

Figure 2 Characterization of AAV1 VLPs expressed in a baculovirus system. (A and B) SDS-PAGE and Western blot analysis against the B1 antibody showing the presence of VP1, VP2 and VP3, respectively. (C) Intact AAV1 VLPs stained by uranyl acetate viewed using a JEOL JEM-100CX II electron microscope. The bar represents 500 Å.

plasmids containing the AAV component insert. Bacmid DNA purified from recombination-positive white colonies was transfected into Sf9 cells using TransIT Insecta reagent (Mirus). Three days post-transfection, media containing baculovirus (pooled viral stock) were harvested and a plaque assay was conducted to prepare independent plaque isolates. Several individual plaques were propagated to passage one (P1) to assay for the expression of the AAV1 capsid genes and a selected clone was propagated to P2 and subsequently amplified to P3 for large-scale virus production.

Production of VLPs in Sf9 Insect Cells

A titered P3 recombinant baculovirus stock (generated as described above) was used to infect Sf9 cells grown in Erlenmeyer flasks at 300 K using Sf-900 II SFM media (Gibco/Invitrogen Corporation) at a multiplicity of infection (MOI) of 5.0 plaque-forming units (PFU) per cell. The cells were harvested at ~ 72 h post-infection (pi), spun down in a Beckman JA-20 rotor at 1090g and resuspended in lysis buffer (50 mM Tris–HCl, pH 8.0, 100 mM NaCl, 10 mM $MgCl_2$, 0.2% Triton X-100) at a final concentration of $\sim 1 \times 10^7$ cells mL^{-1}.

Purification of AAV1 VLPs from Infected Sf9 Cells

The VLPs were released from infected cells by three rapid freeze–thaw cycles in lysis buffer, with the addition of Benzonase (Merck, Darmstadt, Germany) after the second cycle. The sample was clarified by centrifugation at 12 100g for 15 min at 277 K and any resulting pellet was discarded. The cell lysate was pelleted through a 20% w/v sucrose cushion (in 25 mM Tris–HCl, pH 8.0, 100 mM NaCl, 0.3% Triton X-100; buffer A) by ultracentrifugation at 165 000g for 3 h at 277 K in a Beckman Type 70 Ti rotor. This process involved the layering of the sucrose solution at the bottom of a tube containing the clarified cell lysate. The supernatant was quickly discarded after the spin and the pellet was resuspended in ~ 1 mL of buffer A, with 1 mM EDTA added, overnight at 277 K. The sample was further subjected to multiple low-speed spins at 10 000g to remove insoluble material. The clarified sample was loaded on to a sucrose-step gradient (5–40% w/v), prepared by layering ~ 1.5 mL of a sucrose percentage solution into a Beckman ultraclear tube, beginning with the 40% and ending with the 5% fraction at the top. The gradient was spun at 210 000g for 3 h at 277 K in a Beckman SW 41 Ti rotor. A visible blue VLP band (illuminated by a light source) in the 20% sucrose layer was extracted (with a syringe needle) and dialyzed into 50 mM Tris–HCl, pH 8.0, 15 mM NaCl at 277 K to remove sucrose.

AAV1 VLP was also further purified, following a sucrose gradient, using IEX. As described by Zolotukhin *et al.*[75] for recombinant AAV1 vectors, a 5 mL HiTrap Q column (Pharmacia) was equilibrated at 5 mL min^{-1} with five column volumes of buffer B (20 mM Tris–HCl, pH 8.5, 15 mM NaCl), then 25 mL of buffer C (20 mM Tris–HCl, pH 8.5, 500 mM NaCl), followed by 25 mL of buffer B using a peristaltic pump. The AAV VLP-containing fractions were then diluted 1:1 with buffer B and applied to the column at a flow rate of $3\,mL\,min^{-1}$. After the sample had been loaded, the column was washed with 10 column volumes of buffer B and the sample was eluted with buffer C on a Pharmacia ATKA FPLC system, and 0.5–1 mL fractions were collected.

The purity and integrity of the viral capsids were monitored using SDS–PAGE and negative-stain electron microscopy (Figure 2), respectively. The sample was generally buffer-exchanged at 5000g using Amicon Ultra filters (Amicon Ultra-15, 10 kDa molecular weight cutoff; Millipore) into 100 mM HEPES–NaOH, pH 7.3, 50 mM $MgCl_2$, 0.03% NaN_3 and 25% glycerol and concentrated to a final concentration of $\sim 10\,mg\,mL^{-1}$ for X-ray crystallographic studies or into a different buffer and another desired concentration as appropriate for the study to be undertaken.

5 Summary

The production and purification of virus capsids in quantities suitable for structural characterization provides a means for functional annotation of numerous virus systems. A number of these viruses can be produced using different expression systems and the majority of the viral capsids produced are amenable to purification by two well-developed methods, ultracentrifugation and column chromatography.

6 Acknowledgments

This work was supported in part by NSF grant MCB-0718948 and NIH R21 AI81072341.

References

1. J. J. Greene, in *Recombinant Gene Expression*, ed. P. Balbás and A. Lorence, Humana Press, Totowa, NJ, 2004, p. 3.
2. P. Balbás and A. Lorence (eds), *Recombinant Gene Expression*, Humana Press, Totowa, NJ, 2004.
3. R. Zhao *et al.*, *Structure*, 1996, **4**, 1205.
4. A. T. Hadfield *et al.*, *Structure*, 1997, **5**, 427.
5. D. J. Filman, M. W. Wien, J. A. Cunningham, J. M. Bergelson and J. M. Hogle, *Acta Crystallogr., Sect. D*, 1998, **54**, 1261.
6. C. R. Parrish, *Virology*, 1991, **183**, 195.
7. L. Govindasamy, K. Hueffer, C. R. Parrish and M. Agbandje-McKenna, *J. Virol.*, 2003, **77**, 12211.
8. A. L. Llamas-Saiz *et al.*, *Acta Crystallogr., Sect. D*, 1997, **53**, 93.
9. E. Tsitoura *et al.*, *J. Gen. Virol.*, 2002, **83**, 561.
10. E. Tsitoura, U. Georgopoulou and P. Mavromara, *Curr. Gene Ther.*, 2006, **6**, 393.

11. N. Frenkel, *Curr. Gene Ther.*, 2006, **6**, 277.
12. C. J. McCormick, D. J. Rowlands and M. Harris, *J. Gen. Virol.*, 2002, **83**, 383.
13. V. K. Ward *et al.*, *Proc. Natl. Acad. Sci. USA*, 2007, **104**, 11050.
14. D. Porro and D. Mattanovich, *Methods Mol. Biol.*, 2004, **267**, 241.
15. A. Goffeau *et al.*, *Science*, 1996, **274**, 546.
16. J. Reiser, V. Glumoff, M. Kälin and U. Ochsner, *Adv. Biochem. Eng. Biotechnol.*, 1990, **43**, 75.
17. M. A. Romanos, C. A. Scorer and J. J. Clare, *Yeast*, 1992, **8**, 423.
18. M. Janda and P. Ahlquist, *Cell*, 1993, **72**, 961.
19. M. A. Krol *et al.*, *Proc. Natl. Acad. Sci. USA*, 1999, **96**, 13650.
20. I. Alves-Rodrigues, R. P. Galão, A. Meyerhans and J. Díez, *Virus Res.*, 2006, **120**, 49.
21. J. L. Rossi, L. Gissmann, K. Jansen and M. Müller, *Hum. Gene Ther.*, 2000, **11**, 1165.
22. K. Zhao and I. H. Frazer, *J. Virol.*, 2002, **76**, 12265.
23. K. Zhao and I. H. Frazer, *J. Virol.*, 2002, **76**, 3359.
24. H. Naitow, J. Tang, M. Canady, R. B. Wickner and J. E. Johnson, *Nat. Struct. Biol.*, 2002, **9**, 725.
25. B. Kelly, L. King and R. Possee, in *Baculovirus and Insect Cell Expression Protocols*, ed. D. W. Murhammer, Humana Press, Totowa, NJ, 2007, p. 25.
26. M. D. Ayres, S. C. Howard, J. Kuzio, M. Lopez-Ferber and R. D. Possee, *Virology*, 1994, **202**, 586.
27. N. Cheshenko, N. Krougliak, R. C. Eisensmith and V. A. Krougliak, *Gene Ther.*, 2001, **8**, 846.
28. J. L. Vaughn, R. H. Goodwin, G. J. Tompkins and P. McCawley, *In Vitro*, 1977, **13**, 213.
29. R. R. Granados, L. Guoxun, A. C. Derksen and K. A. McKenna, *J. Invertebrate Pathol.*, 1994, **64**, 260.
30. K. L. Hefferon, *J. Mol. Microbiol. Biotechnol.*, 2004, **7**, 89.
31. G. D. Pennock, C. Shoemaker and L. K. Miller, *Mol. Cell. Biol.*, 1984, **4**, 399.
32. G. E. Smith, M. D. Summers and M. J. Fraser, *Mol. Cell. Biol.*, 1983, **3**, 2156.
33. G. E. Smith, M. J. Fraser and M. D. Summers, *J. Virol.*, 1983, **46**, 584.
34. J. M. Vlak *et al.*, *J. Gen. Virol.*, 1988, **69**(Pt 4), 765.
35. R. D. Possee and L. King, in *Baculovirus and Insect Cell Expression Protocols*, ed. D. W. Murhammer, Humana Press, Totowa, NJ, 2007, p. 55.
36. A. S. Belyaev, R. S. Hails and P. Roy, *Gene*, 1995, **156**, 229.
37. A. S. Belyaev and P. Roy, *Nucleic Acids Res.*, 1993, **21**, 1219.
38. P. Roy, *Vet. Microbiol.*, 1992, **33**, 155.
39. P. Roy, M. Mikhailov and D. H. Bishop, *Gene*, 1997, **190**, 119.
40. X. Jiang, M. Wang, D. Y. Graham and M. K. Estes, *J. Virol.*, 1992, **66**, 6527.
41. R. Kirnbauer, F. Booy, N. Cheng, D. R. Lowy and J. T. Schiller, *Proc. Natl. Acad. Sci. USA*, 1992, **89**, 12180.

42. S. Laurent, J. F. Vautherot, M. F. Madelaine, G. Le Gall and D. Rasschaert, *J. Virol.*, 1994, **68**, 6794.
43. M. Urabe, C. Ding and R. M. Kotin, *Hum. Gene Ther.*, 2002, **13**, 1935.
44. A. Prel, G. Le Gall-Reculé and V. Jestin, *Avian Pathol.*, 2008, **37**, 513.
45. E. Hernando *et al.*, *Virology*, 2000, **267**, 299.
46. A. J. Fisher, B. R. McKinney, A. Schneemann, R. R. Rueckert and J. E. Johnson, *J. Virol.*, 1993, **67**, 2950.
47. T. A. Kost and J. P. Condreay, *Trends Biotechnol.*, 2002, **20**, 173.
48. J. A. Fornwald, Q. Lu, D. Wang and R. S. Ames, *Methods Mol. Biol.*, 2007, **388**, 95.
49. A. Lorence and P. Balbás, *Recombinant Gene Expression: Reviews and Protocols*, Humana Press, Totowa, NJ, 2004.
50. R. Bock, *J. Mol. Biol.*, 2001, **312**, 425.
51. N. Grimsley, B. Hohn, T. Hohn and R. Walden, *Proc. Natl. Acad. Sci. USA*, 1986, **83**, 3282.
52. A. Lorence and R. Verpoorte, *Methods Mol. Biol.*, 2004, **267**, 329.
53. H.-L. Liu *et al.*, *Acta Biochimica et Biophysica Sinca*, 2005, **37**, 153.
54. W. Zhang *et al.*, *Virology*, 2001, **279**, 471.
55. L. Santi, Z. Huang and H. Mason, *Methods*, 2006, **40**, 66.
56. S. Cantrell, in *Methods in Molecular Biology*, ed. N. Casali and A. Preston, Humana Press, Totowa, NJ, 2003, **235**, p. 257.
57. A. I. Derman, W. A. Prinz, D. Belin and J. Beckwith, *Science*, 1993, **262**, 1744.
58. W. A. Prinz, F. Aslund, A. Holmgren and J. Beckwith, *J. Biol. Chem.*, 1997, **272**, 15661.
59. P. H. Bessette, F. Åslund, J. Beckwith and G. Georgiou, *Proc. Natl. Acad. Sci. USA*, 1999, **96**, 13703.
60. M. Simcox, A. Huvar and T. Simcox, *Strategies*, 1994, 68.
61. S. Sahdev, S. K. Khattar and K. S. Saini, *Mol. Cell. Biochem.*, 2008, **307**, 249.
62. C. Yanisch-Perron, J. Vieira and J. Messing, *Gene*, 1985, **33**, 103.
63. D. V. Goeddel *et al.*, *Nature*, 1980, **287**, 411.
64. X. S. Chen, R. L. Garcea, I. Goldberg, G. Casini and S. C. Harrison, *Mol. Cell*, 2000, **5**, 557.
65. P. Willingmann *et al.*, *J. Mol. Biol.*, 1990, **212**, 345.
66. R. McKenna *et al.*, *Nature*, 1992, **355**, 137.
67. A. Molla, A. V. Paul and E. Wimmer, *Science*, 1991, **254**, 1647.
68. Y. V. Svitkin and N. Sonenberg, *J. Virol.*, 2003, **77**, 6551.
69. T. Allander *et al.*, *Proc. Natl. Acad. Sci. USA*, 2005, **102**, 12891.
70. N. Brieu, B. Gay, M. Segondy and V. Foulongne, *J. Clin. Microbiol.*, 2007, **45**, 3419.
71. R. Killington, A. Stokes and J. Hierholzer, in *Virology Methods Manual*, ed. B. W. J. Mahy and H. O. Kangro, Elsevier, Amsterdam, 1996, p. 71.
72. J. F. Obijeski, A. T. Marchenko, D. H. Bishop, B. W. Cann and F. A. Murphy, *J. Gen. Virol.*, 1974, **22**, 21–33.

73. M. S. Chapman and M. Agbandje-McKenna, in *Parvoviruses*, ed. J. R. Kerr, S. F. Cotmore, M. E. Bloom, R. L. Linden and C. R. Parrish, Edward Arnold, New York, 2006, p. 107.

74. G. Gao *et al.*, *Proc. Natl. Acad. Sci. USA*, 2002, **99**, 11854.

75. S. Zolotukhin *et al.*, *Methods*, 2002, **28**, 158.

CHAPTER 2

Microscopic Analysis of Viral Cell Binding, Entry and Infection in Live Cells

COLIN R. PARRISH

Baker Institute for Animal Health, College of Veterinary Medicine, Cornell University, Ithaca, NY 14850, USA

1 Introduction

The process of cell infection by animal viruses initiates with binding of one or more receptors on cells. Although some (mostly enveloped) viruses infect cells directly from the plasma membrane, most infect only after being taken up by endocytosis resulting in the virions being enveloped in endosomes, after which the particles are trafficked through various endosomal pathways for varying periods of time. Within the endosome, exposure to low pH, reducing conditions and/or proteases may cause the particles or viral proteins to change their structures and interact with the endosomal membrane, enabling the viral particle or the nucleocapsid or nucleoprotein to be released into the cytoplasm for the subsequent steps of infection. Some viral particles or their nucleic acids and associated components enter the nucleus. Fluorescence microscopy has been a key method for following the cell entry processes of virus particles, allowing virions and their components to be followed during the infectious process. Particles produced within the cytoplasm can also be followed and in some cases the separation of viral proteins or components can be followed in the live cells, along with the viral genome. Many studies have involved fixation of cells at various times after uptake and detection of the viral and cell components by antibody or other staining methods. That approach allowed

RSC Biomolecular Sciences No. 21
Structural Virology
Edited by Mavis Agbandje-McKenna and Robert McKenna
© Royal Society of Chemistry 2011
Published by the Royal Society of Chemistry, www.rsc.org

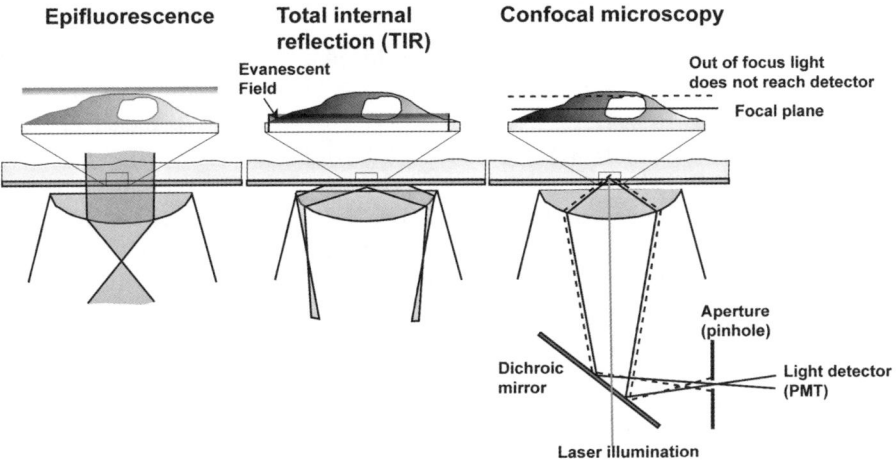

Figure 1 Three principle methods of imaging viruses in live cells: epifluorescence, total internal reflection (TIRF) and confocal microscopy. Highly diagrammatic representations of each imaging method are shown. Each form of imaging detects the fluorescence at different positions in the cells. Epifluorescence images all of the materials in the cells, including those out of the plane of focus. TIRF detects the fluorescence that is close to the coverslip where it is illuminated by the evanescent field of the illumination applied at an angle so that it is refracted by the interface between two different media, ensuring that only materials close to the coverslip are imaged. Confocal imaging detects only the fluorescence of materials that are in one focal plane by blocking the out-of-focus light from reaching the detector (PMT, photomultiplier tube).

many aspects of the infection process to be revealed for different viruses and have allowed many of the basic steps involved to be defined. Recently, various forms of live cell microscopy have allowed the processes to be followed over time, revealing additional dynamic aspects of viral cell entry and other steps in the viral lifecycle, and also showing the variation in the trafficking of different particles or their components within the same cell. Those studies involve following fluorescently labeled particles and cell proteins in real time. There are three light microscopic methods commonly used for imaging of viral proteins in live cells, which are in rapid development with many new applications being developed: widefield fluorescence microscopy (sometimes with deconvolution), confocal microscopy (single or multi photon), and total internal reflection fluorescence (TIRF) microscopy (Figure 1). The technical aspects of the various methods used in live cells have been described in recent reviews.[1–5] Each method has advantages and disadvantages for the analysis of viral binding, entry and trafficking. This chapter reviews how these methods have illuminated the dynamic aspects of cell entry pathways used by different viruses.

2   Endocytosis, Cytoplasmic Transport and Viral Entry

Receptor-mediated endocytic uptake and endosomal trafficking pathways have been defined in general terms for a number of years and for some ligands the pathways are likely to be fairly well understood. However, it is also clear that we have an incomplete understanding of the details of the processes used by many viral particles and their components, even when they use the same receptors as well-defined ligands, such as transferrin. The general properties of endocytic pathways, viral entry and cytoplasmic trafficking have been reviewed[6-11] and details of the analysis of HIV-1 entry pathways in live cells using fluorescence microscopy have also recently been specifically reviewed.[12] In summary, receptor–ligand complexes are taken up from the cell surface through one of a number of endocytic mechanisms, including clathrin-mediated, caveolar-associated and a number of non-clathrin/non-caveolar processes (Figure 2). The complexes then traffic through a limited number of routes, including the Rab5-associated early endosomal compartment (which may be separated into two functional components[13]) or enter a caveolin-associated compartment termed the caveosome when taken up through the caveolae-dependent process.[14] Some ligands recycle rapidly back to the cell surface from the early endosome, whereas others enter the Rab11-dependent recycling endosome and then may recycle more slowly back to the cell surface. Other receptor–ligand complexes enter the degradative pathway, which includes the late endosome or multivesicular endosome and then the lysosome, where the complex is degraded. Intracellular vesicles of various types are transported within the cytoplasm by molecular motors, including dynein and kinesin, which mediate minus- and plus-end microtubular transport, respectively. Viral components may also be transported within the cytoplasm by specifically engaging the same microtubular motors. Dynein is also responsible for the transport of protein aggregates or denatured proteins that are destined for degradation through aggresomal or proteosomal processes, and some viruses or their components may also be transported by that non-specific mechanism.

3   Virus Labeling for Fluorescence Experiments – Allowing Tracking of Viral Particles, Components and/or Nucleic Acids

There are various methods for labeling viral particles with one or more fluorescent tags that allow them to be followed on the cell surface or within live cells. For enveloped viruses fluorescent lipids can be incorporated into the viral envelope by incubating the purified virus with dye, allowing it to be incorporated by exchange and diffusion. Dyes that have been used in these studies include 1,1′-dioctadecyl-3,3,3′,3′-tetramethylindodicarbocyanine (DiD) and related dyes, octadecylrhodamine B chloride (R18) and pyran. Such membrane labeled viruses can then be examined in microscopic studies, where the virus is followed within the cell and the fluorescent properties of the particle are

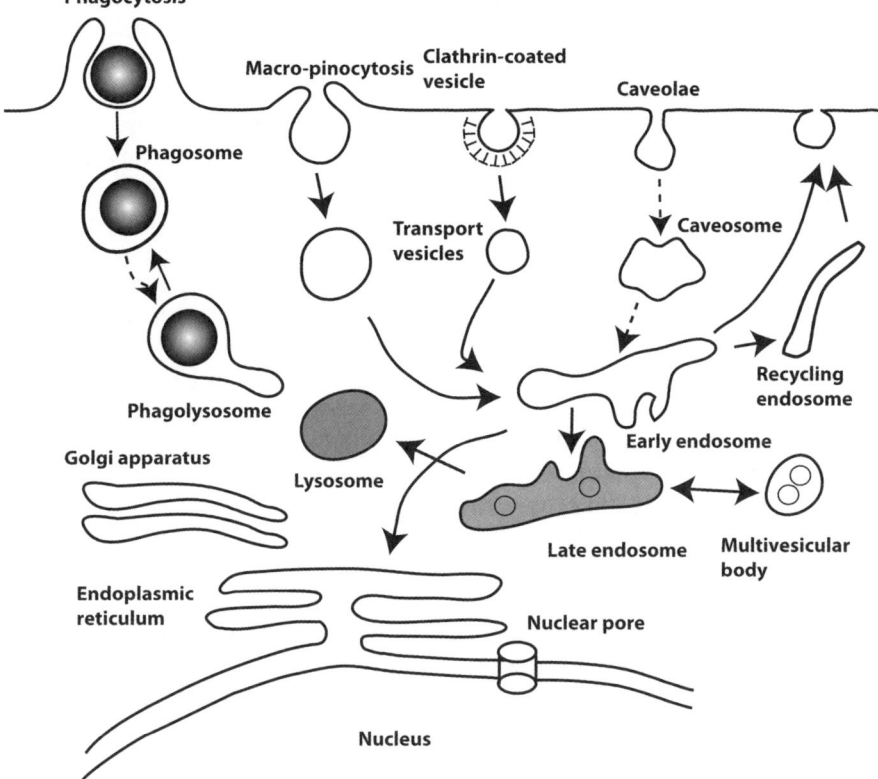

Figure 2 A simplified diagram of the endocytic and vesicle trafficking pathways that can be used by viruses during infection of mammalian cells. Sorting of proteins within the endocytic system is accompanied by increasing acidification of endosomes as they move down microtubules towards the perinuclear region. Incoming viruses or viral components may be sorted through one or more of any of these pathways, depending on the virus, the receptor and the specific cell involved.

monitored over time. One approach commonly used has been to label the virus to a level where the dye is self-quenched and then fusion of the labeled viral membrane with the endosomal membrane or on supported lipid bilayers results in a pulse of increased fluorescence that can be specifically monitored.[15,16]

Other dyes can be chemically conjugated to label the surface-exposed proteins of the viruses, including fluorescein, and also Cy, Alexa or Atto dyes. Carefully balancing the level of labeling is necessary to ensure that the label does not significantly interfere with the viral functions to be examined. Some dyes, such as fluorescein or modified Cy dyes (CypHer), are sensitive to low pH and the low pH of the endosome therefore reduces the fluoresence of the label on the virus; this property may also be used to measure the pH of the environment that the particle is exposed to within the cell.[13,17]

For many viruses, fluorescent proteins such as green fluorescent protein (GFP) and its various spectral variants and also other fluorescent proteins can be fused to one or more of the viral proteins or to a cellular protein that is incorporated into the virus, resulting in metabolically labeled particles that can be followed within the cell during both production and trafficking, and after purification those can be used in binding and entry assays. Many different examples have been reported, including the GFP–VP26 capsid protein of herpes simplex virus and the GFP–gag protein of HIV or other retroviruses.[18–21] In the studies outlined below, GFP variants or other fluorescent proteins are used in most cases and all appear similar in terms of their properties for live cell imaging experiments and the choices are mostly associated with the need for specific colors in the studies. Modified GFP forms are available that can be photoactivated or which have short half-lives and these can be used in timing experiments to monitor the distribution of the protein over specific periods. The use of split GFP molecules allows the association between proteins *in vivo* to be followed in live cells by monitoring the fluorescence where one protein contains part of the GFP and another contains the remainder of the protein, so that when the carrier proteins bind the GFP assembles to produce an intact fluorescent protein.[22]

Another labeling method for application in live cell studies is the use of tetracysteine tags, which can be labeled with bis-arsenical fluorescein derivatives (FLAsH) that can diffuse into the cell and specifically label the tagged viral components.[23,24]

Labeling of the viral genome or of genome-associated proteins can also be used to monitor the progress of the DNA or RNA either alone or along with protein components. Examples include the GFP-fused nucleoprotein protein of rabies virus (with separate labeling of the membrane), the integrase protein of HIV and histones of polyomaviruses, which allow the tracking of the genome within the cell in real time.[25–27] The genome can sometimes be directly labeled by adding nucleic acid-interchelating dyes to the virions that label the nucleic acid by diffusing into the capsid. Labeling of poliovirus RNA was accomplished by growing the virus in cells in the presence of the Syto82 and after purification the capsids were then labeled with Cy5, so that the locations of the capsid proteins and the viral RNA could be followed separately after uptake into cells (Figure 3).[28] In these studies the RNA and protein could be identified simultaneously for many particles and the viral RNA was released from the capsids within endosomes soon after uptake and close to the cell surface, with the uptake being dependent on active processes, including the actin cytoskeleton.

4   Receptor Attachment and Cell Entry

The process of viral entry initiates with attachment of the particles to one or more cellular receptors. Receptors used by different viruses include many forms of glycans, such as polysaccharides, and also various glycoproteins or

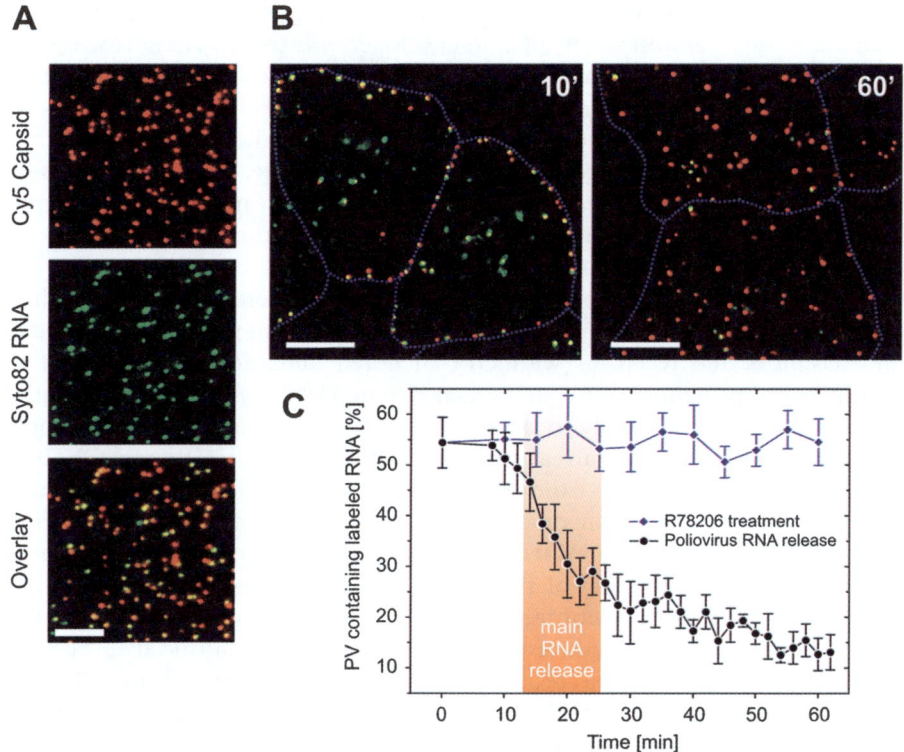

Figure 3 An example of the analysis of viral entry into cells using live cell analysis, revealing the dynamics of the system and the ability to follow the different fates of the viral capsid and the RNA which have different labels. (A) Dual-labeled PV genomic RNA, labeled with Syto 82 (green) co-localizes with the Cy5 labeling of the viral capsid (red). Scale bar indicates 5 μm. (B) Imaging of RNA release in live HeLa cells at the single virus particle level. Cells infected with dual-labeled PV and imaged 10 min and 60 min post-infection. Scale bar indicates 10 μm. (C) PV RNA release kinetics in HeLa cells. The cells were infected with PV at multiplicity of infection (MOI) = 1. The fraction of Cy5-positive particles containing Syto82 signal was detected and counted at different time points post-infection. In contrast to the R78206 negative control (R78206 specifically binds PV capsid and inhibits conformational change and RNA release), untreated dual-labeled PV releases RNA efficiently. From Figure 1 of reference 28, with permission.

glycolipids. Some viruses appear to bind and infect cells through a single functional receptor that leads directly into the infectious pathway, whereas others use multiple receptors that bind simultaneously to different viral proteins or bind in sequence. Capsid binding to a low-affinity but high-density receptor such as heparin sulfate proteoglycan (HSPG) or sialic acids may also serve to attach the virion to the cell, allowing engagement of a second receptor, generally a glycoprotein, to mediate infectious entry. The entry event may involve fusion with the plasma membrane in the case of some enveloped viruses or uptake through endocytosis and fusion or release from within the endosome for many enveloped and most or all non-enveloped viruses. In some cases, viral host or tissue association occurs due to specificity for particular glycans such as 3-*O*-sulfated forms of heparin sulfate or sialic acid forms or linkages, such as *N*-acetyl- or *N*-glycolylsialic acids or α2–3- or α2–6-linked terminal sialic acids, whereas other viruses bind to sialic acids in either linkage. The sialic acids used for binding and infection may be attached to glycoprotein or to glycolipids or both, and the requirement for specific additional receptors may be difficult to determine in the presence of the binding glycan.

Glycoprotein receptors for viruses fall into many different classes, including type I and type II transmembrane proteins, membrane-embedded transporters and receptors and receptors with GPI-anchored tails. Some viruses bind to multiple glycoprotein receptors in sequence or in parallel; for example, the HIV gp120 protein binds initially to the CD4 receptor, triggering a conformational change to allow binding to a second receptor, either CCR4 or CXCR5, depending on the virus and the cell that it is binding. Adenoviruses initially bind to the JAM-A receptor on cells through the knob on the end of the viral fiber and then engage the αVβ1 integrin through the penton base molecules, which also induce signaling of that receptor.[29] The form of the receptor, its location on the plasma membrane, efficiency of mediating endocytosis and in some cases its ability to induce signaling within the cells all need to be considered when examining the process of viral entry and infection of cells. For example, transferrin receptors bind viruses from at least three different families (parvoviruses, hantaviruses and retroviruses) and these are located in the non-raft regions of the plasma membranes, and when binding their normal ligand, transferrin, enter cells through clathrin-mediated entry pathways. In contrast, charged polysaccharides such as HSPG would be a less mobile receptor on the surface of cells and do not engage specific intracellular mechanisms, so they may be less efficient in leading viruses into specific endocytic pathways.

5   Membrane Association and Cell Surface Movement and Uptake

The binding and movement of cell surface receptors or other events that occur at or close to the plasma membrane can be observed specifically using confocal microscopy by focusing on the top or bottom cell surface. This allows a fluorescently labeled virus to be tracked in real time as it binds to the cell receptor, moves on the cell surface and then is either released or enters the cell. TIRF allows particles to be followed only when they are within a short distance of the cell substrate or coverslip and therefore detects virus that has diffused into the space below the cell and associated with the undersurface of the cell. The sample is therefore illuminated at an oblique angle so that the light is refracted from the coverslip and the evanescent wave produced by the refracted light only illuminates the sample that is within 100 nm of the surface. Fluorescent molecules that are on or near the surface of the cells are specifically illuminated, including cellular proteins linked to fluorescent proteins or labeled viral particles, as shown for murine polyomavirus virus-like particles binding, moving on the surface and entering cells that express GFP–clathrin or GFP–caveolae (Figure 4).[30] Similar studies have been reported for human papillomavirus 16 (HPV) moving on the surface of cells, which shows four modes of movement before endocytosis.[31] Using this method, particles can be observed to enter the cell by endocytosis and once they move into the cell beyond the evanescent field their signal is lost. As a demonstration of the resolution of the method, TIRF was used to monitor the stages of fusion of influenza with membranes *in vitro* and identified partial fusion (hemifusion) and full fusion of the virus membrane with the membrane bilayer.[32] In each case, TIRF specifically allows the observation of surface events and avoids the problem of out-of-focus fluorescence that can be encountered using other methods.

The mechanism of endosomal uptake can also be explicitly defined using approaches where the cells expressing fluorescently tagged markers of endocytosis such as Rab proteins (*e.g.* Rab5, Rab7 and Rab11, which can be used as markers for the early, late and recycling endosomes, respectively), EPS15, clathrin light chain or caveolin (among others) can be used along with labeled particles for observing uptake of the virus along with the labeled cell

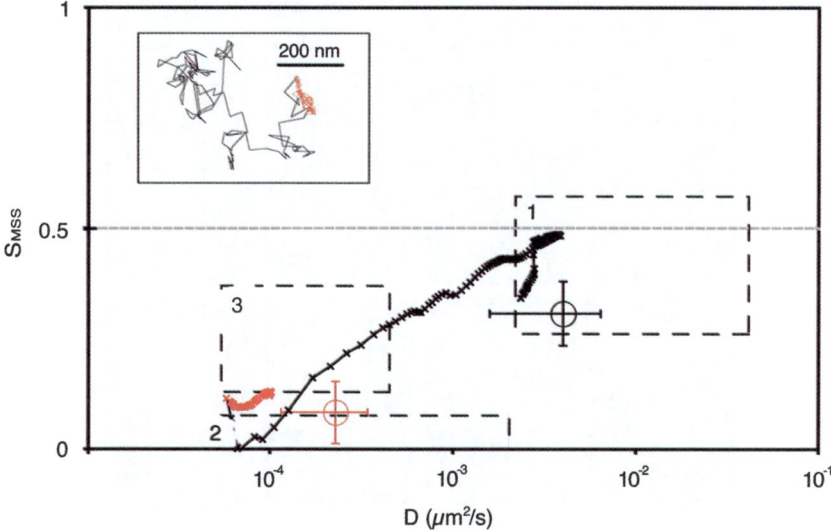

Figure 4   An example of the ability to follow the dynamic movement of virus (polyomavirus) particles on the surface of cells followed by total internal reflection fluorescence (TIRF) microscopy. The movement of individual particles is shown, where S_{MSS} represents the degree on non-linear motion and D = diffusion coefficient. Particles show both slow and fast trafficking that occurs, followed by the trapping of the particles at sites of endocytic uptake. Confined and mobile phases of surface movement were observed and the average movement for the confined (red) and the mobile (black) part of the trajectories (*n* = 10) are shown. From reference 30, with permission.

component (Figure 5). In addition, the use of constitutively active or dominant negative mutants of cellular proteins can allow additional manipulation of the endosomal pathways for visualization.

For some viruses examined in cell culture, attachment occurs to receptors displayed on filopodia that extend from the cell and which contain bundles of actin filaments. Examples include avian leukosis virus, canine parvovirus, HPV16 and HPV31, among others.[21,31,33] There may be directional trafficking of the bound particles towards the cells in a process that can involve myosin-2 and actin or through the retrograde flow of the actin filaments as they polymerize near the tips of the filopodia. The retrograde transport moves the particles at between 1.6 and $3.5\,\mu m\,s^{-1}$.[21,31] The filopodia-specific effects are most readily demonstrated by examining cells that express GFP–actin or that were injected with labeled actin, which labeled the filopodia so that they could

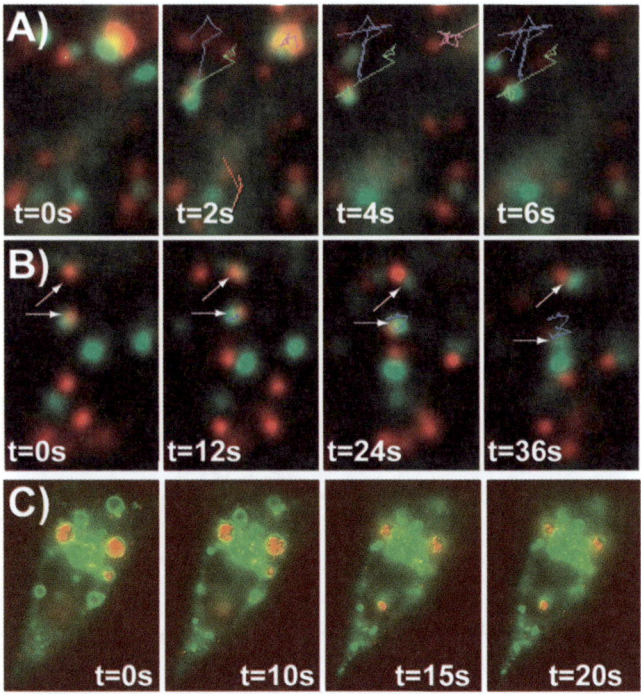

Figure 5   An example of analysis of an virus entering live cells, showing the localization of the virus along with different cellular components over time after entry. The association of Alexa 594-labeled capsids (red) with Rab5–GFP (green) after endocytosis into cells. Time lapse frames show the co-localization and co-movement of Alexa 594-labeled capsids with wild-type Rab5–GFP in feline cells at (A) 15 and (B) 80 min. (C) CRFK cells expressing constitutively active Rab5–GFP (green) contain large vesicles that accumulate capsids (57 min after uptake). Tracks of co-localized particles and vesicles are shown, and white arrows highlight co-localized virus and Rab5. From Figure 6 of reference 46, with permission.

be visualized (Figure 6). In some cases movement of the fluorescent actin within the filopodia can be followed and where the receptors engage the myosin-actin transport mechanisms this can be seen in the similarity of the viral particles and the cytoskeleton. Some receptors that bind viruses to filopodia do not move the particles in an actin–myosin-dependent manner, but the dynamic extension and retraction of the filopodia still results in viruses accumulating at the cell body. The regions of the cell at the base of the filopodia allow rapid endocytosis and so the virions enter the normal intracellular vesicular trafficking pathways. A number of newly produced viruses are also released from cells along filopodia or cytonemes which connect cells to each other and in some cases the virus may induce the formation of the cell extensions, to enhance the cell–cell transmission of the viruses (Figure 7).[33–35]

6 Receptor Signaling and Endosomal Uptake

Many different mechanisms of endosomal uptake from the cell surface have been described, including clathrin-mediated uptake, caveolae-mediated uptake and various non-clathrin–non-caveolae-mediated uptake processes, as well as active uptake through induction of macropinocytosis after cell ruffling (see below). The non-clathrin–non-caveolar pathways are still relatively poorly understood but most lead to the same locations in the cells as the uptake through other routes, but the caveolar uptake used by the SV40 virus and echovirus capsids may lead to a distinct compartment that has been termed the caveosome, and also trafficking to the endoplasmic reticulum.[36,37]

The multivalent binding of virus particles results in clustering of receptors on the cell surface and in many cases this leads to receptor signaling. Some viruses, including poxviruses and different strains of adenoviruses, induce signals that cause ruffling or blebbing of the cell membrane, leading to active viral uptake through fluid phase endocytosis or macropinocytosis (Figure 8).[38–40] A primary sign of many signaling processes associated with endocytosis is the clustering or polymerization of actin in the vicinity of the virus binding site, which enhances the efficiency of endocytosis. This can be detected directly in cells expressing fluorescent actin. As an example, when SV40 binds to receptors in the plasma membranes of cells, there is a transient breakdown of actin stress fibers, followed by actin recruitment to the virus-loaded caveolae, resulting in an actin tail where one end of the tail is associated with site of uptake (Figure 9).[41]

7 Tracking Endosomal Trafficking of Particles within Live Cells

Having entered the cell, viruses or their components follow variants of the normal routes of ligand trafficking through the early endosome to late endosome and then the lysosome, into the recycling pathways (where the viral particles may or may not be recycled) or to the endoplasmic reticulum or Golgi

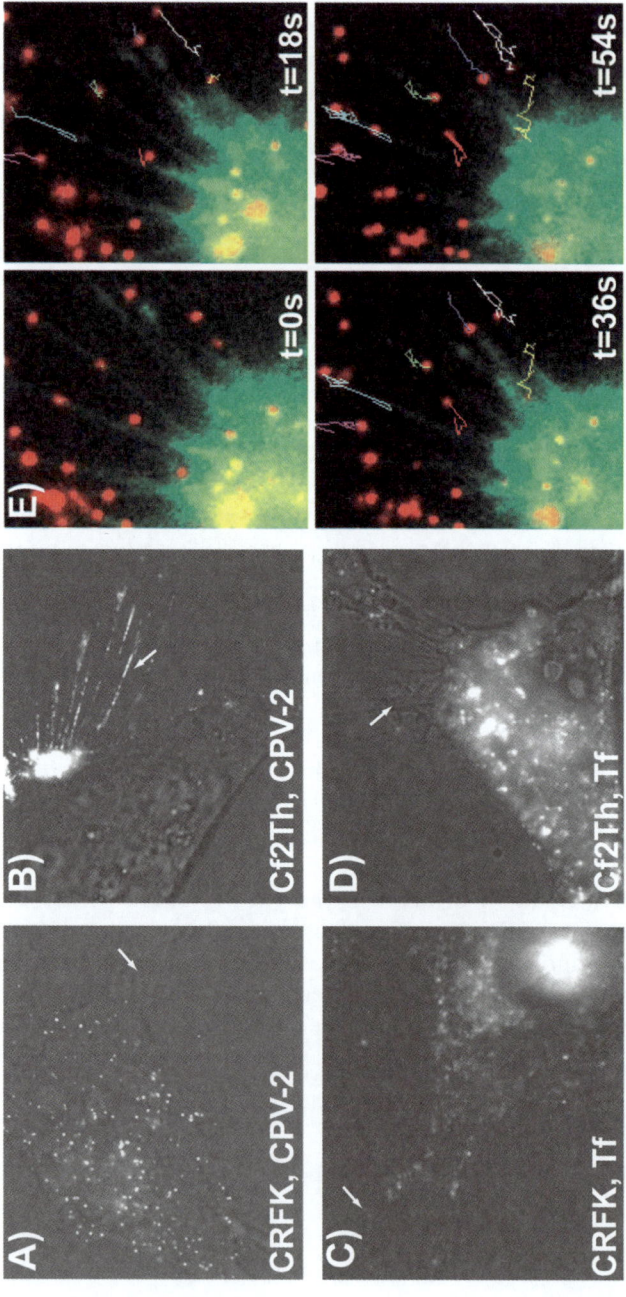

Figure 6 An example of fluorescently labeled canine parvovirus (CPV) particles analyzed on cell filopodia of live cells, showing the dynamic aspects of the process. Alexa 594-labeled CPV capsids incubated with (A) feline or (B) canine cells for 5 min at 37 °C then observed immediately after. Alexa 594–transferrin was also incubated with (C) feline and (D) canine cells under the same conditions. The white arrows highlight filopodia without virus or Tf bound, and the black arrow in (B) shows virus concentrating on the filopodia of canine cells. (E) Time lapse frames showing CPV particles (red) bound to filopodia of Cf2Th cells containing microinjected Alexa 488–actin (green). The tracks show particle movement on the filopodia; the yellow arrow indicates one particle moving towards the cell at the same rate as filopodial retraction. From Figure 4 of reference 46, with permission.

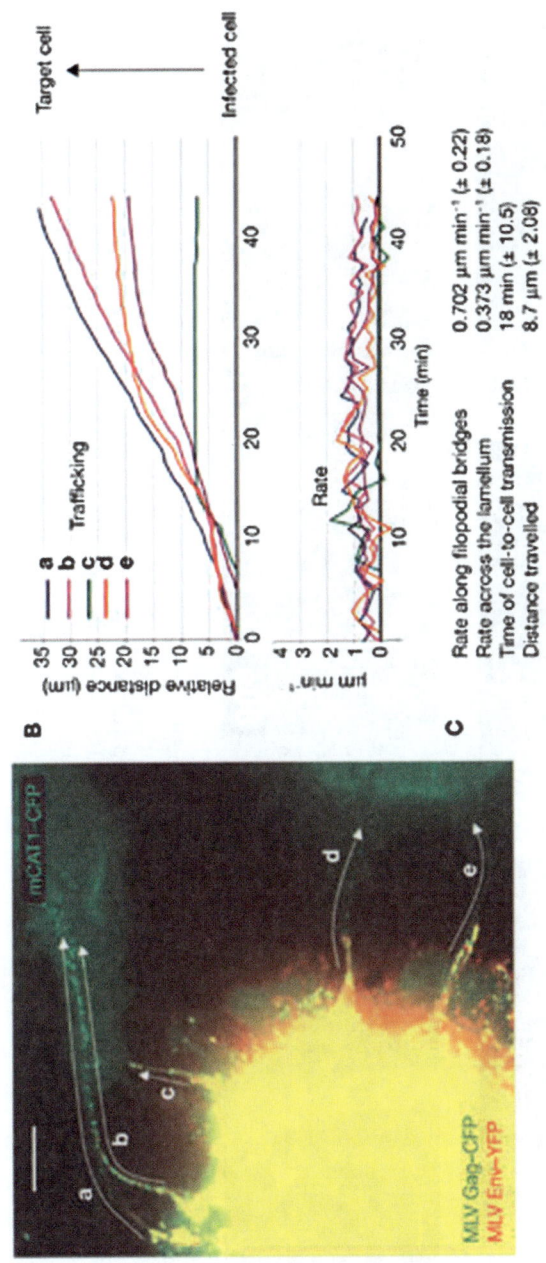

Figure 7 An example of the analysis of movement of fluorescently labeled viral particles labeled with two colors along with the labeled viral receptor (mCAT1) from cell to cell along cell–cell bridges (cytonemes) using live cell fluorescence microscopy. (A) Cos-1 cells generating infectious MLV labeled with Gag–CFP (green) and Env–YFP (red) were co-cultured with target XC cells expressing mCAT1–CFP (green). To illustrate the overall movement of viruses from cell to cell, 22 frames of a time-lapse movie were superimposed. Arrows indicate the paths of five viral particles (a–e) undergoing transmission. (B) Single-particle tracking of particles a–e [shown in (A)] moving from the infected cell towards the non-infected target cells. (C) Average rates of particle movement, average transmission time and average distance traveled for 117 MLV particles undergoing cytonemal transmission. From Figure 1 of reference 34, with permission.

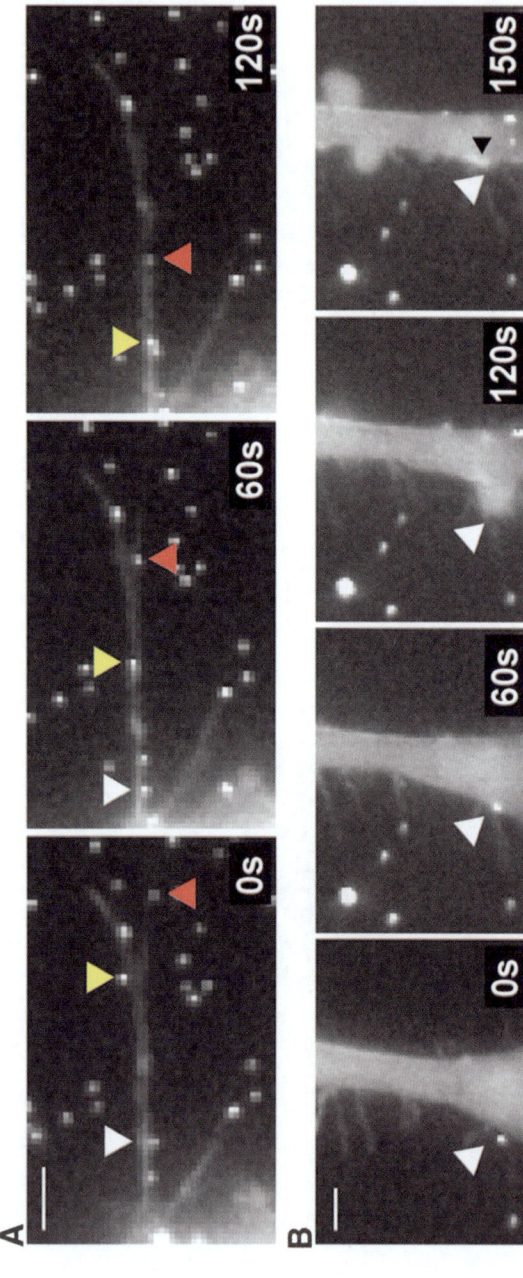

Figure 8 The use of live cell imaging with fluorescently labeled virions and actin GFP in the cells, to follow the changes in the cell membrane and underlying cytoskeleton after vaccinia virus attachment. The figure shows virion movement and membrane perturbation during MV entry in two time lapse series. (A) Enhanced yellow fluorescent protein (EYFP)–CORE–vaccinia viruses were added to cells expressing GFP–actin. Arrowheads highlight virions. (B) As in (A). The virion of interest is indicated by the white arrowheads, showing the blebbing of the membrane and changes in the actin structure. The actin patch at the site of bleb collapse is indicated by a black arrowhead. Scale bars, 2 μm. From Figure 1 of reference 38, with permission.

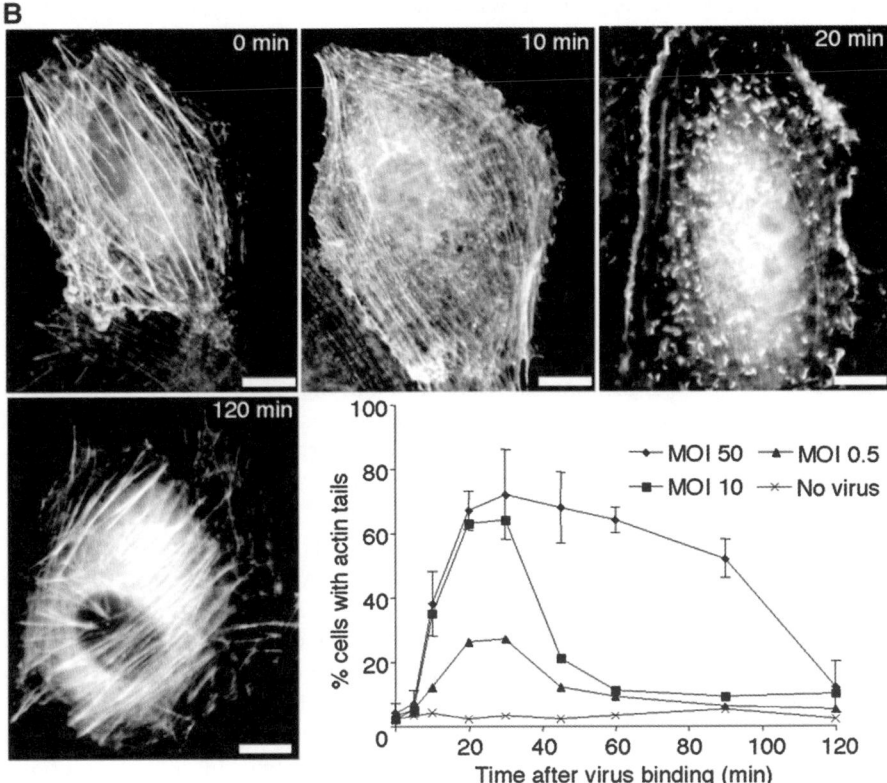

Figure 9   The analysis of SV40 binding and entry into live cells and the effects on the actin cytoskeleton. Cells expressing GFP–actin were incubated with labeled SV40 particles and showed that, initially, most actin is present in stress fibers. Then, actin foci appear and, subsequently, actin tails. The number and intensity of stress fibers are correspondingly reduced (20 min). After further incubation, actin tails disappear and stress fibers reappear (120 min). At lower multiplicity of infection (MOI), the actin cytoskeleton changes are less dramatic. Scale bars, 10 μm. From Figure 2 of reference 41, with permission.

apparatus (Figure 2). To follow the locations of viruses in the endosomal systems, a variety of intracellular markers that can be labeled with fluorescent markers.[6,42,43] Among the many markers that have been examined in live cells, the Rab proteins have been particularly useful. There are >100 Rab proteins within vertebrate cells and those are small GTPases that mediate membrane recruitment and retention within different endosomal or other cellular compartments.[44] These mostly function normally when conjugated with GFP or other fluorescent proteins and both dominant-negative and constitutively active forms of the proteins can be prepared or their expression can be suppressed by siRNAs. Commonly used markers include Rab5 as a label for the early endosome, Rab7 associated with the late endosome and multivesicular endosomes and Rab11 associated with the recycling compartments (Figure 5). Many endosomes can occur in tubular–vesicular forms where different portions of the extended membrane structures are associated with different Rab proteins and transitional forms of some endosomes can be marked by more than one Rab protein (*e.g.* as reported by Lakadamyali *et al.*[13]). The use and analysis of these markers can be complex, as over-expression of the wild-type proteins itself can alter the membrane distribution in the cells and modify the trafficking dynamics of the endosomal system. Co-localization of virus particles with particular markers allows the locations of the virus to be identified. However, the methods used for confirming true co-localization and residence can be difficult in some cases. Many different endosomal vesicles accumulate in the pericentriolar region around the microtubule organizing center. Confocal microscopy allows a higher spatial resolution and more accurate co-localization by reducing the amount of overlapping label that is detected.

For many vesicular compartments, further proof of the association is obtained by showing co-movement of the viral particle fluorescence and that of the endosomal marker protein.[45,46] Movement of intracellular vesicles is commonly seen in live cell studies and the vesicles containing virus particles or components may be moved around within the cytoplasm on microtubules by molecular motors dynein and kinesin or by myosin-2 associated with transport on actin microfilaments. The vesicles may move in a reversible fashion as motors engage or disengage from the microtubules, or they can change direction as they engage different molecular motors. This movement can be readily observed in live cells and labeled viral particles can be tracked as they move in vesicles within the vesicular network. To determine the involvement of the cytoskeleton in the endosomal trafficking, the cells may also express GFP–tubulin or GFP–actin to label cytoskeletal components.

8   Low pH, Membrane Fusion and Other Plasma Membrane or Intra-vesicular Events

Other properties of the virions can also be determined using microscopic approaches, including following the pH of the compartments that the virus enters and also the time of fusion or entry of the virion into the cytoplasm. Labels that are sensitive to low pH include fluorescein and CypHer, which are quenched by the low pH. The pH is more accurately determined from the ratio of the fluorescence of the pH-sensitive dye and another that fluoresces at a different wavelength that is not pH sensitive (*e.g.* Texas Red, Cy or Alexa dyes).[47]

As described above, fusion of enveloped virus membranes with the cellular membrane can also be observed directly by live cell microscopy and in some cases that can be directly correlated with the compartment that the particle is in or the pH of that compartment. For example, the entry and fusion of influenza was directly examined by labeling the membrane of purified virus with the fluorescent dye R18, at levels where it was partially self-quenched, and then

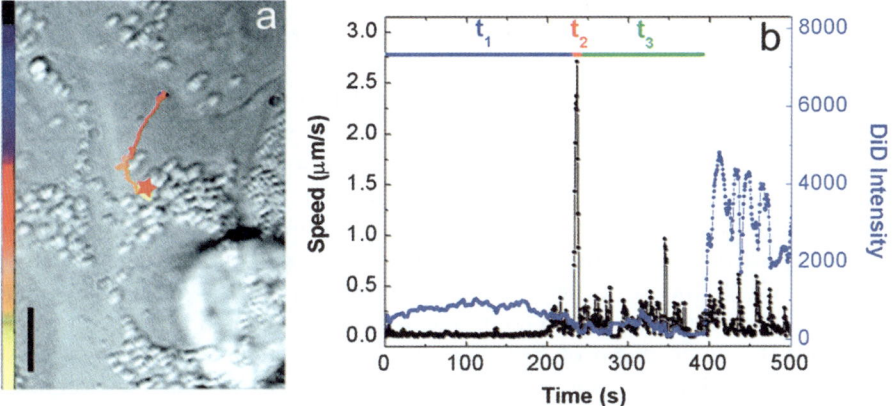

Figure 10 An example of the ability to track the movement and fusion of enveloped virus particles in cells, showing the tracking the transport and fusion of individual influenza viruses. (A) The trajectory of a DiD-labeled virus inside a cell. The color of the trajectory codes time with the colored bar indicating a uniform time axis from 0 s (black) to 500 s (yellow). The red star indicates the fusion site. (Scale bar: 10 μm.). (B) Three stages of movement were defined and are shown as the time trajectories of the velocity (black) and the DiD fluorescence intensity (blue) of the virus. t_1, t_2 and t_3 are the durations of stages I, II and III, respectively. Stage II movements include the rapid unidirectional translocation from the cell periphery to the perinuclear region. Stage I is then defined as the period before this transient motion and stage III is defined as the period after stage II but before fusion. From Figure 2 of Reference 15, with permission.

following that fluorescence into cells.[15] When fusion occurred, the mixing of the viral and endosomal membrane allowed the dye to diffuse into the endo-somal membranes, giving an increase in intensity of the fluorescence upon dequenching of the dye, identifying the time and location of fusion event (Figure 10).

In vitro the dynamics of the fusion process can be dissected further, into the initial hemifusion of the outer leaflets of the two membranes, and this may go on to give a complete fusion or sometimes that can reverse to give dissociation of the virus and the cell membranes. By combining this monitoring with labels on intracellular compartments, the specific site of the membrane fusion event could be monitored.[16,32]

9 Trafficking of Viral Components Within the Cytoplasm – Role of the Cytoskeleton in the Direct Movement of the Viral Components

After the release of the virus nucleoproteins into the cytoplasm, the particle may be able to replicate in that location within the cell or may be transported to other sites within the cytoplasm or to the nuclear pore from where the entire particle or the genome and sub-components can enter the nucleus. The cytoplasm is structured and relatively viscous and particles the sizes of viruses or their nucleocapsids would not be expected to diffuse at significant rates. Active transport of virus nucleocapsids has been reported for herpesviruses, adenoviruses and retroviruses,[20,48–50] and this is also likely for other viruses.[10] Trafficking of herpesvirus capsids or other components, or of other viruses within axons is also seen, with movement between the cell body and the axonal termini (sometimes in both directions), over distances from centimeters up to meters.[51] GFP-labeled particles can be followed within the cytoplasm of the live cells when the viral or cell components are labeled with fluorescent molecules and the involvement of specific molecular motors determined by the use of particular inhibitors; for the dynein motors, these include the dominant interfering version of the dynactin subunit dynamitin, a component of the dynein structure.[19,48] Viral microtubular trafficking of free particles is saltatory and often bidirectional, resulting from the engagement and release of dynein or kinesin motors to the microtubules within the cell. The transport process can also be reconstructed *in vitro* by the assembly of microtubules, cytoplasmic components and a suitable energy source; for example, with herpesvirus capsids the trafficking was observed directly on microtubules on microscope slides.[52]

10   Nuclear Transport and Entry

Many viral genomes enter the nucleus for their replication and this process can also be followed using either widefield or confocal microscopic analysis. The nuclear periphery or the nuclear pores may be labeled with GFP-fused lamin B receptor, lamin A/C protein or nuclear porins, respectively, allowing the virus components to be localized in the region of the nucleus. Many studies have involved the analysis of cells that are fixed and stained for viral or cellular proteins, but some studies that examined the transport of viral proteins into or within the nucleus of live cells have been reported.[23,53] In many cases viral proteins show a balance between the import and export processes and understanding the true distribution requires the analysis of cells where each process is blocked-for example Crm-dependent export can be blocked by leptomycin B1 and import blocked by microinjection of the lectin wheat germ agglutinin or by antibodies against nuclear pore proteins such as nucleoporin.[54]

11   Summary and Conclusions

Following the trafficking of virus particles on and within live cells, along with the analysis of cellular components, has revealed many new details of the infectious entry pathways of viruses. These include the spatial and temporal dynamics of the entry processes, including the fast and slow movement of particles both inside and outside the cells and the specific routes followed by individual particles.

Despite concerns that the virions being followed by microscopy might be in non-productive pathways, in general where direct assays of the role of individual steps in infection have been conducted these appear to be functional for virus infection. Newer imaging technologies are being developed, including higher resolution microscopy, faster image processing, additional fluorescent proteins to allow multicolored imaging and methods for detecting co-localization through FRET analysis or labeling particles or their components within cells. These will no doubt give more information in the future.

References

1. A. D. Hoppe, S. Seveau and J. A. Swanson, *Cell Microbiol.*, 2009, **11**, 540.
2. T. Zal, *Adv. Exp. Med. Biol.*, 2008, **640**, 183.
3. T. Yanagida, M. Iwaki and Y. Ishii, *Philos. Trans. R. Soc. Lond. B*, 2008, **363**, 2123.
4. P. Watson, A. T. Jones and D. J. Stephens, *Adv. Drug Deliv. Rev.*, 2005, **57**, 43.
5. D. J. Stephens and V. J. Allan, *Science*, 2003, **300**, 82.
6. M. N. Seaman, *Cell. Mol. Life Sci.*, 2008, **65**, 2842.
7. E. M. Damm and L. Pelkmans, *Cell. Microbiol.*, 2006, **8**, 1219.
8. J. Gruenberg and F. G. van der Goot, *Nat. Rev. Mol. Cell. Biol.*, 2006, **7**, 495.
9. D. Perrais and C. J. Merrifield, *Dev. Cell.*, 2005, **9**, 581.
10. P. L. Leopold and K. K. Pfister, *Traffic*, 2006, **7**, 516.
11. M. Marsh and A. Helenius, *Cell*, 2006, **124**, 729.
12. E. M. Campbell and T. J. Hope, *Trends Microbiol.*, 2008, **16**, 580.
13. M. Lakadamyali, M. J. Rust and X. Zhuang, *Cell*, 2006, **124**, 997.
14. L. Pelkmans, T. Burli, M. Zerial and A. Helenius, *Cell*, 2004, **118**, 767.
15. M. Lakadamyali, M. J. Rust, H. P. Babcock and X. Zhuang, *Proc. Natl. Acad. Sci. USA*, 2003, **100**, 9280.
16. L. Wessels, M. W. Elting, D. Scimeca and K. Weninger, *Biophys. J.*, 2007, **93**, 526.
17. I. Le Blanc *et al.*, *Nat. Cell Biol.*, 2005, **7**, 653.
18. T. H. Ch'ng, P. G. Spear, F. Struyf and L. W. Enquist, *J. Virol.*, 2007, **81**, 10742.
19. K. Dohner, K. Radtke, S. Schmidt and B. Sodeik, *J. Virol.*, 2006, **80**, 8211.
20. D. McDonald *et al.*, *J. Cell Biol.*, 2002, **159**, 441.
21. M. J. Lehmann, N. M. Sherer, C. B. Marks, M. Pypaert and W. Mothes, *J. Cell Biol.*, 2005, **170**, 317.
22. S. Cabantous, T. C. Terwilliger and G. S. Waldo, *Nat. Biotechnol.*, 2005, **23**, 102.
23. N. Arhel *et al.*, *Nat. Methods*, 2006, **3**, 817.
24. N. J. Arhel and P. Charneau, *Methods Mol. Biol.*, 2009, **485**, 151.
25. Y. Klingen, K. K. Conzelmann and S. Finke, *J. Virol.*, 2008, **82**, 237.
26. A. Albanese, D. Arosio, M. Terreni and A. Cereseto, *PLoS ONE*, 2008, **3**, e2413.

27. S. Bernacchi, G. Mueller, J. Langowski and W. Waldeck, *Biochem. Soc. Trans.*, 2004, **32**, 746.
28. B. Brandenburg *et al.*, *PLoS Biol.*, 2007, **5**, e183.
29. U. F. Greber, *Cell. Mol. Life Sci.*, 2002, **59**, 608.
30. H. Ewers *et al.*, *Proc. Natl. Acad. Sci. USA*, 2005, **102**, 15110.
31. M. Schelhaas *et al.*, *PLoS Pathog.*, 2008, **4**, e1000148.
32. D. L. Floyd, J. R. Ragains, J. J. Skehel, S. C. Harrison and A. M. van Oijen, *Proc. Natl. Acad. Sci. USA*, 2008, **105**, 15382.
33. J. L. Smith, D. S. Lidke and M. A. Ozbun, *Virology*, 2008, **381**, 16.
34. N. M. Sherer *et al.*, *Nat. Cell Biol.*, 2007, **9**, 310.
35. R. Dixit, V. Tiwari and D. Shukla, *Neurosci. Lett.*, 2008, **440**, 113.
36. L. Pelkmans, J. Kartenbeck and A. Helenius, *Nat. Cell Biol.*, 2001, **3**, 473.
37. V. Pietiainen *et al.*, *Mol. Biol. Cell*, 2004, **15**, 4911.
38. J. Mercer and A. Helenius, *Science*, 2008, **320**, 531.
39. O. Meier *et al.*, *J. Cell Biol.*, 2002, **158**, 1119.
40. B. Amstutz *et al.*, *EMBO J.*, 2008, **27**, 956.
41. L. Pelkmans, D. Puntener and A. Helenius, *Science*, 2002, **296**, 535.
42. E. V. Vassilieva and A. Nusrat, *Methods Mol. Biol.*, 2008, **440**, 3.
43. H. Cai, K. Reinisch and S. Ferro-Novick, *Dev. Cell*, 2007, **12**, 671.
44. B. L. Grosshans, D. Ortiz and P. Novick, *Proc. Natl. Acad. Sci. USA*, 2006, **103**, 11821.
45. E. M. Damm *et al.*, *J. Cell Biol.*, 2005, **168**, 477.
46. C. E. Harbison, S. M. Lyi, W. S. Weichert and C. R. Parrish, *J. Virol.*, 2009, **83**, 10504.
47. M. Lakadamyali, M. J. Rust and X. Zhuang, *Microbes Infect.*, 2004, **6**, 929.
48. K. Dohner *et al.*, *Mol. Biol. Cell*, 2002, **13**, 2795.
49. S. A. Kelkar, K. K. Pfister, R. G. Crystal and P. L. Leopold, *J. Virol.*, 200, 78, 10122.
50. H. Mabit *et al.*, *J. Virol.*, 2002, **76**, 9962.
51. G. A. Smith, S. P. Gross and L. W. Enquist, *Proc. Natl. Acad. Sci. USA*, 2001, **98**, 3466.
52. G. E. Lee, J. W. Murray, A. W. Wolkoff and D. W. Wilson, *J. Virol.*, 2006, **80**, 4264.
53. T. O. Ihalainen *et al.*, *Cell Microbiol.*, 2007, **9**, 1946–59.
54. M. Yamada and H. Kasamatsu, *J. Virol.*, 1993, **67**, 119.

CHAPTER 3

Probing Viral Capsids in Solution

BRIAN BOTHNER AND JONATHAN K. HILMER

Department of Chemistry and Biochemistry, Montana State University, Bozeman, MT 59717, USA

1 Introduction

A great deal of what we know about viruses has come from the analysis of capsid structure. Early structural information came from electron micrographs and X-ray diffraction patterns which displayed the shape and general quaternary organization of particles. Then, with the maturation of X-ray crystallography, the elegance and intricacies of viral capsids were revealed. Icosahedral capsids represent some of the most stunning structural models produced and even non-scientists are intrigued by their symmetrical beauty. The symmetry and size of icosahedral particles also made them good subjects for cryo-electron microscopy (cryo-EM) and numerous technical advances were driven by structural virology. In one respect, the structural models may have been too convincing, because many virologists were persuaded to think of the protein capsid only as the rigid shell depicted in the static images. However, even as the structural models were shaping the way virologists and structural biologists thought about the particles, biochemical evidence was accumulating to suggest that, in solution, there was more to these structures than met the eye.

Capsid proteins are responsible for an array of functions critical for completion of a virus lifecycle. These include particle assembly, intracellular transport, genome protection and release and, in the case of non-enveloped viruses, receptor binding. It is now clear that protein dynamics have an essential role in each of these steps. The most obvious indication that capsids are active

RSC Biomolecular Sciences No. 21
Structural Virology
Edited by Mavis Agbandje-McKenna and Robert McKenna
© Royal Society of Chemistry 2011
Published by the Royal Society of Chemistry, www.rsc.org

structures comes from the dramatic protein rearrangements which can occur after assembly and before the particle becomes fully infective. These maturation-associated changes have been caught in sequential still-life models for a number of viruses. Each step in the progression represents a distinct capsid form that can be isolated and often has unique physical properties. A second more subtle, yet equally important, form of dynamics exists in particles at each step along this progression. Unlike the large-scale maturation-induced changes that can be kinetically trapped, the second mode of dynamics is a solution-phase equilibrium process and cannot be captured by structural models. Equilibrium dynamics are important for each of the functions mentioned above and change throughout the maturation process. The two categories of dynamics differ in the physical characteristics of the dynamic motion, and also the techniques that can be used to study them. This chapter focuses on viruses with icosahedral capsids, beginning with discussions of quaternary dynamics and solution-phase equilibrium dynamics. It concludes with an overview of techniques that have been used to study virus particles in solution, the information that can be obtained from such experiments and future directions.

2 Quaternary Dynamics

Large-scale quaternary rearrangement of subunits in icosahedral capsids is generally associated with maturation events or swelling and contraction induced by solution conditions.[1] These changes can involve major alterations of capsid size or geometry, but are not necessarily accompanied by large changes in secondary or tertiary elements. The capsid as a whole undergoes a symmetric transformation, with radial translocation and/or subunit rotation, creating the net effect of a larger or smaller capsid that still retains its overall symmetry and general features (Figure 1). In a biological context, these are generally one-way events triggered by packaging of nucleic acids,[2] trafficking solution conditions[3,4] or receptor binding.[5] Although it may be possible in some cases to reverse the process,[4] the steps are not necessarily populated as an equilibrium in solution. In many cases, the one-way nature of quaternary transitions serves as a regulatory gateway for structural maturation that coincides with key events in the viral lifecycle. Because the particles can become kinetically trapped in a particular form, quaternary dynamics have been well characterized using classical structural techniques.[1] This has permitted detailed comparison of the pre- and post-transition structures and in some cases the transition itself has been directly observable.

Maturation-associated Dynamics in a Bacteriophage

Icosahedral capsids often have a spherical form after assembly, adopting their final quasi-equivalent form upon maturation. One of the most interesting and best detailed characterizations of this process involves the bacteriophage HK97. HK97 is a dsDNA λ-like coliophage with a T = 7 capsid possessing a

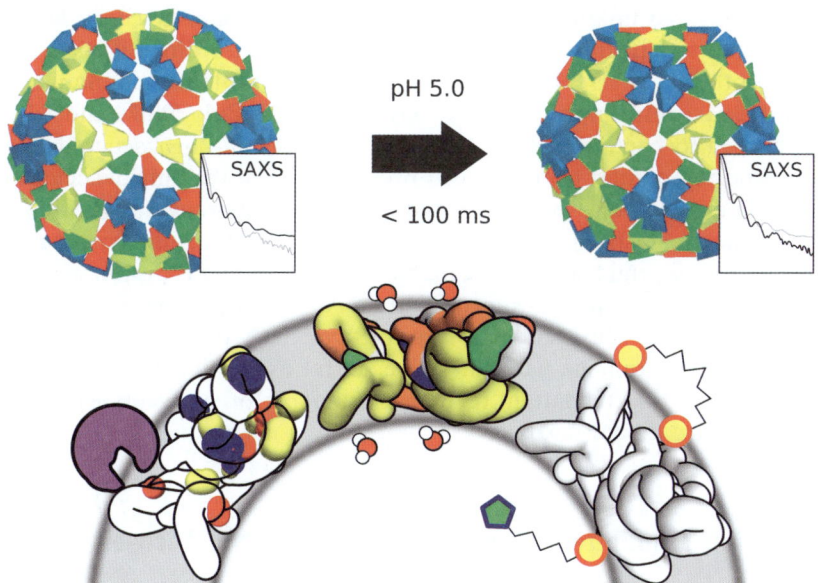

Figure 1   Quaternary and localized equilibrium dynamics. Top: maturation of the *Nudaurelia capensis* omega virus VLP. The contraction and shifting of the subunits reduce the diameter by 15% and occur very rapidly, within 100 ms. The transition was studied by SAXS (data shown in insets as the scattering vector), which revealed substantial changes in the radial density distribution. Adapted from Canady *et al.*[4] Bottom: common methods of probing localized equilibrium dynamics: a cartoon structure of HRV14 VP1–4 overlaid with three different data sets. On the left, observed cleavage sites from limited proteolysis are mapped to the transparent structure. The first cleavages are red (5 min) followed by yellow (10 min), then blue (60 min). Data from Lewis *et al.*[27] At center, HX experiments probe smaller scale dynamics of HRV14. The first labeled sites are blue, followed by green, yellow, orange and red (time-scale ranging from 12 s to > 30 hours). Data from Wang and Smith.[61] On the right, chemical labeling reactions measure dynamics at approximately the same amplitude and rate as proteolysis and HX. Chemical probes can be doubly reactive with variable linkers to measure distance constraints or functionalized with fluorescent tags to detect single reactions.

portal structure in place of one pentamer.[6] Progression from the initial Prohead form to the final mature Head II has been followed *in vitro* by expressing the major capsid protein GP5 and the viral protease GP4. In the mature form of the capsid, GP5 proteins are topologically linked together, creating protein chain-mail.[7] Maturation begins with digestion of the N-terminal 103 amino acids by the viral protease. The subsequent quaternary rearrangement involves subunit rotation and radial expansion (with the diameter going from 450 to 650 Å).[3,8] This dramatic rearrangement is also accompanied by a thinning of the capsid cross-section and a transition from a nearly spherical shape to a much more angular polyhedron. Once expanded, the protein shell uses autocatalytic cross-linking between Lys and Asn side-chains to create hexamers and pentamers that are concatenated together, but not directly cross-linked. The final product is a remarkably thin, yet robust, protein shell. Similar Pro-head to Head transitions have been described for P22, Phi6, Phi29 and T7,[1] and is not limited to bacteriophages, as members of the *Adenoviridae* and *Herpes-viridae*[9] undergo similar transitions. However, HK97 is the only particle known to use this cross-linked chain-mail and the precise choreography required to generate the final structure makes this a jewel in structural virology. The mature capsid is a rigid structure, capable of withstanding the estimated 60 atm of pressure exerted by the packaged DNA.[6,7] As such, the dynamics which were important for assembly and maturation have been quenched, but because HK97 is a bacteriophage with an active portal assembly, the mature capsid does not need to participate in receptor binding or other events which might require conformational flexibility.

Maturation-associated Dynamics in Small RNA Viruses

A second recurring theme in maturation-associated dynamics is the use of an autocatalytic proteolysis event to trigger maturation. For small RNA viruses of the *Picornaviridae*, *Nodaviridae* and *Tetraviridae*, assembly of the procapsid positions the subunits such that autohydrolysis of the protein chain occurs. Cleavage of the subunit acts as a molecular switch, preventing a reversal of the assembly process allowing access to local energy minima not populated by the procapsid. This event has been described with some detail for Flock House virus (FHV)[10] and is a requirement for generation of infective particles.[11] Cleavage of the alpha capsid protein in FHV produces a 363 amino acid beta-protein, which comprises the capsid shell and cellular receptor binding region and the 44 residue gamma-peptide which is situated inside the capsid shell next to the RNA in the structural models. Both *in vivo* and *in vitro* experiments with FHV have shown that the gamma-peptide (found in noda- and tetraviruses) is involved in membrane penetration/disruption,[12] analogous to the functional role of VP4 in picornaviruses.[13]

The tetraviruses provide an interesting example of the relationship between quaternary dynamics and autohydrolysis. *Nudaurelia capensis* omega (NωV) is the best studied member of this family of T = 4 viruses that only infect members of the Lepidoptera. In the case of the tetraviruses, pH can be used to induce a transition from the procapsid to a smaller mature capsid: the structures of both forms of NωV have been determined at moderate resolution.[14-16] In addition to a radial contraction of 16%, subunit rotation and tertiary changes in the internal helical region occur in the transition. These rearrangements can be reversed by simply raising the pH, provided that the autohydrolytic cleavage has not proceeded beyond 15% of the subunits.[16] The driving force for structural rearrangement is electrostatic and it appears that pH may be the trigger for maturation *in vivo* also, based on recent data showing that infection by NωV induces apoptosis and a decrease in intracellular pH in insect midgut cells.[17] As mentioned above, NωV uses a capsid protein processing event that is similar to FHV in which the C-terminal region is hydrolyzed by an asparagine side-chain-mediated attack on the peptide backbone.[10] A mutant form of NωV in which the catalytic function has been abrogated can repeatedly transition between forms, conclusively demonstrating that hydrolysis disconnects the driving force for the procapsid-capsid transition.[18] As a model system, NωV is significant because it was the first virus for which the conformational change between a capsid intermediate and the mature form were monitored in real time. This was initially accomplished using solution small-angle X-ray scattering.[4] Subsequently, a number of biophysical techniques have been applied to characterize both forms and the transition, including fluorescence,[19] chemical reactivity,[20] proteolysis[21] and FT-IR.[21] Considering the size and complexity of a $T = 4$ capsid, the 240 subunits react surprisingly rapidly upon pH reduction from 7.5 to 5.0, completing the rearrangement within 100 ms. Hydrolysis occurs much more slowly, having a half-life of hours.[16]

Structural Transitions

One of the unifying themes of quaternary dynamic transitions is their use as a delineator between distinct structural states. These states often have dramatic differences in structural stability and the end product of the transition is highly tailored for its environment. In the case of HK97, the end product is a highly robust capsid capable of packaging large amounts of DNA. Viruses which lack a portal assembly require a capsid with the functional and structural flexibility to dock with the host and release the genome using only the capsid shell subunits. Such functionality can be conferred by making the end product of maturation a metastable state: a local energy minimum robust enough to serve as a delivery vehicle and yet primed to release its contents upon the right environmental signal. Two examples of mature metastable particles are polio[22] and NωV.[21] In many cases, metastability can be inferred from available data on the infection process without biophysical confirmation. One such example involves the puzzling case of parvoviruses, a group of small single-stranded DNA viruses with a non-enveloped $T = 1$ icosahedral capsid. Parvoviruses all use receptor mediated endocytosis for internalization. Lacking an envelope that would allow membrane fusion, they gain entry into the cytoplasm by deploying a phospholipase domain.[23] Structural models clearly indicate that this domain resides on the inside of the capsid.[24] Release of the 130 amino acid domain occurs after receptor binding and endosomal acidification without particle disruption.[25] Heating can be used as a surrogate trigger, again without particle disruption. Analyses of the structural models reveal no pores or channels that could accommodate such a large domain. Mutational analysis[23,26] and post-release cryo-EM data identify regions that affect translocation and are altered by the process, respectively; however, where and how a folded globular domain of ~ 15 kDa extricates itself from the capsid interior is still a mystery.

3   Solution-phase Equilibrium Dynamics

Whereas maturation events largely involve quaternary rearrangements, some-times with tertiary structural triggers, solution-phase dynamics is characterized by rapid equilibrium motion with localized perturbations in the tertiary structure of the subunits (Figure 1). These motions have been described as 'breathing',[27] but the motion may involve only a subset of the subunits or capsid population and there is no evidence that it is a symmetric transition. Due to the rapid equilibrium, it is not possible to isolate populations of just one state and crystal packing forces or reduced temperatures quench these motions, severely limiting the applicability of X-ray crystallography and cryo-EM as viable study tools. Instead, a variety of solution-phase techniques have been applied to detect and quantify equilibrium dynamics. The technical limitations and difficulty in performing these assays has hampered a full characterization of equilibrium structural dynamics in capsids.

Virus Particles are Dynamic

Dynamic protein regions in assembled particles were first identified in NMR experiments on the plant virus cowpea chlorotic mottle virus (CCMV).[28] The observed dynamics were on the time-scale of 1–10 ns and were associated with the N-terminus of the capsid protein. The number of dynamic side-chains dramatically increased in capsids without RNA and it was proposed that RNA induced the formation of internal alpha-helices. This suggested to the authors that the N-terminal domain must be important for assembly of infectious particles. A different form of rapid dynamics was observed for picornaviruses. In poliovirus, antibodies raised against intact particles actually recognized a domain that was on the internal surface of the static structural model.[13,22] Further studies on poliovirus using antibodies demonstrated that the externa-lization of internal domains on VP1 and VP4 could be reversed and involved regions important for cell entry.[29–31] Antiviral drugs that increased the thermal stability of capsids, such as the hydrophobic WIN compounds (discussed in Chapter 16), are known to prevent transition out of the metastable phase in picornaviruses.[32] These observations for poliovirus and other picornaviruses led to the concept of the mature particle as a metastable structure, as discussed above. Later computational studies using molecular dynamics suggested that the increased stability had an entropic basis, diminishing the entropy gain of uncoating.[33,34]

Proteolysis and Mass Analysis

Initially, the extent of the dynamics responsible for the reversible exposure of internal domains was not fully appreciated. A striking experiment that changed this involved the use of limited proteolysis and mass spectrometry. It had recently been demonstrated that this combination of a standard biochemical technique with a powerful analytical platform could be used to investigate capsid protein dynamics.[35] Time-dependent analysis of a protease reaction using mass spectrometry provided precise identification of the cleavage sites and qualitative kinetic data on the solution-phase dynamics of the capsid. Researchers in Siuzdak's group at The Scripps Research Institute showed that the dynamic motion of the internal VP4 protein in human rhinovirus made it more cleavage-accessible than regions on the exterior of the capsid. Importantly, by using proteases as a probe, the whole capsid was screened for dynamics without bias towards any specific site. When carried out in the presence of the WIN stabilizing compounds, the proteolysis experiments showed a loss of the dynamic 'breathing' motion.[27] These results definitively established that icosahedral capsids exist as an ensemble of conformations in solution, only one of which is captured in the structural models.

The application of proteolysis and mass spectrometry to the study of capsid protein dynamics was initially unintentional. Intrigued by the potential of identifying antigenic peptides on the surface of virus particles, Bothner and co-workers exposed FHV, a member of the *Nodaviridae*, to proteases and identified the released peptides using MALDI and electrospray mass spectrometry. Surprisingly, the very first peptides to be generated mapped to the interior of the capsid, positioned next to the RNA.[35] Following extensive control experiments to assure that a subpopulation of disrupted particles was not the source of the released peptides, it was accepted that nanometer-range, reversible protein motion was responsible for the exposure of internal domains on the capsid surface and subsequent protease-mediated cleavage. This indicates that FHV particles are present as an ensemble of conformers in solution and is consistent with the fact that transient exposure of the internally located amphipathic gamma-peptide on the particle surface mediates interaction with host cell membranes.[12]

In addition to their contribution to the understanding of the maturation process in icosahedral capsids, tetraviruses have proven to be an interesting model system for the study of solution-phase properties. As described above, their T = 4 capsids undergo an autocatalytic cleavage during maturation that is very similar to what occurs in FHV. The original structural model of NωV had distinct similarities to FHV with respect to the arrangement of gamma-peptides around the fivefold axes. It was therefore reasoned that the gamma-peptide would also be highly dynamic and exposed to the capsid surface. However, the proteolytic susceptibility and chemical reactivity were much less than those seen in FHV.[21] Subsequent refinement of the X-ray structure revealed an additional set of helices contributed by the N-termini of the protein, creating a previously undetected interaction with the gamma-peptides at the fivefold axes.[14] In order for the gamma-peptides to transiently sample external positions at the fivefold axes, a strong set of interactions would need to be broken. The work with NωV and *Helicoverpa armigera* stunt virus (HaSV) demonstrates the power of solution-based approaches in cases where structural data are not available or may lead to incorrect conclusions.

4 Methods for Studying Viruses in Solution

Equilibrium-type dynamic motions within viral capsids are inherently difficult to study, as it is not possible to kinetically trap conformational isoforms for static structural studies. When a population can be found in a nearly homogeneous state, it will typically be the low-energy form that dominates, whereas it is the higher-energy, 'activated' form which is of most interest as the functional species. Experimentally, methods need to be specific enough to resolve the difference between the two states and sensitive enough to provide accurate measurements despite a heavy population bias for one form over the other (Figure 2). Although these challenges have by no means been met in entirety, a wide variety of approaches have proven successful for detecting capsid protein dynamics in solution.

Due to their large size and high degree of symmetry, the application of standard biophysical approaches for studying protein dynamics in viral capsids has been challenging. For example, NMR, electron paramagnetic resonance (EPR) and Förster resonance energy transfer (FRET) are standard methods used to investigate protein dynamics in mono- and multimeric proteins. With respect to the large supramolecular complexes that virus particles are, only NMR has made a significant contribution. However, this has not stopped researchers from trying and steady advances are being made using both standard biophysical techniques and creative new approaches (Table 1). Regardless of the approach, certain precautions must be taken to assure the validity of the results. Foremost is the assurance of a homogeneous population of intact particles. Depending on the experiment, even a few percent of unassembled

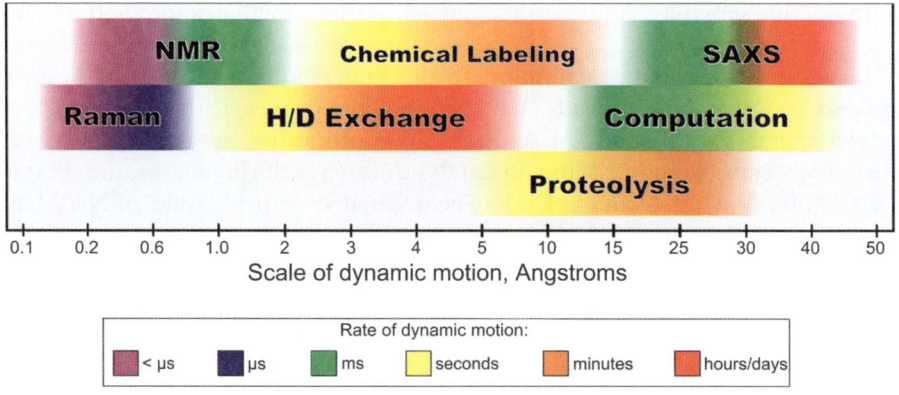

Figure 2   Time-scale and amplitude of protein dynamics that can be measured using different techniques. The horizontal axis indicates the scale of the dynamic motion that can be probed by each technique in nanometers. Some methods, such as Raman spectroscopy, may be combined with other tools such as HX to change their usefulness. Each method also measures a particular rate of dynamics (see color key). The time and amplitude boundaries of the techniques shown here are not necessarily from inherent limitations, but indicate past applications of the techniques specifically for the study of viral capsids.

Table 1 Techniques for studying virus particles in solution.

Method	Information available	Drawbacks	Benefits
NMR	Structure, detection of HX time-scale and range of motion	Peak broadening with particle size Difficulty assigning residues Requires concentrated samples	Can be highly specific Bias towards dynamic regions
Raman	Secondary structural elements, detection of HX	Difficult assigning site(s) Low signal-to-noise ratio	Does not requires concentrated protein Can probe low-frequency motions
FRET	Distance constraints between groups	May require site-directed mutation Signal assignment is difficult	Highly specific Accurate distance measurements Time-scale with fluorescence lifetime experiments
Hydrogen–deuterium exchange (HX)	Backbone dynamics at a resolution of several residues	Quantitative experiments are technically difficult Difficult to assign sites for protein isoforms of varying length	High spatial resolution Time-scale data are possible Native proteins and protein mixtures
Chemical reactivity	Local chemical environment	Limited to availability of reactive side-chains	Highly specific Many probes available
SAXS	Quaternary structure, real-time data on changes	Requires synchrotron light source	Can give very low-resolution structures: size and approximate shape
SANS	Low-resolution structures with compositions	Requires neutron source and detector: few facilities available	Differentiates between lipids, nucleic acids and protein
Computational simulation	Limited only by computational resources	Viral particles push the limits of computational scale Requires known structure	Avoids physical limitations of experimental setup Emergent field with exponential growth
Proteolysis	Backbone dynamics at a resolution of several residues	Protein sequence limits possible enzymes Simultaneous cleavages complicate kinetic interpretation	Provides detailed kinetics Probes larger dynamic motion Does not require known protein structure Tolerant of solution conditions

protein could seriously bias the results. Establishing sample quality can be a non-trivial task, requiring size-exclusion chromatography, light scattering and gradient centrifugation, along with a well-established understanding of capsid stability. Meeting these requirements is a necessity that may rule out some experimental techniques, due to protein concentration limitations or solution conditions. Even when a capsid is stable under the required conditions, the experiment itself may have destabilizing effects. Small-angle X-ray scattering (SAXS), resonance-based and covalent labeling experiments can be disruptive and have the potential to alter the dynamics, if not destabilize the capsid itself.

Spectroscopy

Historically, NMR, Raman and small-angle scattering techniques have been the dominant spectroscopic approaches to study virus capsids in solution. This should not be taken to imply that these techniques are the only viable ones: FRET, EPR and other methods all have potential advantages, limited primarily by the ability to produce appropriately labeled capsids. Considering the challenges of detecting small sub-populations of highly dynamic regions, additional sensitive techniques such as paramagnetic relaxation enhancement should not be overlooked as potential strategies for examining solution-phase dynamics.

NMR

As discussed previously, NMR has been instrumental in showing that regions of viral capsids have dynamic character in solution.[28] The primary advantage of NMR over other analytical methods is the resolution and content of the data: it is theoretically possible to determine the position and motion of the protein backbone with single-residue precision. However, limits in current instrumentation make this impractical for intact viral capsids, due in part to slow tumbling in solution which dramatically speeds relaxation, crowded spectra because of the large number of atoms and the necessity for isotope-enriched protein concentrations in the millimolar range. NMR does have an advantage in that it selectively detects dynamic regions. Because of the lack of signal from well-ordered regions in a slow-tumbling capsid, any regions with a large degree of localized motion generate a very unique signature that immediately stands out from the background. With successful assignment, the signals from dynamic regions can be tracked in response to solution conditions, maturation state or other perturbations. The early experiments with CCMV used this approach to examine the dynamics at the N-terminus which were quenched in the presence of packaged RNA. More recently, the latest advances in NMR technology have been applied to similar problems to detect dynamic regions in the course of HK97 maturation.[36]

Icosahedral capsids such as CCMV or HK97 continue to present technical challenges to NMR analysis, but another category of viral capsid is much more amenable to such experiments. Filamentous capsids, which are composed of a very large copy number of fairly small monomeric subunits, can be spatially aligned along their axis, either via a magnetic field or with solid-state crystals. Pulsing schemes designed to take advantage of the resultant dipole–dipole interaction can generate high-resolution structures, without the need for particle tumbling and regardless of the large particle size.[37] One of the interesting observations from such work is the extreme lack of dynamics in some filamentous capsids: in the case of the Pf1 bacteriophage, order parameters for alpha-carbons approached the upper limit of 1.0, indicating a very static structure. For comparison, crystalline glycine needs to be cooled to –45 °C to reach such a limit. However, despite the overall high degree of order, some of the side-chains displayed rapid mobility. Residues deep in the interior of the virion were highly dynamic, despite being in proximity to the packaged DNA. This finding contrasts with opposite observations for many icosahedral viruses, in which the presence of nucleic acids decreases the dynamics of residues in the local proximity and sometimes generally throughout the capsid.[38,39]

As with all experiments for detecting dynamic motion, NMR spectroscopy imposes a particular set of criteria for interpreting the results. The time-scale of motion that can be detected by NMR varies depending on the protein and the exact experiment, but in general it involves motion in the microsecond range or faster. Although structural transitions that complete in milliseconds could be considered rapid in the context of maturation and large-scale equilibrium associated motions, they would be on the slow side for most NMR experiments.

Raman

Raman spectroscopy provides several key features that are invaluable to the study of viral capsids in solution. By probing the vibrations of specific bonds, Raman spectroscopy reports on the local environment for each particular signal. Unlike NMR, Raman spectroscopy is far more tolerant of dilute protein solutions and it does not suffer from the scattering effects of large assemblies as does circular dichroism spectroscopy. Raman spectra can be predicted for specific secondary structure elements, allowing approximate assignment of observed spectra to the corresponding helices or sheets and although the single-dimension spectra can become very crowded with overlapping signals, digital difference comparisons between two conditions can be used to isolate unique signals. Side-chains can also generate unique peaks, especially for aromatic amino acids and cysteine.

As a specific probe of secondary structure and local environment, Raman spectroscopy has been used extensively to study the effects of capsid assembly and maturation events.[40–42] For the P22 capsid, maturation from the procapsid to the capsid form produced a substantial number of changes in the side-chain signals without altering the distribution of secondary structure.[41] This suggested that the maturation involved quaternary translation of relatively intact domains in the capsid expansion. Due to its sensitivity to bond vibration,

Raman spectroscopy is also sensitive to hydrogen–deuterium exchange, making it a very specific detection system for time-resolved proton exchange reactions, discussed below. As with NMR spectroscopy, direct Raman measurements depend on the sum population in solution and the time-scale of Raman 'dynamics' is very fast (on the order of bond motion), which necessitates the use of a coupled kinetic probe such as deuterium exchange to detect slower motions.

Small-angle X-ray Scattering (SAXS)

Both Raman and NMR spectroscopy can be highly specific for certain regions in a virus capsid and both can give information on the local environment of the signal in question. Unfortunately, it can also be very time consuming to collect enough spectra to gain sufficient signal-to-noise ratio. In contrast to these methods, SAXS represents an orthogonal approach. The solution-phase complement to X-ray diffraction, SAXS measures the distribution of X-rays scattered at low angles ($< 10°$) from the incident beam. This provides information on the distribution of density within the sample directly from solution. The resulting reconstruction is low resolution (~ 25–$100 \, \text{Å}$), but can be collected in milliseconds. This rapid data collection allows SAXS to be used in a time-dependent manner to study maturation events and capture transient intermediate structures.[4] This ability was critical in the study of HK97 expansion between Prohead II and Head I: data collected at 1 min intervals demonstrated a biphasic transition with two isosbestic points on an intensity–resolution plot.[8] At the present time, the resolution limits for SAXS restrict its ability to detect smaller localized and equilibrium dynamics, but continuing improvements to synchrotron light sources are expanding the scope of this technique. Smaller scale light sources with high brilliance are also now available for local installation.

Small-angle Neutron Scattering (SANS)

The general application of small-angle scattering can also be used with a collimated neutron beam, known as small-angle neutron scattering (SANS). In addition to being less destructive than SAXS, this technique is of particular interest to the study of viral particles due to the sensitivity of SANS to hydrogen, which is nearly silent in SAXS experiments. Hydrogen nuclei interact efficiently with the incident neutrons, giving the solvent a substantial and distinctive scattering length density, which can be controlled according to the ratio of H_2O to D_2O in the solvent. Protein in the sample also scatters, but with a different length density as a function of the average elemental composition. Likewise, nucleic acids and lipids both have unique scattering signatures. Because these scattering length densities have different responses to the presence of deuterium, by performing measurements in a series of H_2O/D_2O ratios, the individual contributions from proteins, nucleic acids and lipids can be discerned. This technique is known as contrast variation and it has been successfully applied to solution-phase structural studies of MS2 bacteriophage.[43] Kuzmanovic *et al.* found that in solution, the RNA is compacted within a radius of 83 Å, despite the fact that the inner radius of protein is 115 Å: the area between the capsid and RNA was composed of 81% water. Separate studies were performed to investigate the effects of the A protein, a 44 kDa RNA-binding protein which is present as only a single copy within the capsid shell.[44] Recombinant viruses without the A protein formed thicker shells (~ 34 *vs* 24 Å) compared with normal MS2 viruses, but the outer diameter of the viruses and capsids remained nearly the same.[44] The values obtained from these solution-phase measurements are subtly different from measurements of MS2 crystal structures: the crystal structures retain the same protein shell thickness, but with a slightly reduced overall diameter, presumably due to crystallization conditions or packing forces. More importantly, when comparing crystal structures obtained with the A protein with those without, no difference can be seen, which illustrates the value of the solution-phase studies, despite the reduced resolution of such methods.

Computation

By far the most recent development in the study of capsid dynamics is the application of computational simulations and network and graph theory. Only recently has computing hardware advanced to the point where meaningful results can be obtained for systems of any magnitude: in 2006, an all-atom simulation of satellite tobacco mosaic virus (STMV) for 13 ns represented the first such work of its kind.[45] The results of that simulation indicated that the RNA core would be self-stable without the protein capsid, which matched previous experimental results. Likewise, the calculations indicated that the empty capsid would be unstable, whereas the RNA-filled capsid would not, in agreement with experimental observations. Less than a year after that initial virus simulation, a 70S ribosome was simulated with approximately double the number of atoms.[46]

Despite technological progress, the short time-scales and limited size of all-atom simulations remain a bottleneck for practical simulations of more viral systems. To circumvent this limitation, one approach is to reduce the complexity of the system, a technique known as coarse-graining. There are several schemes to achieve this goal: the protein structure and its network of connections can be represented by a series of point masses. In the most common form of simulation, known as normal mode analysis (NMA), the connections are simulated as a series of springs and the lowest-mode harmonic distortions represent the large-amplitude, low-frequency motions possible in the capsid structure. These harmonic modes have been compared with the experimentally observed quaternary dynamics of HK97, CCMV and other viruses with relatively good agreement.[47,48] NMA has also been used to distort high-resolution crystal structures to fit into a lower density cryo-EM map, while maintaining the lowest energy conformation possible.[49] Computational tools such as NMA have invaluable use as a modeling framework upon which to put experimental results. One of the most intriguing developments is the combination of coarse-grained simulations and graph theory to understand unique spectroscopic features and shared characteristics of highly symmetric viral capsids.[50,51] Elastic wave theory and an amorphous isotropic bond polarizability model have been used to predict the extreme low-frequency Raman spectra for M13.[52] The results predicted an axial distortion that was found to be in good agreement with experimentally collected data. Another study used group theory to analyze the Caspar–Klug viruses from the VIPER database with weighted subunit interactions.[53] Surprisingly, a low-energy plateau that extended through 24 modes before dramatically increasing was common across capsid architectures. One explanation is that the plateau provides a theoretical basis for the metastable, dynamic structures that have been repeatedly observed in a variety of experimental studies.

Labeling Experiments

Spectroscopic approaches to the study of viruses in solution have the advantage of being relatively non-destructive, but regardless of the precise technique there are recurring problems of sensitivity, signal assignment and a lack of control over the scale of dynamics being measured. Alternatively, the high-energy dynamic state of a viral capsid can be differentiated from the ground state by means of a permanent covalent modification. The general scheme involves the careful labeling or cleavage of the capsid under kinetically controlled conditions. After the labeling phase has been completed, an analysis step (or multiple steps) is employed to quantitate the course of the reaction. The three variants of this approach are discussed in detail below.

Chemical Reactivity

Chemical cross-linking is the process of reacting a small chemical probe with an engineered or native reactive group on the body of a viral capsid. In the case of a cross-linking reagent with two reactive termini, the distance between two amino acid side-chains can be inferred from the length of the linker region. A selection of cross-linking reagents, ranging in size from 8 to 25 Å with a variety of reactive groups, are commercially available. Key uses of cross-linking are to map subunit interfaces when the structure in not known and to investigate solution dynamics when structural models are available. Analysis normally involves the use of proteolysis and mass spectrometry to identify the specific residues involved. P22 and HIV-1 CA systems have been probed in this manner. With HIV-1, the ability of a linker to form a bridge between Lys70 and Lys182 in CA provided evidence that the N domain of one subunit is in close contact with the C domain of another.[54] For the P22 system, the use of cross-linking reagents with different lengths allowed inter-subunit distances between two lysine residues (Lys183–Lys183) to be determined in relation to a known intra-subunit lysine distance (Lys175–Lys183).[55]

Chemical labeling experiments which measure the rate of site-directed labeling reactions are another way to probe capsid dynamics. In place of distance constraints within the structure, this method provides information on the chemical environment and solvent exposure at a particular site. Using this technique, Bothner, Taylor and others probed the reactivity of the specific capsid regions including the internal gamma-peptides of tetra- and nodaviruses and were able to show distinct differences between T = 4 procapsids and capsids and T = 4 and T = 3 mature capsids, consistent with other data indicating that the solution properties are different.[20,21] Similar techniques were used to show that FHV virus-like particles (containing heterologous cellular RNA) and wild-type particles have dramatically different dynamic properties in solution even though they are crystallographically identical.[38] In addition to the rate of chemical modification, the maximum stoichiometry of labeling also provides information regarding the structure of capsids in solution. The maturation of the NωV capsid is a very rapid and dramatic contraction, with a decrease in diameter of 16% within 100 ms.[4] This conformational change involves a transition from a highly fenestrated structure to a much more tightly packed capsid. Labeling experiments on the two forms of the protein with fluorescein derivatives showed that the maximum extent of labeling was far greater for the expanded procapsid than the capsid: up to ~ 800 lysine or cysteine (depending on the reactive chemistry used) were labeled on the procapsid.[20] In comparison, the capsid form of the protein could react with only ~ 100 dye molecules per capsid.

Hydrogen–Deuterium Exchange

A powerful technique for investigating protein dynamics is hydrogen–deuterium exchange (HX). This approach makes use of the varying rates of exchange for amide or side-chain protons with solvent, depending on their local environment and participation in hydrogen bonding. Upon dilution or buffer exchange of a protein (containing normal environmental ratios of ^{1}H:^{2}H) into ^{2}H$_2$O (D$_2$O), the protein protons will begin to equilibrate with their surroundings. In practice, the rate of exchange varies from nearly instantaneous to effectively zero over the course of months, depending on the stability and solvent exposure at a specific site. Detection of the exchange can be accomplished via spectroscopic methods such as NMR or Raman and this approach has been used to study capsid subunit interactions, folding and conformational changes.[40,41,56] However, recent application of HX has focused heavily on mass spectrometry to detect the mass shift of deuterium incorporation. After exchange, the proteins are cleaved with pepsin to generate peptides so that changes can be assigned to a localized area. The number of sites that have exchanged is measured as the shift in mass difference before and after exposure to D$_2$O. This approach requires small quantities of material and is less subject to the size scaling issues discussed above for NMR. The only real limitation to the study of large complexes using HX–MS is that as the number of different proteins in a sample increases, it becomes more difficult to assign exchange data to a specific site. Virus capsids, with their use of multiple copies of the same protein abrogate this issue, at least with respect to low T number capsids. However, HX is an intrinsically reversible process and the phenomenon of back-exchange (^{2}H to ^{1}H) during proteolysis, separation and detection can complicate data analysis: in some cases up to 30% of the deuterons are lost in this second phase of analysis. Back-exchange can be minimized by controlling pH, reducing temperature and minimizing the time of analysis, but these complications make intact protein HX–MS an attractive option due to the lack of post-labeling processing for experiments which do not necessitate specific assignment of exchanging regions. Due to the flexibility with which HX experiments can be conducted, there is a wealth a excellent studies using this approach, many of which have contributed greatly to our understanding of viral capsid structures. Our discussion of this topic is only a brief selection to highlight some of the features of this experimental method; more examples of HX applied to viruses can be found elsewhere.[57–61]

As opposed to spectroscopic studies, one of the key features of HX–MS is the ability to work at low concentrations and in the presence of a hetero-geneous mixture of proteins. This tolerance to mixtures of proteins was used to probe the dynamics of Ø29 scaffolding protein gp7 in the free and procapsid-bound forms.[62] A combination of intact protein and proteolyzed fragments was analyzed, allowing characterization of the H–L–H domain which interacts with the coat protein. Furthermore, HX–MS provided con-firmation for the multiple concentric layers of gp7 implied by the cryo-EM reconstructions, based on the presence of two population of gp7 with different exchange rates.

The HX experiments conducted on HRV14 provide a good example of the application of HX data to a system with a known structure, placing dynamics measurements directly into the context of classic static structures.[61] As with the studies on Ø29 gp7, post-exchange proteolysis was used to improve the reso-lution of assignments: ~80% of the protein sequence was measured *via* 90 unique peptides and rates of exchange were tracked for all peptides. To achieve this level of coverage only 20 pmol of protein per sample was required, which is a benefit of the sensitivity of HX–MS. In contrast to the exchange curves for VP1, VP2 and VP3, which had varying regions of protection and dynamic motion, VP4 was almost uniformly dynamic and highly exchanged after only 2.5 min. These data support other experimental data such as limited proteo-lysis, which indicates that VP4 is a very dynamic region (see Figure 1). How-ever, the dynamics of VP4 as indicated by HX–MS have a subtle nuance: because the threshold for deprotection in HX is so low, it is impossible to evaluate the full limit of the dynamic motion. Limited proteolysis of the same region provides evidence that the dynamic behavior of VP4 is not just limited to solvent exposure on the interior of the capsid, but rather that it undergoes substantial translocations.

Kinetic Hydrolysis

Quantitative measurements of protein dynamics are critical if we are to understand the thermodynamic nature of conformational changes, how they are coordinated within the supramolecular structure of a capsid and how they relate to the lifecycle of a particular virus. To date, only a single example exists in which equilibrium and rates of conversion have been determined for a megadalton complex. The hepatitis B virus capsid has been instrumental as a model system to understand viral capsid assembly and has served as the foundation for substantial theoretical models of supramolecular complex assembly. A member of *Hepadnaviridae*, the core protein forms an unusual dimer composed almost entirely of alpha-helices. The dimer cannot be separated without denaturing conditions and 120 copies assemble to form the T = 4 capsid. Due to a well-established strong hysteresis to disassembly, intact capsids can be studied under the same conditions as the dimer. This feature was exploited recently by Bothner's group to measure quantitatively the dynamics of dimer and capsid forms using kinetically controlled proteolysis (kinetic hydrolysis).[63] Because the assembled and unassembled protein can be studied under the same solution conditions, dynamics associated with protein dimer can be separated from emergent dynamic properties resulting from assembly of the capsid. Using SDS-PAGE to measure rates of intact protein degradation and peptide mass mapping to identify the sites of hydrolysis, it was shown that the C-terminal region of the capsid protein Cp149 was dynamic in both forms. By performing assays across a range of protease concentrations, kinetic curves of HBV digestion were obtained which detailed both the rate of exposure from the closed to the open conformation, and also the equilibrium between those states in solution. This approach works because whereas the sequence specificity is a function of the protease, the rate of hydrolysis at a particular site is highly dependent upon accessibility by the protease, as mediated by the local backbone dynamics. Based on a two-state model using the structure from X-ray crystallography as the low-energy state, the transition has a lifetime of approximately 2 s and about three subunits per capsid are in the open (or high-energy) conformation at any time. By docking enzymes to the HBV surface, it was estimated that a translocation of $>13 \, \text{Å}$ from the location in the crystal structure is required to reach the open, cleavage-accessible conformation. A surprise finding was that the protein in the dimer and capsid forms was thermodynamically distinct (Figure 3). Although no substantial differences were determined for opening rate, the equilibrium between open and closed forms had an opposite temperature dependence when comparing dimer and capsid. The dynamic site is also in close proximity to the binding location of HAP compounds that have demonstrated antiviral activity.[64] Together these results have interesting implications for the role of dynamics in HBV assembly and the specific targeting of dynamic regions with antiviral agents. The surprising behavior of the HBV system illustrates the power of solution-phase measurements of dynamic protein motion. Future applications of this technique to other systems will undoubtedly illuminate trends for viral

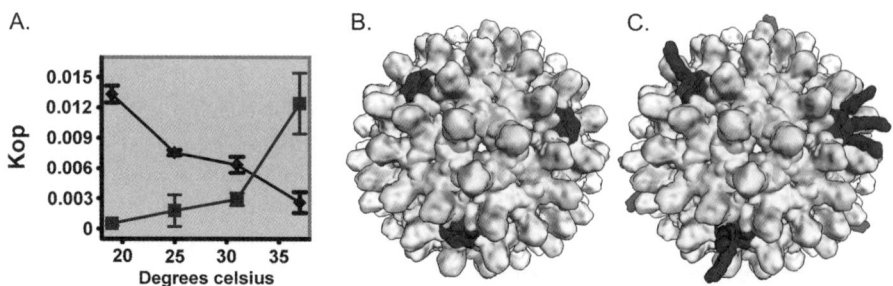

Figure 3   Dynamics of HBV capsid as probed with proteolysis. (A) The equilibrium between open and closed forms of the C-terminus for dimer (black diamonds) and capsid (gray squares). Data from Hilmer *et al.*[63] (B) Conformation of the capsid as seen in the crystal structure: the C-terminus is shown as dark gray. (C) Estimate of unfolding required for trypsin cleavage. The C-termini cluster around the fivefold and quasi-sixfold centers of symmetry. Model shown has randomized 25% open conformation.

capsids in general and serve to better connect our understanding of structure and function.

5   Summary

Biochemical and biophysical investigations of virus particles in solution are critical to understanding their functional properties. The array of functional demands that are placed on capsid proteins requires numerous structural calisthenics to be performed, often in the context of a delicate balance between assembly and disassembly. Together with structural models, information on the location and extent of capsid dynamics provides a basis for linking structure to function with greater detail than either approach alone. A number of significant questions remain to be addressed, including the role of mutations that alter dynamics on viral fitness and whether dynamics is a symmetric or asymmetric property. The latter point has implications for receptor binding, cell entry and genome release, all of which remain poorly characterized in non-enveloped viruses. The significance of this information goes beyond the basic biology of viruses. Dynamic regions are a relatively untapped target for antiviral therapy and viruses are excellent model systems for studying allostery and dynamics in supramolecular complexes. In addition, viruses are being used and developed as bioinspired nanomaterials, with applications from gene delivery to nanowires. As scientists seek next-generation nanomaterials that can actively respond to selected stimuli, a thorough understanding of the dynamic properties of capsids will be important.

References

1. A. C. Steven, J. B. Heymann, N. Cheng, B. L. Trus and J. F. Conway, *Curr. Opin. Struct. Biol.*, 2005, **15**, 227.

2. P. E. Prevelige Jr, ed. *The Bacteriophages*, 2nd edn, Oxford University Press, New York, 2005.
3. J. F. Conway *et al.*, *Science*, 2001, **292**, 744.
4. M. A. Canady, H. Tsuruta and J. E. Johnson, *J. Mol. Biol.*, 2001, **311**, 803.
5. C. E. Fricks and J. M. Hogle, *J. Virol.*, 1990, **64**, 1934.
6. W. R. Wikoff *et al.*, *Science*, 2000, **289**, 2129.
7. R. L. Duda, *Cell*, 1998, **94**, 55.
8. R. Lata *et al.*, *Cell*, 2000, **100**, 253.
9. J. B. Heymann *et al.*, *Nat. Struct. Biol.*, 2003, **10**, 334.
10. A. Zlotnick *et al.*, *J. Biol. Chem.*, 1994, **269**, 13680.
11. A. Schneemann, R. Dasgupta, J. E. Johnson and R. R. Rueckert, *J. Virol.*, 1993, **67**, 2756.
12. A. Janshoff, D. T. Bong, C. Steinem, J. E. Johnson and M. R. Ghadiri, *Biochemistry*, 1999, **38**, 5328.
13. M. Chow, R. Yabrov, J. Bittle, J. Hogle and D. Baltimore, *Proc. Natl. Acad. Sci. USA*, 1985, **82**, 910.
14. C. Helgstrand, S. Munshi, J. E. Johnson and L. Liljas, *Virology*, 2004, **318**, 192.
15. S. Munshi *et al.*, *J. Mol. Biol.*, 1996, **261**, 1.
16. M. A. Canady, M. Tihova, T. N. Hanzlik, J. E. Johnson and M. Yeager, *J. Mol. Biol.*, 2000, **299**, 573.
17. M. Tomasicchio, P. A. Venter, K. H. Gordon, T. N. Hanzlik and R. A. Dorrington, *J. Gen. Virol.*, 2007, **88**, 1576.
18. D. J. Taylor, N. K. Krishna, M. A. Canady, A. Schneemann and J. E. Johnson, *J. Virol.*, 2002, **76**, 9972.
19. K. K. Lee, J. Tang, D. Taylor, B. Bothner and J. E. Johnson, *J. Virol.*, 2004, **78**, 7208.
20. D. J. Taylor *et al.*, *Chem. Commun. (Cambridge)*, 2003, 2770.
21. B. Bothner *et al.*, *Virology*, 2005, **334**, 17.
22. J. M. Hogle, *Annu. Rev. Microbiol.*, 2002, **56**, 677.
23. G. A. Farr, L. G. Zhang and P. Tattersall, *Proc. Natl. Acad. Sci. USA*, 2005, **102**, 17148.
24. M. Agbandje, R. McKenna, M. G. Rossmann, M. L. Strassheim and C. R. Parrish, *Proteins*, 1993, **16**, 155.
25. S. F. Cotmore, M. D'Abramo Jr, C. M. Ticknor and P. Tattersall, *Virology*, 1999, **254**, 169.
26. F. Sonntag, S. Bleker, B. Leuchs, R. Fischer and J. A. Kleinschmidt, *J. Virol.*, 2006, **80**, 11040.
27. J. K. Lewis, B. Bothner, T. J. Smith and G. Siuzdak, *Proc. Natl. Acad. Sci. USA*, 1998, **95**, 6774.
28. G. Vriend, M. A. Hemminga, B. J. M. Verduin, J. L. Dewit and T. J. Schaafsma, *FEBS Lett.*, 1981, **134**, 167.
29. M. Roivainen, A. Narvanen, M. Korkolainen, M. L. Huhtala and T. Hovi, *Virology*, 1991, **180**, 99.

30. M. Roivainen, L. Piirainen, T. Rysa, A. Narvanen and T. Hovi, *Virology*, 1993, **195**, 762.
31. Q. Li, A. G. Yafal, Y. M. Lee, J. Hogle and M. Chow, *J. Virol.*, 1994, **68**, 3965.
32. M. P. Fox, M. J. Otto and M. A. McKinlay, *Antimicrob. Agents. Chemother.*, 1986, **30**, 110.
33. D. K. Phelps and C. B. Post, *J. Mol. Biol.*, 1995, **254**, 544.
34. D. K. Phelps and C. B. Post, *Protein Sci.*, 1999, **8**, 2281.
35. B. Bothner, X. F. Dong, L. Bibbs, J. E. Johnson and G. Siuzdak, *J. Biol. Chem.*, 1998, **273**, 673.
36. B. R. Szymczyna, L. Gan, J. E. Johnson and J. R. Williamson, *J. Am. Chem. Soc.*, 2007, **129**, 7867.
37. S. J. Opella, A. C. Zeri and S. H. Park, *Annu. Rev. Phys. Chem.*, 2008, **59**, 635.
38. B. Bothner *et al.*, *Nat. Struct. Biol.*, 1999, **6**, 114.
39. A. T. Da Poian, J. E. Johnson and J. L. Silva, *J. Biol. Chem.*, 2002, **277**, 47596.
40. R. Tuma, J. K. Bamford, D. H. Bamford and G. J. Thomas Jr, *Biochemistry*, 1999, **38**, 15025.
41. R. Tuma, J. H. Bamford, D. H. Bamford, M. P. Russell and G. J. Thomas Jr, *J. Mol. Biol.*, 1996, **257**, 87.
42. P. E. Prevelige Jr, D. Thomas, K. L. Aubrey, S. A. Towse and G. J. Thomas Jr, *Biochemistry*, 1993, **32**, 537.
43. D. A. Kuzmanovic, I. Elashvili, C. Wick, C. O'Connell and S. Krueger, *Structure*, 2003, **11**, 1339.
44. D. A. Kuzmanovic, I. Elashvili, C. Wick, C. O'Connell and S. Krueger, *J. Mol. Biol.*, 2006, **355**, 1095.
45. P. L. Freddolino, A. S. Arkhipov, S. B. Larson, A. McPherson and K. Schulten, *Structure*, 2006, **14**, 437.
46. K. Y. Sanbonmatsu and C. S. Tung, *J. Struct. Biol.*, 2007, **157**, 470.
47. M. K. Kim, R. L. Jernigan and G. S. Chirikjian, *J. Struct. Biol.*, 2003, **143**, 107.
48. F. Tama and C. L. Brooks III, *J. Mol. Biol.*, 2002, **318**, 733.
49. F. Tama, O. Miyashita and C. L. Brooks III, *J. Struct. Biol.*, 2004, **147**, 315.
50. E. C. Dykeman, O. F. Sankey and K. T. Tsen, *Phys. Rev. E*, 2007, **76**, 011906.
51. K. M. ElSawy, A. Taormina, R. Twarock and L. Vaughan, *J. Theor. Biol.*, 2008, **252**, 357.
52. K. T. Tsen *et al.*, *J. Biomed. Opt.*, 2007, **12**, 024009.
53. F. Englert, K. Peeters and A. Taormina, *Phys. Rev. E*, 2008, **78**, 031908.
54. J. Lanman *et al.*, *J. Mol. Biol.*, 2003, **325**, 759.
55. S. Kang, A. M. Hawkridge, K. L. Johnson, D. C. Muddiman and P. E. Prevelige Jr, *J. Proteome Res.*, 2006, **5**, 370.
56. R. Tuma, H. Tsuruta, J. M. Benevides, P. E. Prevelige Jr and G. J. Thomas Jr, *Biochemistry*, 2001, **40**, 665.

57. T. E. Wales and J. R. Engen, *Mass Spectrom. Rev.*, 2006, **25**, 158.
58. J. Lanman *et al.*, *Nat. Struct. Mol. Biol.*, 2004, **11**, 676.
59. J. Lanman and P. E. Prevelige Jr, *Curr. Opin. Struct. Biol.*, 2004, **14**, 181.
60. B. Suchanova and R. Tuma, *Microb. Cell Factories*, 2008, **7**, 12.
61. L. Wang and D. L. Smith, *Protein Sci.*, 2005, **14**, 1661.
62. C. Y. Fu and P. E. Prevelige Jr, *Protein Sci.*, 2006, **15**, 731.
63. J. K. Hilmer, A. Zlotnick and B. Bothner, *J. Mol. Biol.*, 2008, **375**, 581.
64. C. R. Bourne, M. Finn and A. Zlotnick, *J. Virol.* 2006, **80**, 11055.

CHAPTER 4

Three-dimensional Structures of Pleiomorphic Viruses from Cryo-Electron Tomography

ALASDAIR C. STEVEN, GIOVANNI CARDONE, CARMEN BUTAN*, DENNIS C. WINKLER AND J. BERNARD HEYMANN

Laboratory of Structural Biology, National Institute of Arthritis and Musculoskeletal and Skin Diseases, National Institutes of Health, Bethesda, MD 20892, USA

1 Introduction

The past two decades have seen remarkable progress in structural studies of viruses that possess icosahedral capsids, stemming from technical advances in cryo-EM[1,2] and X-ray crystallography.[3] Among other insights, it has emerged that protein subunits with some half-dozen different folds assemble into icosahedral shells;[4] that there are capsids with previously undocumented T-numbers;[5,6] and that capsids exhibit diverse variations on the themes of quasi-equivalence and non-equivalence.[7,8] Some capsids have been found to be highly dynamic, undergoing massive conformational changes during maturation and/or cell entry. On a functional level, it has transpired that some viruses encapsidate enzymes as well as genomes and some capsids afford compartments for specific activities as well as serving as passive containers.

---

*Current address: Department of Biological Chemistry and Molecular Pharmacology, Harvard Medical School, 240 Longwood Ave., Boston, MA 02115

RSC Biomolecular Sciences No. 21
Structural Virology
Edited by Mavis Agbandje-McKenna and Robert McKenna
© Royal Society of Chemistry 2011
Published by the Royal Society of Chemistry, www.rsc.org

The high order of symmetry implicit in icosahedral geometry has been instrumental in these structural analyses; in X-ray diffraction studies, it has enabled phase determination by exploitation of non-crystallographic symmetry; and three-dimensional reconstructions of cryo-electron micrographs have been facilitated both by the 60-fold reduction, compared to asymmetric particles, of the angular range that must be searched to determine viewing geometries and the commensurate benefit to be derived from averaging, considered on a per-particle basis. However, many viruses do not have icosahedral capsids but confine their genomes in other less regular structures, and viruses that have icosahedral capsids also have other components, *e.g.* packaged genomes, that do not conform to this symmetry. For these viruses and these viral components, there has been a relative dearth of three-dimensional structural information.

In considering departures from regularity, we may consider three progressively more extreme stages: asymmetry, polymorphism, and pleiomorphism. Particles that lack symmetry but assume the same structure in each assembled particle may be tackled, albeit more laboriously, by the same approaches of cryo-EM with "single particle analysis" (SPA) and X-ray crystallography. We take polymorphism to refer to a discreet set of related structures that may, for instance, represent topologically distinct closures, *e.g.* tubular and polyhedral foldings of the same hexagonal lattice. Although polymorphic structures may not crystallize, their repetitive structures are potentially amenable to image averaging strategies in cryo-EM. Pleiomorphic structures represent the most challenging specimens; different particles may contain different numbers of their constituents arranged in a continuum of stochastically variable spatial distributions. Here, crystallography is not an option and direct interpretation of electron micrographs, even those of very high quality, is subject to ambiguity arising from the co-projection of many layers of molecules.

The recently implemented visualization technique of cryo-electron tomography (cryo-ET)[9] is capable of producing three-dimensional density maps of individual macromolecular complexes – even pleiomorphic ones – in their native hydrated states.[10,11] (Although cryo-ET is a branch of cryo-EM, for clarity, we use the term cryo-EM only for studies of vitrified specimens that involve no tilting or only minimal tilting of the specimen). Cryo-ET is emerging as a tool of great promise not only for structural investigations of viruses, but also for the study of sub-viral particles and of viruses interacting with antibodies, receptors, or larger cell-related complexes. In this chapter, we review the basic principles of cryo-ET; discuss the resolution currently attainable, with particular reference to viral applications; briefly summarize work of this kind published to date; and discuss in further detail some studies that point to the presence of some host cell-derived proteins as integral components of pleiomorphic viruses, as well as cryo-ET observations pertaining to cell entry mechanisms in several viral systems.

2 Cryo-electron Tomography: How It's Done

Cryo-ET operates on the same basic principle as the well known clinical imaging modality of CAT (computerized axial tomography) in that projection images are recorded of the same object as viewed from different directions, and these data are then combined in calculating a three-dimensional density map. However, there are three differences in terms of practical implementation: (i) the relationship between the specimen (patient) and the recording device; (ii) the range of viewing angles that can be covered; and (iii) the alignment of projections. In CAT, the specimen (patient) remains stationary and the recording device rotates to different viewing positions, whereas in cryo-ET, the recording device (an electron microscope) remains stationary and the specimen is rotated. During a CAT scan, the data can be collected through a full angular range (180°), whereas in cryo-ET, conventional specimen holders can only be tilted through a range of $\pm 70°$ or less before the specimen projection along the line-of-sight becomes prohibitively thick. The practical upshot of the truncated range of views is that resolution in the tomogram is anisotropic – the so-called "missing wedge" effect (see §5.3). The alignment of the CAT scanner is good enough to allow direct reconstruction of tomograms, whereas the nominally eucentric electron microscope gives only an approximate mutual alignment of the projections, and these data require further processing to ensure their correct registration.

The key development underlying cryo-ET as a practical procedure (see Figure 1 for a typical workflow) has been development of protocols for the semi-automatic or automatic acquisition of tilt series at very low electron dose per projection.[12,13] The most advanced of these use predictive tracking of the area-of-interest, thereby keeping the specimen approximately centered with minimal cost of additional irradiation. Computational alignment of the projections is mostly done using colloidal gold particles as fiducial markers,[14,15] although the alignment can also be done without markers.[16,17] Once a satisfactory alignment is obtained, the tomogram is calculated using one of a variety of reconstruction algorithms – *e.g.*.[18,19] Because of the low doses used in cryo-ET, the signal-to-noise ratio of tomograms is low. Several denoising algorithms have been developed to remove noise and enhance contrast for visualization and interpretation.[20–23]

3   Resolution in Cryo-ET

Resolution in tomograms is a complex issue depending on multiple factors, and comprehensive discussion goes beyond the scope of this article; nevertheless, several general points can be made. First, the *maximum resolution achievable-* given noise-free data and a complete (180°) tilt series-depends on both the tilt-angle increment ($\Delta\theta$, in degrees) and specimen thickness (D) according to the relationship.[24,25]

$$\text{res}_{max} = \pi D(\Delta\theta/180)$$

In practice, this limit will not be reached on account of noise and other considerations but it provides a useful bound. In studies of isolated viruses, tilt increments between 1° and 2° are typically used. For a virus 150 nm in

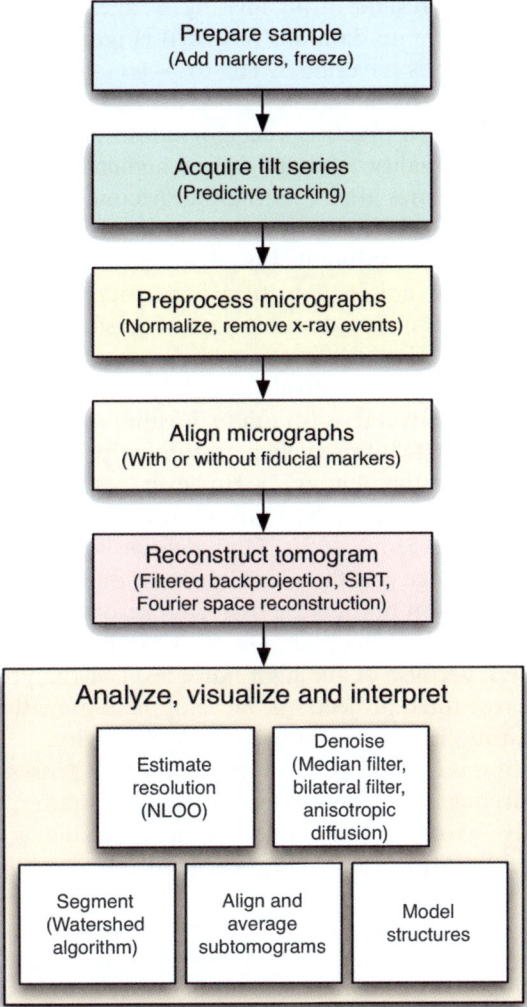

Figure 1   Sequence of operations in cryo-ET.

diameter, this limits resolution to 2.6 nm (1°) or 5.2 nm (2°). This restriction may, in principle, be extended by reducing the tilt increment, $\Delta\theta$. However, with vitrified specimens, there may be little to be gained because: (1) if the electron dose per projection is kept fixed, radiation damage increases and may become resolution-limiting; or (2) if the electron dose per projection is reduced to maintain a fixed total dose (say, 70 electrons/Å^2), the projections become noisier and therefore more difficult to align with high precision, potentially impairing the resolution of the tomogram.

Second, resolution in ET is anisotropic, being higher in-plane than in the third dimension. Anisotropy arises from the incomplete angular range covered in the tilt series which results in a "missing wedge" in the specimen's 3-D Fourier transform, where no data are recorded (Figure 2). It has been calculated that, for a tilt series covering $+60°$ to $-60°$, the effect of the missing information may be approximately described as reducing the resolution in the z-dimension by $\sim 40\%$. In practice, the shortfall in z-resolution may be even greater, as projection quality tends to fall at higher tilt-angles. In principle, collecting a second tilt series after rotating the specimen in-plane by $90°$ ("dual tilt series") can mitigate this effect, by reducing the missing wedge to a missing pyramid. However, this procedure incurs either increased radiation damage (in the second tilt series) or noisier data (if a fixed total dose is divided equally between the two tilt series), and dual tilt data acquisition has yet to enter into widespread use in cryo-ET.

Third, cryo-ET tends to use relatively high defocus values, e.g. $4\,\mu m$ to$10\,\mu m$, as compared to cryo-EM where $\sim 0.7\,\mu m$ to $2.5\,\mu m$ is typical. These conditions enhance low resolution features, which facilitates projection alignment and reduces the noise level in the tomogram. However, resolution is usually limited to the spacing of the first zero of the contrast transfer function (CTF),[26] which falls at a spacing of $(3.5\,nm)^{-1}$ for $5\,\mu m$ defocus or $(5.0\,nm)^{-1}$ for $10\,\mu m$ defocus, for data recorded at $200\,keV$. Computational correction procedures for recovering information beyond the first CTF zero are now in routine use in cryo-EM. However, such corrections are considerably more challenging proposition in cryo-ET because of the high noise level of the projections and the focal gradients across tilted projections. Development of methods to effect such corrections constitutes an area of intense current activity.

Fourth, resolution is limited by noise. In addition to possessing signal out to a certain spatial frequency, a tomogram (or, for that matter, any image) must have a sufficiently favorable signal-to-noise ratio for that signal to be interpretable.[27] Noise in tomograms may be reduced to a limited extent by

---

Figure 2   Anisotropic resolution and the "missing wedge". The schematic at top left conveys the collection of projections at successive tilt-angles. Specimen thickness and mechanical constraints limit the angular range to $\pm 70°$ or so. In calculating the tomogram, these projections are combined computationally. According to the central section theorem, each projection contributes a central slice through the 3D Fourier transform of the object (gray planes in the schematic at top right). Because of the truncated tilt-range, there is a zone in the transform for which no information is obtained – the so-called "missing wedge" (red). Bottom: the effect of this missing information on real-space density sections is illustrated, using a cryo-EM reconstruction of the HBV T $= 4$ capsid at $9\,\text{Å}$ resolution.[58] Sections in-plane $(x - y$; bottom row) and perpendicular to it $(x - z$; upper row) are shown at full resolution, i.e. no missing wedge, at left. The same sections after computationally removing wedges of $40°$ (corresponding to acquisition range of $\pm 70°$), and $80°$ (acquisition range of $\pm 50°$) are shown at center and on right. Although not immune to "missing wedge" artifacts, the x – y sections are relatively unaffected, contrary to the x – z sections that are strongly affected. Bar, $10\,nm$.

electron beam source
sample
detector
missing
wedge
z
x
y
x

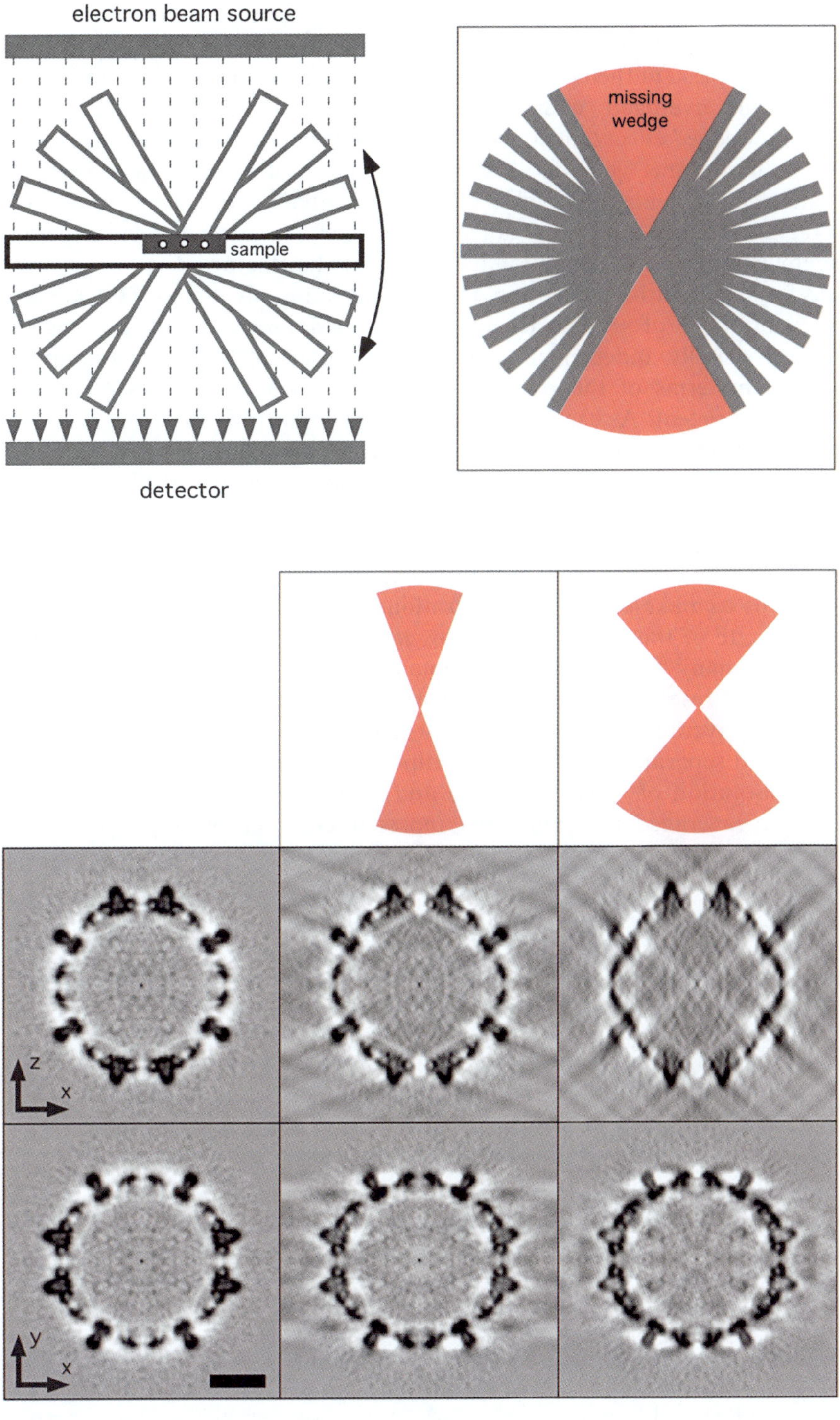

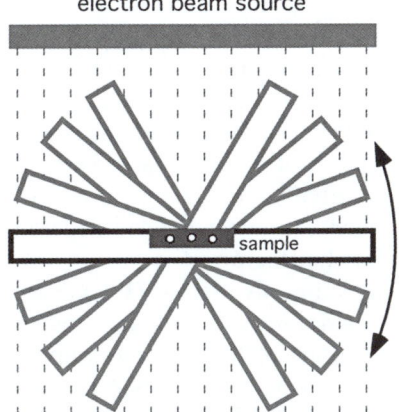

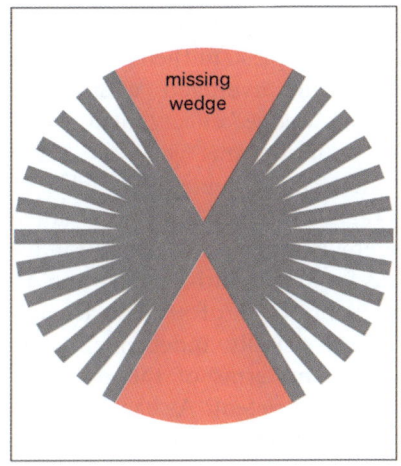

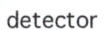

filtering[20,21] and, more powerfully, by averaging. For averaging to be possible, the object of interest has to be identifiable in the primary tomograms, from which its copies are then excised and the resulting subtomograms are mutually aligned and averaged, with due attention paid to bias introduced by the "missing wedge" effect (see above). Averaging brings important advantages. As the subtomograms generally present the object of interest in a variety of orientations, the missing wedge effect is essentially annulled. Moreover, as averaging suppresses noise, it is feasible to work at somewhat lower defocus, thereby extending the resolution as limited by that factor.

As far as measuring resolution, the criteria commonly used in SPA are based on the principle of self-consistency: reconstructions are calculated separately (but not, usually, independently) from two subsets of data, and resolution is specified in terms of the smallest spacing to which the reconstructions are deemed consistent. As each tomogram of a pleiomorphic specimen is one-of-a kind, such an approach is not applicable. Nevertheless, some quantitative resolution criteria are now available.[28–30] Also, conventional SPA resolution criteria such as the Fourier Shell coefficient[18] are applicable to averaged datasets.

It should be evident from the foregoing discussion that it is difficult if not hazardous to make blanket generalizations about resolution in cryo-ET at the present state of the art. With this *caveat*, we offer the following opinions: (i) given the many factors that may potentially be limiting, quantitative measurement of resolution is desirable; (ii) the resolution of primary tomograms and that of averaged subtomograms – which may be significantly higher-should be assessed separately; and (iii) for primary tomograms of pleiomorphic viruses, a resolution of ∼5 nm in-plane and 7.5 nm in the third dimension, and 2.5 nm for averaged subtomograms, represent considerable accomplishments.

4 Features of Pleiomorphic Viruses

The number of published cryo-ET studies on pleiomorphic viruses is growing quite rapidly (Table 1). In Figure 3, we present illustrative cryo-ET data on three pleiomorphic viruses – herpes simplex virus 1 (HSV-1; Figure 3A), Rous sarcoma virus (RSV; Figure 3B), and influenza A virus (Figure 3C). Each virus exhibits substantial variation in size, albeit within a distinctive range. All three are enveloped, with outwards-protruding glycoprotein spikes.

HSV-1 has ∼800 spikes per virion, quite closely packed.[31] They are of eleven different kinds in varying molar ratios,[32] four of which are involved in membrane fusion activity during cell entry. A major goal of current interest is to map their distributions over the viral surface, identify eventual colocalizations, and correlate tomographic information on the structures of complete membrane-embedded spikes with high resolution information on partial or complete ectodomains.[33,34] Influenza virus has two kinds of spikes: hemagglutin (HA, the fusagen), and neuraminidase (NA-Figure 4). There are about 400 spikes per virion (∼85% HA and ∼15% NA) that are densely packed except in

Table 1 Pleiomorphic viruses studied by cryo-electron tomography.

Virus	*Reference*
Herpesviruses	
Herpes simplex virus type 1	[31]
Herpes simplex virus type 1 (A-capsids)	[59]
Herpes simplex virus (urea-extracted capsids)	[60]
Kaposi's sarcoma-associated herpesvirus	[61]
Murine gamma herpesvirus-68	[62]
Retroviruses	
Moloney murine leukemia virus	[63]
Human immunodeficiency virus (mature)	[64–70]
Human immunodeficiency virus (immature)	[68, 71]
Rous sarcoma virus	[36]
Other viruses	
Vaccinia virus	[72]
Influenza A virus	[35]
Cystovirus f12	[73]
Bunyavirus	[74]
Porcine reproductive and respiratory virus	[75]
Coronavirus (mouse hepatitis virus)	[76]

occasional bare patches.[35] The minority NA spikes are not randomly distributed but tend to cluster. On RSV virions, spikes are generally much sparser, varying from < 5 to ~ 120 per virion.[36] There is a single viral glycoprotein, Env, the fusagen, but spikes are also observed that have distinctly different morphologies, as discussed in the following section.

Influenza virus and RSV, but not HSV-1, have layers of matrix proteins lining their envelopes (Figures 3 & 4). That of influenza virus is clearly resolved from the membrane, whereas in RSV tomograms of similar resolution, these two layers are so tightly apposed that they are not resolved. In both viruses, there are gaps in the matrix layer: indeed, some influenza virions appear to entirely lack a matrix layer.[35,37] The latter observation points to an alternative budding mechanism.

The three viruses differ markedly in their nucleocapsids. HSV-1 has a $T = 16$ icosahedral shell, containing its 152-kbp genome of dsDNA. In contrast, RSV capsids, like those of other retroviruses, are polymorphic and exhibit a variety of foldings – tubular, bi-conical, and irregular polyhedral – of a hexagonal lattice.[38–40] Its spacing, 9.5 nm, is large enough to have been detected in the tomograms (~ 5.5 nm resolution, in-plane) but the contrast of the basic repeat is too low for it to be seen. However, the second order of this repeat at ~ 4.5 nm is detectable by cryo-EM of capsid fragments.[36] Tomographic data on the three-dimensional forms of individual capsids allows lattice models to be built (Figure 5). Influenza virus has eight segments of single-stranded RNA, that combine with N-protein and the viral polymerase in filamentous ribonucleoprotein particles (RNPs). Key questions of long standing are whether each

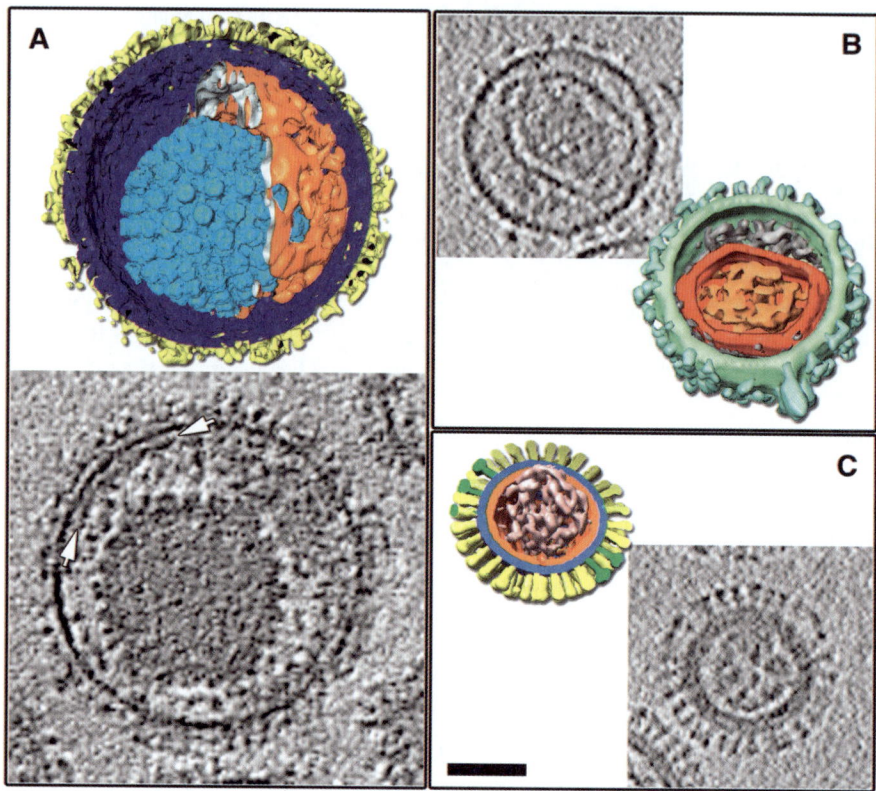

Figure 3 Tomographic reconstructions of three pleiomorphic viruses, at the same scale. In each case, a cutaway view of the virion interior is shown in surface rendering, together with a central thin (~ 1.5 nm) section, shown in gray-scale. All three panels: bar, 50 nm. (A) HSV-1 has an icosahedral T = 16 capsid (light blue), encased in tegument (orange), an irregular proteinaceous structure, and surrounded by its envelope, a lipid bilayer is which the viral glycoproteins are embedded. The inner surface of the envelope is colored dark blue and the glycoprotein spikes, yellow. In the section, the filamentous tegument density (arrow) tracking the inner surface of the envelope may be an actin filament. The lipid bilayer is perceived as a single layer of continuous density. In-plane resolution, ~ 65 Å. (B) RSV virion with a polyhedral capsid (red) containing density that presumably represents its diploid ssRNA genome and associated proteins (orange). The lipid bilayer of its envelope is lined with an unresolved layer of matrix protein. Note the greater thickness of this envelope compared to its counterpart in HSV (*cf.* A and B). The Env spikes are in pale green. Although this virion is relatively well endowed with spikes, they are sparser than on HSV-1 (A) or influenza virus (C). There is also material between the RSV capsid and the envelope (gray), at least some of which is unassembled capsid protein subunits. In-plane resolution, ~ 60 Å. (C) Influenza A virus (strain X-31) has its genomic material in the form of multiple ribonucleoprotein (RNP) complexes (pink) which have flexible filamentous shapes. In its envelope, the bilayer (blue) is resolved from the matrix protein layer (orange) – see the section. The two viral glycoproteins, hemagglutinin (HA) and neuraminidase (NA), are distinguishable, HA ectodomains are shown in yellow, and those of NA in green. In-plane resolution, ~ 55 Å. For further details, see Figure 4.

virion acquires one copy of each RNP and, if so, how such selectivity is accomplished. Current cryo-ET data provide only partial answers: on one hand, some virions clearly contain fewer than eight RNPs, and some virions have entirely different RNPs in the form of single solenoids (cryo-ET data support earlier observations by other kinds of EM – *e.g.*,[41,42]). An immediate priority is to improve the resolution of the tomograms and/or the efficacy of segmentation procedures in order to be able to discriminate and, potentially, to average different classes of intraviral RNPs.

5 Stowaways or Conscripts? Host Cell Proteins in Virus Particles

One paradigm of structural virology has been that virus components are produced by the expression of viral genes and, although assembly takes place in a crowded intracellular environment, cellular molecules are efficiently excluded from nascent virions. (There are rare exceptions to this rule – such as the assembly of cellular histones into the chromatinized genomes of papovaviruses and the incorporation of cellular lipids, together with viral glycoproteins, into the envelopes of many viruses). Two arguments that rationalize this trend of confining assembly to virally encoded components are: (i) as has long been recognized, there do not appear to be any host proteins that naturally form capsids and which could be appropriated as viral building-blocks; and (ii) any cellular proteins needed by a virus to establish a productive infection should be already available in the host cell, so that it is unnecessary to import them from the host cell of the previous cycle. However, if – for example – a cellular protein were to be needed during infection but is not normally present in sufficient amount or if it would have to be chemically modified for its role in the viral replication, these would represent valid reasons for the molecule be delivered in prefabricated form by an infecting virion.

The experimental basis for this paradigm ("viruses are assembled from viral gene products") lies in the identification as such of all major bands on gel electropherograms of purified virus preparations, and in the plausible assignment of features in density maps calculated from X-ray diffraction or cryo-EM data. However, if virions were to incorporate, in moderate amounts, a sampling of host cell components, these molecules would not show up as prominent bands in SDS-PAGE analysis, and it is generally difficult to determine whether faint gel bands represent *bona fide* viral constituents or contaminants. And if a given host cell-derived intruder were to be irregularly or variably distributed within viral particles, it would not show up in symmetrized density maps calculated from cryo-EM or X-ray crystallographic data. One strength of cryo-ET is its potentiality for detecting in individual virions, components whose occupancies are too low for them to be detected by biochemical analysis and too variable to be visualized in conventional structural analyses.

The criteria for detectability in cryo-tomograms of viruses are that the molecule in question should be sufficiently large and distinctive in shape that it may be reliably identified, even given the limited resolution of tomograms. A growing number of observations by cryo-ET[31] and other techniques[43,44] suggest that the cytoskeletal protein, actin, becomes incorporated into virions of herpes simplex virus and other viruses. Actin is abundant in the cytoplasms of all eukaryotic cells. In HSV virions, filaments of appropriate diameter for actin have been seen in the tegument, a capacious compartment lying between the inner surface of the viral envelope and the outer surface of the icosahedral capsid (*e.g.*, Figure 3A). The tegument accommodates an estimated 20 viral proteins,[32] some of which appear to be required early in infection. The functional connotations of the presence of actin filaments in HSV (or other viruses) are not yet clear but we may speculate on some possibilities: for instance, they may provide a scaffold for the binding of tegument components; or they may function as a spacer that prevents the HSV envelope-which is not backed by a layer of matrix protein-from closing too tightly round the nucleocapsid, thereby preventing incorporation of a sufficient complement of tegument proteins.

In a cryo-ET study of the RSV, we encountered numerous examples of molecules protruding from the viral membrane that are apparently host-derived[36]

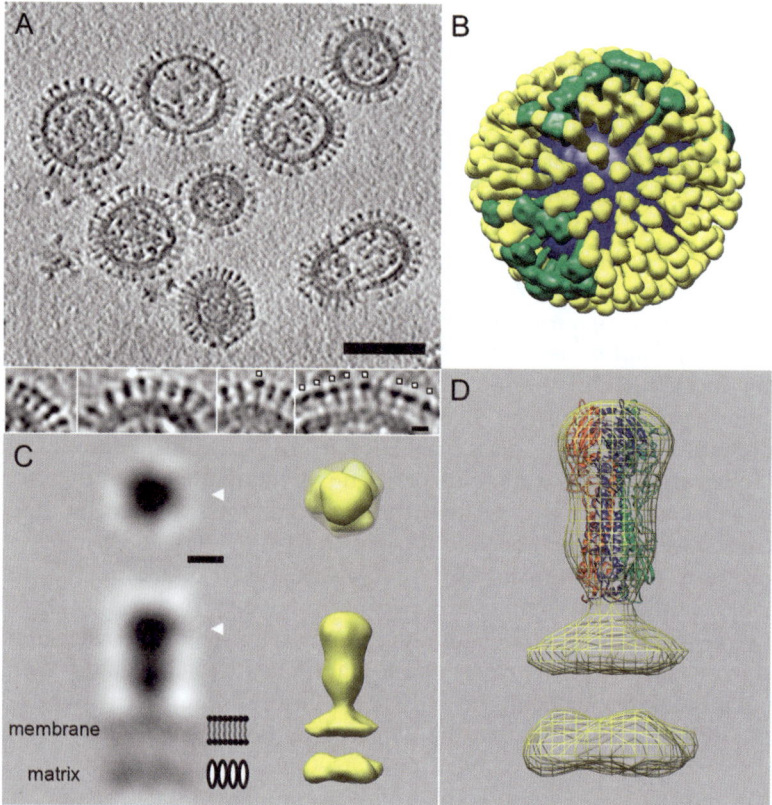

(Figure 6). Most RSV spikes are 7 nm high and 8.5 nm in diameter (Figure 6, right columns), and presumably represent trimers of the viral glycoprotein, Env. The number of Env spikes per virion is quite variable and correlates with the morphology of the capsid contained, but can be 100 or more for virions with closed polyhedral capsids.[36] They are quite sparsely distributed over the virion surface (Figure 3B), a property that facilitates shape perception, and we have detected at least six different kinds (Figure 6). The simplest explanation of their provenance is that they were fortuitously incorporated into nascent virions at the lipid raft-associated sites where budding took place. If a particular host cell membrane protein were to be required to render a virion infectious, it should be

---

Figure 4   (A) Tomographic slice through a field of influenza virions (reproduced from Harris *et al.*, 2006). Bar, 100 nm. The perceived variations in size are a genuine expression of pleiomorphy, not the result of sectioning vicions at different levels. Most virions are roughly spherical but the one at bottom right is filamentous. The two kinds of glycoprotein spikes, trimeric HA and tetrameric NA, may be discriminated in the tomogram. In longitudinal section, HA has a bi-lobed "peanut" shape and is attached to the membrane by a short stalk, whereas NA is slightly longer overall and its globular distal domain is connected to the membrane by a long thin stalk (hence, "lollipop" morphology). At the foot of panel A is a gallery of longitudinal spike sections. Some NAs are marked with square symbols. Bar, 10 nm. In transverse sections, HA and NA may be discriminated by their cross-sections – respectively, triangular and square (data not shown). (B) Distribution of glycoprotein HA (yellow) and NA (green) spikes over the surface of an influenza virus particle. The bilayer is in blue. Spikes were identified as HA or NA in the tomogram and the corresponding densities replaced with molecular models. The two models were obtained by placing the crystal structures of the respective ectodomains on uniform density rods of appropriate length and thickness, and band-limiting both models to 4.5 nm resolution, which approximates the in-plane resolution of the tomogram. There are about seven times as many HA spikes as NA spikes. NA spikes tend to cluster in small patches, an arrangement that may promote adhesion of virions to host cells. Overall, the spikes are packed densely but not in a regular array, and there are occasional patches of exposed bilayer. (C) Density map of a complete HA trimer, obtained by tomographic averaging, 4 nm resolution. 550 individual HA spikes were extracted as subvolumes from tomograms acquired at –4 µm defocus, aligned, and averaged (G. C., unpublished results). At left: (*lower panel*), longitudinal cross-section, showing continuity of density of the ectodomain with the membrane and its disposition relative to the underlying layer of matrix protein. It is likely that the small HA endodomain has some affinity for the matrix protein, as gaps in the matrix protein layer tend to coincide with spike-free patches of membrane. *Top panel*, transverse section through spike at level marked by white arrowhead in lower panel. Both sections are 0.8 nm thick. Bar, 5 nm. Right, surface renderings of the axial (upper) and side (lower) views of the averaged HA trimer and adjacent patches of membrane and matrix protein layer. (D) The crystal structure of the HA ectodomain in its neutral pH (prefusion) conformation (PDB code: 2hmg) fits snugly into the molecular surface defined by the averaged tomogram. The three subunits are in red, blue and green, respectively.

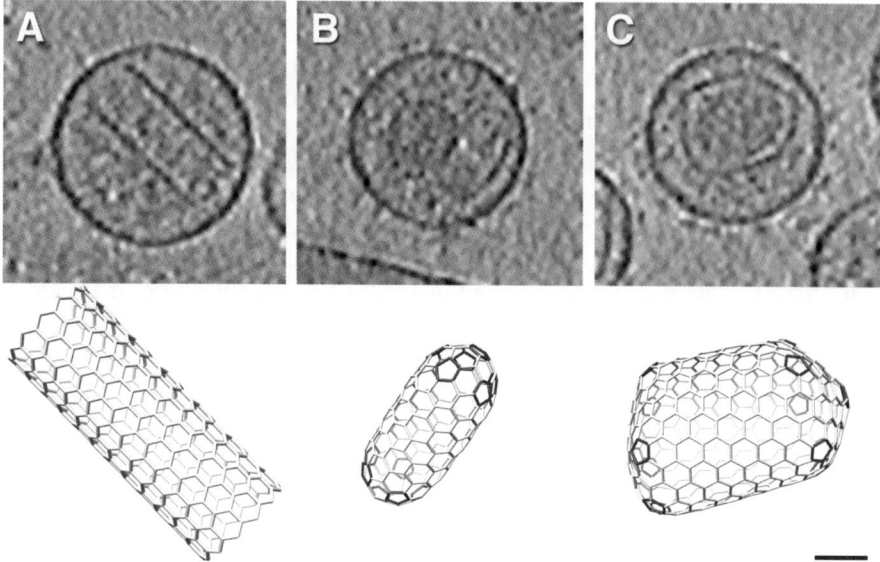

Figure 5   Fullerene lattice models of three RSV capsids visualized by cryo-ET. Bar, 20 nm.

present in most if not all virions, whereas each of the extraneous spikes that we have observed was found only in a minority of virions (see Figure 6 legend). Thus, pending further information, it appears likely that their presence in virions is fortuitous and functionally inconsequential.

Returning to the issue of whether assembling viruses generally eschew host cell proteins, this may represent a design principle on which pleiomorphic viruses diverge from smaller viruses with uniquely specified structures. The internal volume of the latter viruses is quite limited, mandating that the space available to accommodate the genome (and hence, genome size) should be utilized with maximum efficiency by excluding host proteins and nucleic acids and packing the genome to high density. For pleiomorphic viruses, this appears to be less of an issue: for example, the mean packing density of RSV RNA (two 9.6 kb genomes in a roughly spherical space, 105 nm in diameter) is some 30-fold lower than that of poliovirus (one 7.5 kb RNA genome in a spherical space, 24 nm in diameter) or 12-fold lower if only the interior volume of a capsid such as that shown in Figure 3B is considered. It is consistent with pleiomorphic viruses being less precise in how they assemble virions that they should also be less discriminating in which components they incorporate. The examples that we have seen of host cell-derived viral components have been membrane proteins with substantial ectodomains (Figure 6). Internal proteins are less easy to detect as they tend to be smaller and closer-packed. In this context, it appears quite possible that the HSV tegument may also contain some host proteins, as may the corresponding compartment between the capsid

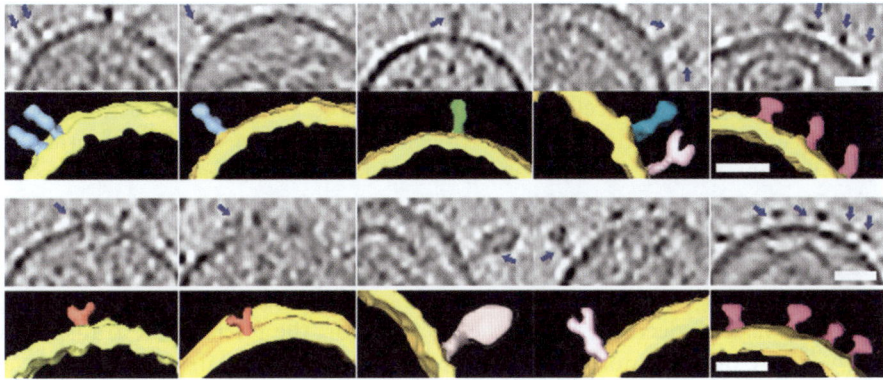

Figure 6 Gallery of morphologically distinct spikes observed in cryo-tomograms of RSV. A complete section through a virion is shown in Figure 3B. Shown here, at higher magnification, are partial sections (rows 1 and 3), showing envelope regions in which spikes of various kinds (marked by dark blue arrows) are present. Below, in rows 2 and 4, are surface renderings at slightly higher magnification in which the spikes are color-coded and the lipid bilayer plus matrix protein layer is in yellow. The majority species of spike, presumably Env trimers, are shown in the right-hand columns (magenta). Other spikes – presumably host cell-derived – are as follows: short Y-shaped molecules, probably dimers, with a 6 nm stem topped by a 6 nm-long bifurcation (bottom left, columns 1 and 2, orange; N = 13 observations); a 2-domain protein ∼ 18 nm long with its long axis perpendicular to the membrane (upper left, light blue; N = 7); a molecule of similar length and orientation but with the two domains less clearly separated (upper rows, third column, green; N = 17); a Y-shaped, presumably dimeric, molecule with a 17 nm-long membrane-proximal domain topped by either a 12 nm bifurcation or an 8 nm bifurcation (column 4, both rows, light pink; N = 7); a large (40 nm long) molecule with a "tennis racquet" morphology (lower row, column 3, pink; N = 1); and a "pillar"-like molecule, 23 nm long (dark blue, column 4; N = 1). Apart from Env, none of these molecules have been identified. Primary observations by Butan et al (2008) are supplemented here with additional data. Bar, 25 nm.

and envelope of RSV and other retroviruses, in addition to their complement of unassembled capsid protein.[45,46]

6 Tomographic Visualization of Cell Entry Events in Subcellular Systems

Hitherto, detailed structural information on events taking place inside virus-infected cells has come mainly from EM of thin plastic sections. As these specimens are chemically fixed, dehydrated, and stained, their portrayal of native structure is compromised. Cryo-ET offers a major extension to this line of investigation, given the superior preservation of native structure in vitrified specimens. However, its applicability is limited by specimen thickness: for practical purposes, the current limit is at about 0.5 µm, which restricts "cellular" cryo-ET studies to the thinnest of prokaryotes and peripheral regions of flat eukaryotic cells. In the long run, the potential of cryo-ET to visualize the substructure of infected cells will be fully realized only when the technology required to prepare vitrified sections of cells[47] comes into widespread practice. In the interim, some studies are appearing, based on the idea of using cryo-ET to image virions interacting with subcellular systems, *i.e.* with receptors or with receptor-containing vesicles or other suitably thin cellular surrogates. We now briefly summarize a few examples.

The first study of this kind visualized the tailed bacteriophage T5 interacting with lipid vesicles decorated with the T5 receptor, the ferrichrome transporter FhuA.[48] The phage genome, initially tightly compressed in its capsid, was seen to appear at lower condensation, in the interior of vesicles. This transfer was seen to be accompanied by a reorganization of the tail-tip, the part that interacts with receptor.

Synaptosomes are membrane-bound structures produced by homogenization of neurons. They have cell-like properties that render them infectable by viruses such as HSV but are much smaller than intact cells, sufficiently so to allow cryo-ET. Maurer *et al.* (2008)[49] used this system to study the entry pathway of HSV, in which the viral envelope fuses directly with the host membrane. They pieced the resulting observations together into a putative time course to propose a pathway in which successive steps in fusion are followed by detegumentation.

Poliovirus has a genome of ssRNA packed into an icosahedral capsid of known structure.[50] Lacking an envelope, it is thought to transfer the genome into an infected cell through a pore created in the cell membrane.[51,52] The poliovirus receptor (Pvr) has been identified: its ectodomain consists of three tandem Ig domains – D1, D2 and D3-whose structure was recently determined.[53] Cryo-EM studies of poliovirus decorated with Pvr ectodomains pinpointed its binding site on the capsid surface.[54–56] Interestingly, they also revealed that although D1 and D2 are linearly aligned, there is a bend of $\sim 60°$ between D2 and D3 (Figure 7). One consequence of this bend is that if receptors on the surface of a susceptible cell were to bind around a capsid vertex, that pentamer would be brought quite close to the cell membrane – a juxtaposition that might be conducive to creation of a channel. This hypothesis has received support from a cryo-ET study of virions interacting with lipid vesicles to which receptors had been bound.[57] The resulting representation of the virion/vesicle complex confirmed the anticipated interaction with a 5-fold site (Figure 7) and also observed an apparent outswelling of the membrane region proximal to the capsid pentamer.

7 Perspective

Cryo-ET of pleiomorphic viruses and virus-containing complexes is currently at about the same stage as cryo-EM of icosahedral capsids was, approximately

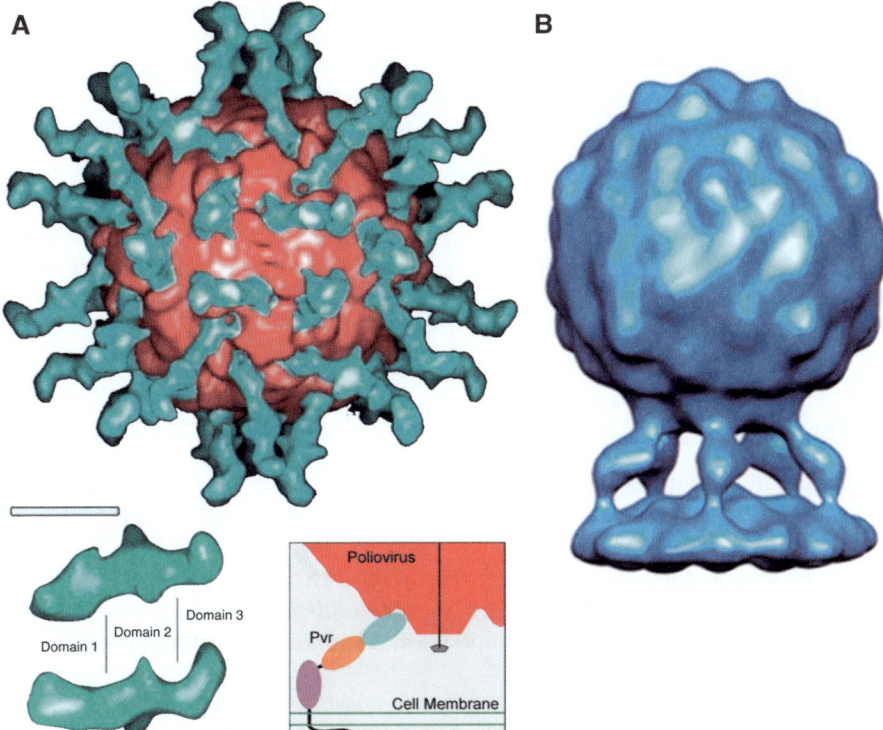

Figure 7 (A) Top: Cryo-EM reconstruction of poliovirus (red) decorated with its receptor ectodomain (turquoise), a monomeric three-Ig-domain molecule – from Belnap *et al.* (2000).[54] Bar, 10 nm. The three domains of the receptor, as excised from this density map, are shown at bottom left; the two prominent protrusions on domain 2 are Asn-linked carbohydrates. Note that domain 3 is rotated by about 60° relative to the axis defined by domains 1 and 2. Bar, 2.5 nm. Bottom right: if domain 3 were to be perpendicular to the membrane surface as shown in the cartoon (domain 3 is violet), domains 2 and 3 would be oriented at about 30° relative to that plane. Binding of multiple receptors around a 5-fold axis would provide a stable attachment bringing the penta-mer, which is thought to be the exit point of the genome during infection, close to the membrane. (B) An averaged tomographic reconstruction of polio-vir-ions bound to receptor-associated lipid vesicles - from Bostina *et al.* (2007)[57]– is consistent with this inferred mode of interaction. Whereas multi-Ig receptors are often drawn as linear stacks whose long axis is at 90° to the plane of the bilayer, the poliovirus receptor is kinked and its long axis at 30° to the bilayer. Other examples of elongated, similarly inclined, spikes have been observed in the envelope of HSV-1 (Figure 3A of Grünewald *et al.*, 2003[31]).

20 years ago, in terms of resolution. We anticipate substantive further technical progress in cryo-ET, with the introduction of more sensitive electronic cameras, improved software for CTF correction and automated segmentation, higher throughput data collection, and maturation of cryo-sectioning, among other

innovations. These technical advances will enhance the resolution achieved in cryo-ET, particularly for averagable components. Whatever the ultimate achievable resolutions turn out to be, the future is bright for the many kinds of investigations now possible in this fertile area of structural virology.

Acknowledgments

We gratefully acknowledge ongoing collaborations with Drs Rebecca Craven (Pennsylvania State University College of Medicine) and Judith White (University of Virginia). This work was supported by the Intramural Research Program of NIAMS and the NIH IATAP Program.

References

1. A. C. Steven and D. M. Belnap, in *"Structural proteomics and its impact on the life sciences"*, ed. J.L. Sussman and I. Silman, World Scientific, Singapore, 2008, 269.
2. Z. H. Zhou, *Curr. Opin. Struct. Biol.*, 2008, **18**, 218.
3. E. Fry, D. Logan and D. Stuart, in *"Methods in Molecular Biology"*, ed. C. Jones, B. Mulloy and M. Sanderson, Humana Press Inc., Totowa, NJ, 1997, 56, 319.
4. M. S. Chapman and L. Liljas, *Adv. Protein Chem.*, 2003, **64**, 125.
5. K. H. Choi, *et al.*, *J. Mol. Biol.*, 2008, **378**, 726.
6. J. T. Huiskonen *et al.*, *J. Virol.*, 2009, **83**, 3762.
7. S. C. Harrison, *Curr. Opin. Struct. Biol.*, 2001, **11**, 195.
8. A. C. Steven *et al.*, *Curr. Opin. Struct. Biol.*, 2005, **15**, 227.
9. V. Lucic, F. Forster and W. Baumeister, *Annu. Rev. Biochem.*, 2005, **74**, 833.
10. K. Grünewald and M. Cyrklaff, *Curr. Opin. Microbiol.*, 2006, **9**, 437.
11. S. Subramaniam *et al.*, *Curr. Opin. Struct. Biol.*, 2007, **17**, 596.
12. D. N. Mastronarde, *J. Struct. Biol.*, 2005, **152**, 36.
13. S. Nickell *et al.*, *J. Struct. Biol.*, 2005, **149**, 227.
14. D. Ress *et al.*, *J. Electron Microsc.*, 1999, **48**, 277.
15. F. Amat *et al.*, *J. Struct. Biol.*, 2008, **161**, 260.
16. H. Winkler and K. A. Taylor, *Ultramicroscopy*, 2006, **106**, 240.
17. D. Castano-Diez *et al.*, *J. Struct. Biol.*, 2007, **159**, 413.
18. G. Harauz and M. van Heel, *Optik*, 1986, **73**, 146.
19. R. Gordon, R. Bender and G. T. Herman, *J. Theor. Biol.*, 1970, **29**, 471.
20. A. S. Frangakis and R. Hegerl, *J. Struct. Biol.*, 2001, **135**, 239.
21. W. Jiang *et al.*, *J. Struct. Biol.*, 2003, **144**, 114.
22. P. van der Heide *et al.*, *J. Struct. Biol.*, 2007, **158**, 196.
23. R. Narasimha *et al.*, *J. Struct. Biol.*, 2008, **164**, 7.
24. R. N. Bracewell and A. C. Riddle, *Astrophys. J.*, 1967, **150**, 427.
25. R. A. Crowther, D. J. DeRosier and A. Klug, *Proc. R. Soc. London, Ser A*, 1970, **317**, 319.

26. J. Frank, *"Electron Tomography"*, 2nd ed., Springer, New York, 2006.
27. M. Unser, B. L. Trus and A. C. Steven, *Ultramicroscopy*, 1987, **23**, 39.
28. P. A. Penczek, *J. Struct. Biol.*, 2002, **138**, 34.
29. G. Cardone, K. Grünewald and A. C. Steven, *J. Struct. Biol.*, 2005, **151**, 117.
30. M. Unser *et al.*, *J. Struct. Biol.*, 2005, **149**, 243.
31. K. Grünewald *et al.*, *Science*, 2003, **302**, 1396.
32. A. C. Steven and P. G. Spear, in *"Structural Biology of Viruses"*, ed. W. Chiu, R.M. Burnett and R.L. Garcea, Oxford Univ. Press, New York, 1997, 312.
33. F. Cocchi *et al.*, *Proc. Natl. Acad. Sci. USA*, 2004, **101**, 7445.
34. E. E. Heldwein *et al.*, *Science*, 2006, **313**, 217.
35. A. Harris *et al.*, *Proc. Natl. Acad. Sci. USA*, 2006, **103**, 19123.
36. C. Butan *et al.*, *J. Mol. Biol.*, 2008, **376**, 1168.
37. B. J. Chen *et al.*, *J. Virol.*, 2007, **81**, 7111.
38. B. K. Ganser *et al.*, *EMBO J.*, 2003, **22**, 2886.
39. B. K. Ganser-Pornillos *et al.*, *J. Virol.*, 2004, **78**, 2545.
40. J. B. Heymann *et al.*, *Comput. Math. Meth. Medicine*, 2008, **9**, 197.
41. N. G. Wrigley, *Br. Med. Bull.*, 1979, **35**, 35.
42. T. Noda *et al.*, *Nature*, 2006, **439**, 490.
43. M. L. Wong and C. H. Chen, *Virus. Res.*, 1998, **56**, 191.
44. M. G. Lyman and L. W. Enquist, *J. Virol.*, 2009, **83**, 2058.
45. J. Lanman *et al.*, *Nat. Struct. Mol. Biol.*, 2004, **11**, 676.
46. J. A. Briggs *et al.*, *Nat. Struct. Mol. Biol.*, 2004, **11**, 672.
47. A. Al-Amoudi *et al.*, *EMBO J.*, 2004, **23**, 3583.
48. J. Bohm *et al.*, *Curr. Biol.*, 2001, **11**, 1168.
49. U. E. Maurer, B. Sodeik and K. Grünewald, *Proc. Natl. Acad. Sci. USA*, 2008, **105**, 10559.
50. J. M. Hogle, M. Chow and D. J. Filman, *Science*, 1985, **229**, 1358.
51. M. T. Tosteson and M. Chow, *J. Virol.*, 1997, **71**, 507.
52. J. M. Hogle, *Annu. Rev. Microbiol.*, 2002, **56**, 677.
53. P. Zhang *et al.*, *Proc. Natl. Acad. Sci. USA*, 2008, **105**, 18284.
54. D. M. Belnap *et al.*, *Proc. Natl. Acad. Sci. USA*, 2000, **97**, 73.
55. Y. He *et al.*, *Proc. Natl. Acad. Sci. USA*, 2000, **97**, 79.
56. D. Bubeck, D. J. Filman and J. M. Hogle, *Nat. Struct. Mol. Biol.*, 2005, **12**, 615.
57. M. Bostina *et al.*, *J. Struct. Biol.*, 2007, **160**, 200.
58. J. F. Conway *et al.*, *Nature*, 1997, **386**, 91.
59. G. Cardone *et al.*, *Virology*, 2007, **361**, 426.
60. J. T. Chang *et al.*, *J. Virol.*, 2007, **81**, 2065.
61. B. Deng *et al.*, *J. Struct. Biol.*, 2008, **161**, 419.
62. W. Dai *et al.*, *J. Struct. Biol.*, 2008, **161**, 428.
63. F. Forster *et al.*, *Proc Natl Acad Sci USA*, 2005, **102**, 4729.
64. J. Benjamin *et al.*, *J. Mol. Biol.*, 2005, **346**, 577.
65. P. Zhu *et al.*, *Proc. Natl. Acad. Sci. USA*, 2003, **100**, 15812.
66. J. A. Briggs *et al.*, *EMBO J.*, 2003, **22**, 1707.

67. G. Zanetti *et al.*, *PLoS Pathog*, 2006, **2**, e83.
68. L. A. Carlson *et al.*, *Cell Host Microbe*, 2008, **4**, 592.
69. P. Zhu *et al.*, *PLoS Pathog.*, 2008, **4**, e1000203.
70. J. Liu *et al.*, *Nature*, 2008, **455**, 109.
71. E. R. Wright *et al.*, *EMBO J.*, 2007, **26**, 2218.
72. M. Cyrklaff *et al.*, *Proc. Natl. Acad. Sci. USA*, 2005, **102**, 2772.
73. G. B. Hu *et al.*, *Virology*, 2008, **372**, 1.
74. A. K. Overby *et al.*, *Proc. Natl. Acad. Sci. USA*, 2008, **105**, 2375.
75. M. S. Spilman *et al.*, *J. Gen. Virol.*, 2009, **90**, 527.
76. M. Barcena *et al.*, *Proc. Natl. Acad. Sci. USA*, 2009, **106**, 582.

CHAPTER 5

Structure Determination of Icosahedral Viruses Imaged by Cryo-electron Microscopy

ROBERT S. SINKOVITS[a,c] AND TIMOTHY S. BAKER[a,b]

[a] Department of Chemistry and Biochemistry, University of California, San Diego, La Jolla, CA 92093, USA; [b] Division of Biological Sciences, University of California, San Diego, La Jolla, CA 92093, USA; [c] San Diego Supercomputer Center, University of California, San Diego, La Jolla, CA 92093, USA

1 Introduction

Three-dimensional (3D) image reconstruction of icosahedral viruses by transmission electron microscopy (TEM) began with the pioneering work on negatively stained samples in Cambridge, UK.[1] This ushered in a new era for virus structure determination and helped lay a firm foundation for subsequent, near atomic resolution X-ray crystallographic studies of viruses such as tomato bushy stunt[2] and southern bean mosaic virus.[3] Crowther *et al.*'s elegant common lines formulation[4] was extremely laborious by today's standards as it required hands-on inspection and analysis of images and their Fourier transforms to identify the orientations of individual virus particles on the TEM support grid. At that time, the ~ 30 Å resolution that could be achieved was limited not by the ability to collect and analyze adequate numbers of images but rather by effects of stain, radiation damage and various distortions to the sample.

The advent of single-particle cryo-electron microscopy (cryo-EM) subsequently revolutionized the use of microscopy for structure determination for

RSC Biomolecular Sciences No. 21
Structural Virology
Edited by Mavis Agbandje-McKenna and Robert McKenna
© Royal Society of Chemistry 2011
Published by the Royal Society of Chemistry, www.rsc.org

viruses and also a wide range of macromolecules and macromolecular complexes.[5,6] In cryo-EM, specimens are preserved in thin layers of vitrified ice, which eliminates the need for negative staining, reduces radiation damage and allows them to be imaged much closer to their native state, albeit at very low contrast. Although structures could now be solved at much higher resolutions, the extremely noisy nature of each particle image necessitated averaging the data from hundreds or even thousands of images to produce reliable 3D reconstructions. As manual processing of large amounts of data became impractical, new approaches were needed to reduce the amount of hands-on intervention. The initial focus was on developing sophisticated software for handling individual steps in the structure determination, but eventually attention turned toward automating the reconstruction from start to finish.

Viruses with a variety of morphologies have been examined by cryo-EM, but those with capsids that have icosahedral symmetry are by far the most common and also often the easiest to study in 3D.[7] They are known to infect hosts from all kingdoms of life and to package a wide range of single- and double-stranded DNA and RNA genomes. Because of the large number of human, livestock, fish and plant diseases caused by these viruses, they have been the subjects of innumerable structural studies, as exemplified throughout this book. The reason why some viruses tend to form icosahedral capsids remains an unsolved mystery, but it is widely argued that evolutionary pressures (*e.g.* genetic economy) led naturally to self-assembling systems comprised of multiple copies of one or a small number of unique subunits.[8]

The basic techniques used to generate cryo-reconstructions of single particles can be straightforwardly applied to icosahedral viruses but, by designing algorithms to exploit their high degree of symmetry, researchers have been able to reach higher resolutions. The symmetry operations that leave the icosahedron invariant result in any non-axial view having 60 equivalent views. This means that we can define an asymmetric unit (ASU) that includes just one-sixtieth of the volume and use this ASU to generate the full icosahedron. We exploit this icosahedral symmetry in two ways. First, when determining the view orientations of the virus particles in our electron micrographs, we only need to consider orientations that are in the ASU rather than the full range of orientations for a 3D object. Second, when a density map is reconstructed from the particle images, each image will make 60 contributions to the reconstruction – one from the assigned orientation in the ASU plus 59 from the symmetry-related orientations. In addition to their symmetry, we also exploit the fact that the icosahedral viruses are roughly spherical and we have developed algorithms that can very efficiently determine reasonable estimates for the orientations of the particle images in the early stages of the reconstruction. We explore all of these points in more detail in Section 3.

The full range of steps necessary for a virus structure determination project is given in Figure 1. Specimen purification has been covered in Chapter 1. Discussions in this chapter are limited to those operations required to go from electron micrographs, acquired on film or CCD camera, to a 3D structure.

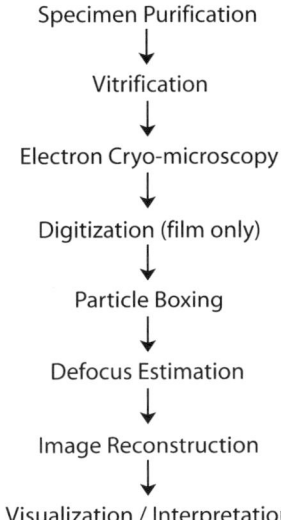

Specimen Purification

↓

Vitrification

↓

Electron Cryo-microscopy

↓

Digitization (film only)

↓

Particle Boxing

↓

Defocus Estimation

↓

Image Reconstruction

↓

Visualization / Interpretation

Figure 1   Primary steps involved in determining and analyzing the 3D structures of biomacromolecules using cryo-electron microscopy. This chapter covers only the steps from digitization through image reconstruction used to study icosahedral viruses.

2   Image Digitization and Preprocessing

Recording Media and Image Digitization

Micrographs are recorded on photographic film or a CCD camera and both have been used to achieve very high-resolution cryo-reconstructions.[9,10] The choice of recording method is often one of personal preference and is influenced by the tradeoff between the ease of use of CCD images and the larger field of view that can be captured with film. Indeed, CCD data are already in a format suitable for boxing particle images but film requires an additional step of digitization. This is normally performed with a flat-bed, scanning micro-densitometer, to produce pixels at a step size of 6–7 μm. For example, an image recorded at a magnification of 50 000× and digitized at 7 μm intervals would generate pixels whose size correspond to $1.4\,\text{Å}$ $(= 7\,\mu\text{m} \times 10^4\,\text{Å}/\mu\text{m}/50\,000)$ in the specimen. A digitized micrograph obtained from scanning an 8×10 cm piece of film at this resolution would have dimensions of approximately 11 400 by 14 300 pixels and contain 155 megapixels. Hence, when each pixel value is represented by a four-byte, floating-point number, the entire micrograph, stored as one file, would consume ∼620 megabytes (0.62 GB) of computer storage space.

Once TEM image data have been recorded and are available in digital form, several preprocessing steps are required before the image reconstruction process can be initiated. Individual virus particles must be identified in the micrographs (see particle boxing), estimates must be made of the defocus levels in order to correct for the effects of the microscope contrast transfer function (CTF)[11] and the image transforms must be scrutinized to make sure that the images are sufficiently free of astigmatism, specimen drift or other artifacts that would degrade the quality of the reconstruction (see defocus estimation).

Particle Boxing

The next step is to identify and window out ('box') individual virus particles. Poor quality micrographs or images of 'bad' particles can always be rejected later, but care should be taken at this stage to reject particles that are over-lapping with other particles or are distorted, deformed, disassembling or show other obvious defects. The size of the box should be large enough so that the boxed images contain a sufficient number of background pixels around the particles. For example, an image of a 500-Å diameter virus such as polyoma or SV40[12] recorded at 50 000× and digitized at 7 μm intervals will require a box size of at least 357 pixels. Hence, a larger box size (perhaps 395 × 395 pixels) would be used to assure that none of the virus particle is missing in the image. Another important reason for providing sufficient padding around the particles is that the use of defocus to enhance image contrast causes an otherwise perfect image to be spread out over a larger area. The density that would have been recorded at a given pixel had the image been acquired in focus is instead replaced by a Gaussian distribution that smears the intensity over neighboring pixels. As a result, the apparently featureless region surrounding the particle image actually contains information that should be used in the reconstruction. This effect is independent of the size of the particle and the extra padding that is required depends only on the value of the defocus.

Defocus Estimation

After particle boxing has been accomplished, the next step is to estimate the defocus level of each micrograph. Unlike the other parameters that are used in the calculation of the CTF correction, the defocus level is not known *a priori* to a sufficient level of accuracy and must be determined from the image data. This is typically achieved through quantitative analysis of the average (incoherent) power spectra of the particle images. A single spectrum is fairly noisy, but averaged spectra generally display a series of concentric 'Thon' rings,[13,14] whose positions are related to the value of the microscope CTF at the time the image was recorded. Various programs[15,16] are used to compute a least-squares fit between a theoretical CTF and the observed rings to determine the defocus level that best agrees with the locations of the nodes in the averaged transform (Figure 2). These programs require as input the known value for the spherical aberration of the microscope objective lens, the accelerating voltage of the electron beam, the pixel size and the averaged power spectra. While estimating the defocus levels, we also have the opportunity to identify artifacts that are not apparent from a simple visual inspection of the micrographs. High levels of

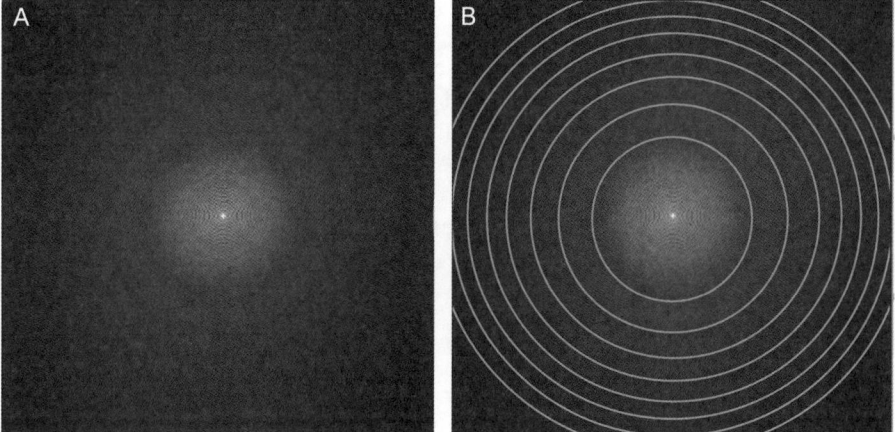

Figure 2 (A) Incoherent average of the Fourier transforms of 100 boxed images of bacteriophage P22 expanded heads. (B) Same as (A) except with nodes in the microscope contrast transfer function highlighted with gray ellipses to illustrate location and presence of slight image astigmatism. Data were acquired at an electron accelerating voltage of 200 kV and an estimated underfocus setting of 1.4 μm.

astigmatism will cause the Thon rings to become elliptical, whereas specimen drift results in loss of intensity in the rings in the direction of the drift.

3 3D Image Reconstruction

Most modern image reconstruction projects, regardless of the symmetry of the system being studied, are based on some type of iterative, model-based refinement method. In its most basic form, the strategy involves determining the origins (*i.e.* common point of reference or center of symmetry) and orientations for a set of particle images by comparing them with projections of a model, then using these aligned images to construct a new, hopefully more accurate model that serves as the starting point for the next iteration.

As expected, an actual implementation is more complex than just described, but these two elements form the core of any iterative reconstruction approach. The first complication is that the quality or resolution of the model must be estimated. There are several mathematically precise definitions for resolution, but here we take it simply to mean the level of detail that can reliably be discerned in the model. The finer detail one is able to discern in a model, the higher is the resolution of that model. For macromolecular structures, low resolution is generally considered to fall somewhere in the ~ 20–50Å range. Here, only gross features of the virus morphology and possibly coarse outlines of the subunits can be distinguished, whereas at very high resolutions ($< 4 \text{Å}$) the tertiary structures of viral protein subunits become visible. Resolution estimates are used to gauge the quality of the model used in refinement and also guide decisions regarding the choice of algorithms and input parameters from one iteration to the next.

Another complication arises because the origins and orientations of the particles are obtained *via* comparison of the particle images to a model, but the model itself is constructed from these same images. We require either a method to assign origins and orientations to the images in the absence of a model or a way of constructing a model without relying on aligned images. This issue is best addressed after first presenting an overview of our automated image reconstruction system.

Iterative Model-based Refinement and Automation

AUTO3DEM is an automation system that we developed to perform icosahedral reconstructions (Figure 3).[17] We specifically highlight features within AUTO3DEM, but two points should be kept in mind. First, the main steps carried out by AUTO3DEM (origin and orientation determination, resolution estimation, model construction) are generic to any iterative, model-based

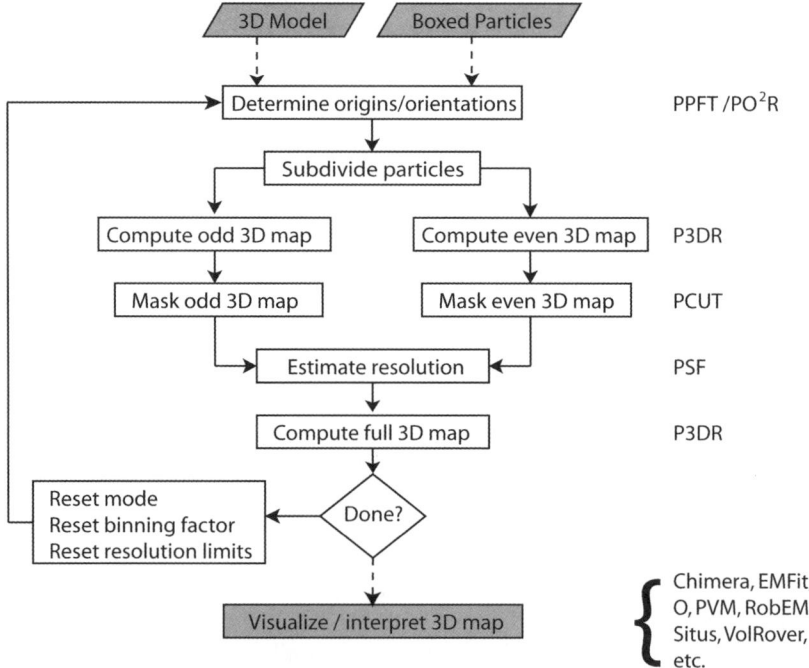

Figure 3   Simplified flow chart of automated image reconstruction process implemented by AUTO3DEM.[17] Shaded boxes represent either input data or steps performed outside of AUTO3DEM. The programs needed at each computational step are listed at the right side of the figure. Dashed and solid lines indicate one-time and iterative operations, respectively. The programs P3DR, PO^{2}R and PPFT impose or assume icosahedral symmetry.

refinement method. Second, other automated systems are available for processing images of particles with icosahedral[18] and lower or no symmetries.[19] The algorithms within AUTO3DEM have been specifically tuned for icosahedral symmetry and the underlying image reconstruction codes have been parallelized and can run on any shared or distributed memory parallel computer.

There are numerous benefits to be realized by automating the image reconstruction process. The most obvious ones are that it relieves the user of many of the repetitive and error-prone steps involved in managing large numbers of data files, setting the parameters for and running multiple programs, interpreting intermediate results and making decisions affecting the overall course of the calculations. Typographic errors may be relatively uncommon at any given step but, over the course of many iterations, even experienced users are likely to make mistakes. Another key advantage of an automation system is that it can be used to capture expert knowledge and make the software more accessible to novice users. Rather than having to understand every detail of the software, less experienced users can rely heavily on the default parameters and generate moderate resolution reconstructions with a minimum of effort. Automation can also reduce the time needed to solve a structure since the delays between the completion of one step and the initiation of the next are eliminated and computer resources are maximally utilized. Finally, the quicker turnaround time made possible through automation enables a researcher to carry out more numerical experiments and reach higher resolutions.

Starting Model/Structure

The iterative refinement process requires an initial model against which a set of particle images can be compared. This starting model does not need to be of very high resolution, but rather just have the correct size and general shape of the structure under investigation. Often a prior reconstruction obtained for a closely related virus can be used, but care must be taken since a size difference of just a few percent can cause the reconstruction process to fail. Geometric models of the proper dimension can also be used, but again are prone to the same problems. An alternative approach that we typically employ is to use the 'random model computation' (RMC)[20] to construct a starting model from a relatively small number of particle images.

A detailed understanding of the RMC is not necessary at this point since the general iterative refinement method only requires that we have a starting map and does not depend on how it was obtained. In addition, the RMC procedure closely follows that used for the general method, but with one small, yet crucial, exception as described later (see Building a Starting Model from Scratch).

Determining Particle Origins and Orientations: Global and Local Refinement

The most computationally intensive and also critical operation in the image reconstruction process is the determination of the five parameters that

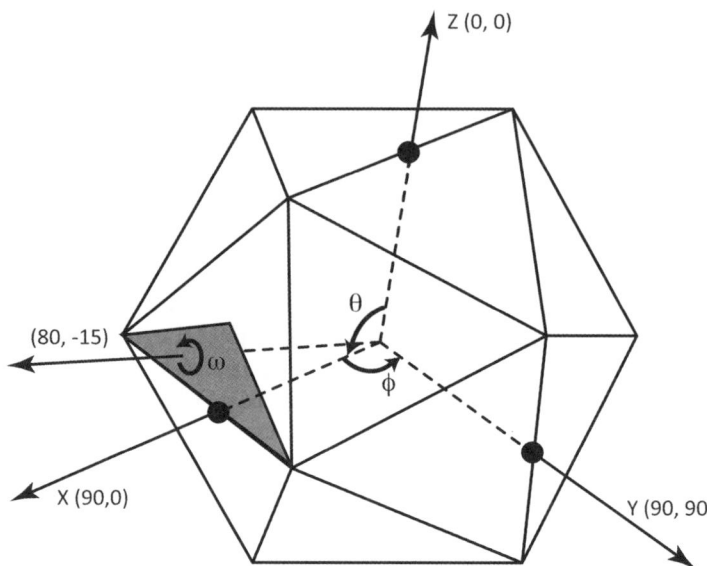

Figure 4   Schematic of icosahedron showing the three angles (θ, φ, ω) that define the view orientations of the icosahedral particles in the electron microscope. The shaded triangle denotes one of 60 equivalent asymmetric units. This unit is bounded by an adjacent pair of fivefold axes ($\theta = 90.0°$, $\varphi = \pm 31.72°$) and an adjacent threefold axis ($\theta = 69.09°$, $\varphi = 0°$). The values of θ and φ for the Cartesian axes and one general view vector appear in parentheses.

characterize each boxed particle image. These are the two coordinates (x, y) that define the origin of the particle relative to the center of the box and the three angles (θ, φ, ω) (Figure 4) that describe the orientation of the particle relative to the electron optical axis of the microscope. Two distinct algorithms are used by AUTO3DEM to carry out this step and in both cases the projections of the model are multiplied by the CTF before comparisons are made with the image data.

The first algorithm is used during the early stages of the reconstruction process, when the model is generally only accurate to and computed at a relatively low resolution (often $\gg 30\,\text{Å}$) and can change significantly from one iteration to the next. To avoid a particular particle being trapped with an orientation assignment that is far from the correct one, we perform a global search of orientation space (*i.e.* all possible orientation angles) for each iteration cycle. This would normally be a very expensive 3D search over the three orientation angles, but the polar Fourier transform (PFT) algorithm[21] subdivides this brute force process into two discrete steps and greatly reduces the computation. In PFT, θ and $|\varphi|$ are estimated first, followed by ω and the sign of φ. The process is made even more efficient by taking advantage of the icosahedral symmetry and restricting the orientation search window to values of (θ, φ) that represent half of the ASU or 1/120th of the total orientation space. (The factor of one half follows from the fact that, in the first part of the algorithm, we only need to determine the absolute value of φ.) Using the default angular step of 0.5°, the PFT program computes a set of 1430 projected views of the model that evenly cover the ASU and then each particle image is compared with every one of these model projection images. If the diameter of the particle is small (*e.g.* <400 Å), we can increase the size of the angular step to 1.0°, in which case only 370 projected views are generated.

The resolution that can be reached at this stage of the reconstruction is variable, but often falls in the range of 15–20 Å and can sometimes reach 10 Å. The time needed for this calculation depends strongly on both the number and size of the particle images and the type of computer hardware being used. We often find that this can be done in a matter of hours when run in parallel mode on a machine with eight dual core Opteron 880 processors running at a clock speed of 2.4 GHz.

In the later stages of the reconstruction process, the program PO^{2}R employs the second algorithm to perform local refinements of the particle origins and orientations.[22] The assumption is made that, at this stage of the processing, the origin and orientation of each particle are relatively close to the true values and that only a very limited region of origin and orientation space, centered about the current values of (x, y) and $(\theta, \varphi, \omega)$, need be searched. The orientation angle search is typically done over a $9\times9\times9$ grid of points with an angular step size typically ranging between 0.1 and 1.0°. The PO^{2}R algorithm has two main advantages over the one used in PFT. First, the computational cost of PO^{2}R does not depend on the size of the angular step, but rather on the number of nearby orientations that are tested. Hence, as the resolution of the model improves, the angular step can slowly be reduced without causing an increase in run time. Second, even when the same angular step sizes are used in PPFT (parallel implementation of the PFT algorithm) and PO^{2}R, the latter generally leads to better alignments and hence higher resolutions since the comparison between the images and projections of the model is performed completely in Fourier space. Of course, both methods are needed since PO^{2}R can only be used after the orientations of the particles are correctly identified within a small region of the ASU.

Image data at full pixel resolution are generally not required during the first few iterations of a reconstruction. The PFT program has the capability of averaging together 2×2 groups of neighboring pixels in the original images to generate sampled or 'binned' images with one-quarter the total number of pixels, thereby reducing the time required for the computations. AUTO3DEM can monitor the progress of the reconstruction and automatically switch from using binned to un-binned data once it detects that the resolution of the model is no longer improving.

The PPFT and PO^{2}R programs can each more accurately determine particle origins and orientations if particles are located close to the center of the image box. For mis-centered, tightly boxed particles, portions of the particle image will be cut off and can never contribute to the 3D reconstruction.

Furthermore, images of particles collected at high defocus contain information ('Fresnel fringes') that can extend a considerable distance beyond the visible edge of the particle. Tight boxing will truncate these fringes, reduce resolution and affect origin and orientation determination. One strategy aimed at achieving higher resolutions is to re-box the particles from the original micrographs using the improved estimates for the particle origins obtained during the reconstruction.

Computing the 3D Reconstruction

The next major step in the reconstruction process is to compute a 3D model from the images of particles whose origins and orientations have been determined. AUTO3DEM uses the program P3DR (parallel 3D reconstruction)[22] to construct a density map of the virus. A density map computationally represents the object as a 3D grid of voxels, each of whose magnitude is proportional to the scattering density within a corresponding small volume of the virus.

Reconstruction of a density map from a series of 2D projection images relies on the well-known projection theorem. This states that the Fourier transform of a projection of a 3D object is equivalent to a central section of the 3D transform of the original object. The implication is that one can easily generate projections of an object from the 3D transform of the object (this operation lies at the heart of the PO^{2}R algorithm) and that the transforms of the particle images can be used to build a map or model of the object once the orientations and origins of the particles imaged are known. This approach is analogous to that used to generate tomographic reconstructions (tomograms), which are obtained by recording a series of images of an object that is rotated systematically ± 60–$70°$ about an axis normal to the electron beam in a microscope.[23] In order to generate a structure that faithfully reproduces the features at all spatial frequencies, the image data must first be corrected for the CTF before performing the back projections. There are several ways of doing this, but the most common are to either correct just the phases[13] or both phases and amplitudes of the Fourier components.[24]

We took advantage of the inherent symmetry of icosahedral viruses when determining particle orientations and limited our search to orientations that fall within a single ASU. We can exploit symmetry again during the reconstruction of the density map by allowing each particle image to make 60 contributions to the reconstruction – one corresponding to the assigned orientation within the ASU and 59 from the symmetry-related orientations.

The quality of the model can be further improved by omitting 'bad' particle images from the reconstruction. The origin and orientation refinement programs, PPFT and PO^{2}R, both output one or more quantitative measures of how well each particle image agrees with the model. AUTO3DEM can use these scores to rank the particles and include just those that lie above a certain threshold or that have scores within a given number of standard deviations of the average score. Particle images can also be eliminated at any time by manually editing the text files that list the particles to be processed.

Although tremendous advantages are realized by assuming icosahedral symmetry when processing images of viruses, the details of structures or substructures will be smeared or averaged out if they are not present in multiples of 60 and arranged with 532 point group symmetry. For example, the nucleocapsids of all herpesviruses are nominally icosahedral but they each possess a small portal complex at one of the vertices.[25,26] Under icosahedral averaging, this component will appear with lower density at all 12 vertices rather than the correct density at a single vertex. More seriously, the structure of such a component as deduced from an icosahedral reconstruction will likely be incorrect since fivefold symmetry would have been imposed regardless of its true symmetry. Icosahedral reconstructions are generally carried out in these cases to solve the capsid to the highest possible resolution and may be followed by lower-symmetry or asymmetric reconstructions to resolve the structure of the non-icosahedral components.[27–29]

Estimating the Resolution of the Reconstruction

An estimate of the level of resolution achieved in the reconstructed virus structure provides an objective gauge of map quality and reliability and helps guide the course of the reconstruction. The Fourier shell correlation (FSC) and phase residual are two measures commonly used to estimate resolution in single-particle reconstructions.[30] Published results often just present or report the FSC since the two tend to assess resolution very similarly. Here, we restrict our discussion to the FSC.

The FSC does not directly measure the resolution of a reconstructed density map, but rather looks at the agreement between a pair of maps. This pair, often referred to as the 'even' and 'odd' maps, is built from image data obtained by dividing the full set of images into two, mutually exclusive sets of equal size. The goal is to determine how well the two maps conform at different spatial frequencies.

An FSC curve plots the correlation coefficients (CCs) between the Fourier transforms of the maps as a function of spatial frequency. Two maps that correlate perfectly (*i.e.* are identical) in a defined band of spatial frequency have a CC = 1.0. Those that exhibit no correlation, yield a CC = 0.0. For real image reconstruction data, the FSC will typically have a value close to one at the very lowest spatial frequencies ($< 1/50\,\text{Å}^{-1}$), indicating that the 'even' and 'odd' reconstructed virus structures are similar in terms of size and overall shape and then eventually drop to zero or lower at high spatial frequencies, suggesting that details within the virus structure at that resolution are unreliable and consist of uncorrelated noise (Figure 5). The spatial frequency at which the FSC first drops below a value of 0.5 is generally considered to be a conservative estimate of the resolution, but this view is not universally accepted and some argue that a cutoff as low as 0.14 is valid.[31] In practice, the FSC curve tends to

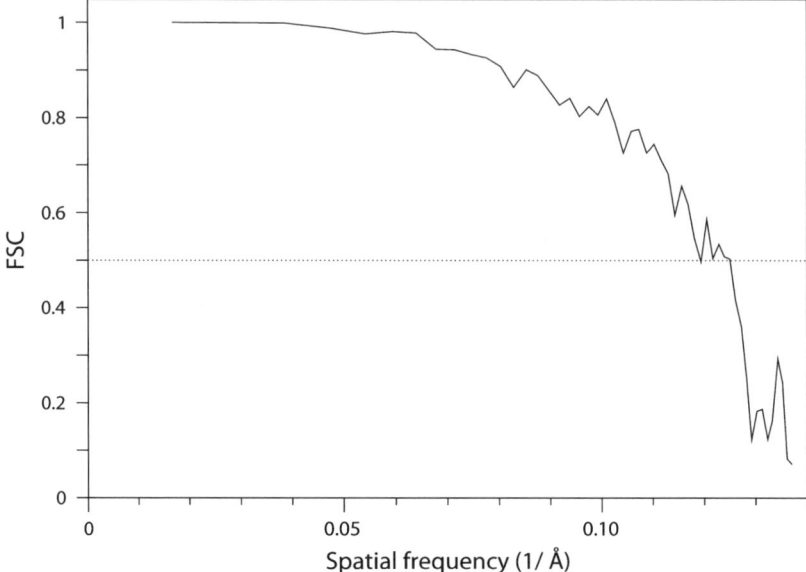

Figure 5   FSC plotted as function of spatial frequency for 3D reconstructions (even
and odd maps) of PsV-F computed from a data set of 2605 boxed particle
images. The spatial frequency at which the FSC drops below a value of 0.5
corresponds to a resolution of 8.1 Å.

drop very rapidly in a narrow region of spatial frequency and the difference in
estimated resolution using these two cutoff values is small. Regardless, the
resolution estimate arising from any FSC-type analysis is simply an estimate
and not a precise determination. It merely offers the researcher some level of
confidence that structural features can be reliability interpreted and correlated
with other relevant data.

The disordered components of the virus, such as the genome, internal or
external lipid bilayers and, in some cases, portions of the major capsid protein
that contact the nucleic acid or bilayer are often solved to a much lower
resolution than the icosahedral capsid. Consequently, a more accurate estimate
of the resolution achieved in the 3D reconstruction is obtained when most of
the disordered regions are computationally excised from the pair of maps
before the FSC curve is determined. Inner and outer radii, r_1 and r_2, are selected
to identify the bounds of the highly ordered region of the virus. All density
values within these radial limits of the 3D maps are unmodified, but those that
lie outside the limits are down-weighted smoothly to zero. That is, a Gaussian,
radial falloff is applied to all densities at $r < r_1$ and $r > r_2$. Such weighting avoids
sharp discontinuities in the density maps at $r = r_1$ and $r = r_2$ that would give rise
to artifacts in the Fourier transforms and lead to artificially high CCs and
overestimation of the resolution achieved.

For most projects, the FSC provides a reliable estimate of the resolution, but
its utility begins to break down as reconstructions are attained at resolutions

beyond about 7–8 Å. In these instances, a more useful although qualitative estimate of resolution is obtained through careful examination of fine features in the map. For example, at about 6–10 Å and sometimes even lower resolution, α-helices in proteins appear as smooth, tube-like density features ('cylinders' or 'tubes') with a diameter of ~6–7 Å. When the resolution of a reconstruction improves to better than 6 Å, the density corresponding to these secondary structure elements transforms from a smooth cylinder morphology into one with more helical character. At about 4–5 Å resolution the helical pitch is unmistakable in cryo-EM density maps.[32,33] It is only at even higher resolution (4 Å or better) that the separation between adjacent β strands in sheets or bulky side chains of amino acid residues can be resolved in proteins.[9,10]

Building a Starting Model from Scratch: the Random Model Computation

The iterative, model-based refinement process requires a starting model to obtain initial estimates for the origins and orientations of the particles. Often we can start with the 3D map of a closely related virus that has been obtained *via* microscopy (single particle image reconstruction) or crystallography (X-ray diffraction of single crystals). However, when a completely new virus structure is being examined, we need some way to bootstrap the refinement process. In our reconstruction scheme, we generally use AUTO3DEM to construct an *ab initio* model using the RMC.[20]

The crux of the RMC approach is to construct a density map from a small number of particle images for which random orientations are assigned and whose origins are set to the center of the box. Although this initial 3D map will bear no resemblance to the actual structure being solved except for appropriately representing the size and symmetry of the virus, it often serves as an effective seed for the image reconstruction process. A rigorous explanation for how this technique succeeds has not been reported, but we believe that the random model contains just enough signal (features consistent with genuine structure) to jump-start the iterative process. This view is also consistent with our observation that the RMC works best when a modest number (generally <200) of images is used. Use of an entire data set of particle images tends to result in a relatively featureless, spherically symmetric model and the refinement process fails to converge and yield a reliable reconstruction.

The RMC works best for viruses whose structures include prominent features such as ridges, arches and protrusions (see, *e.g.*, Section 4), but is less reliable for particles with smoother profiles. In the latter case, multiple random models can be constructed and the one that leads to the best low-resolution model, as measured by the average value of the FSC over a fixed range of spatial frequencies, is selected as the starting point for the full reconstruction using all of the image data.

AUTO3DEM is just one of many image reconstruction packages that now employ similar, random model methods. For example, EMAN[34] and the helical reconstruction package IHRSR[35] offer functionality similar to that embedded in AUTO3DEM.

For completeness, we point out, but with no further explanation, that there are additional ways of constructing low-resolution, starting models from a set of particle images with unassigned origins and orientations. These include the random conical tilt method and angular reconstitution.[13,14] Both of these methods have proven to be very powerful and have utility for examining lower symmetry particles and also those that are asymmetric (the ribosome being the classic example). The reader is encouraged to become familiar with these methods and with the extensive software packages in which they are implemented. These include SPIDER,[36] IMAGIC[37] and SPARX.[38]

Hand Determination

TEM records 2D projections of 3D specimens and handedness information is lost in the images. Hence there is a 50% chance that the reconstruction will be the mirror image of the correct structure. The absolute hand can often be identified from experiments in which pairs of images are collected from the same sample, tilted at two different orientations relative to the electron beam.[31,39] If tilt experiments prove to be inconclusive in distinguishing which enantiomer of a structure is correct, it may be possible to determine hand from the reconstruction itself (see below). When the hand cannot be determined by tilt experiments or direct visualization, the researcher should clearly communicate that the choice of hand is arbitrary for the structure.

For many icosahedral viruses, the subunits that form the capsid are grouped into trimers, pentamers or hexamers that are clearly distinguishable at resolutions of 30 Å or even lower. These multimeric units are further arranged on a regular lattice that can either be symmetric (*e.g.* T = 1, 3, 4, 9, 12, 16, *etc.*) or skewed (*e.g.* T = 7, 13, 19, 21, *etc.*). Since the latter possess a definite handedness, the correct hand for the reconstruction of a virus with a skewed lattice can be inferred by comparison with the lattice of related viruses for which the hand is already known.

Determination of hand is more difficult for capsids that possess a symmetric lattice, but is possible if related capsid structures of known hand have been studied and the individual subunits are clearly resolved in the reconstruction. This is often possible for reconstructions at a resolution of 15 Å or better since both the capsid subunits and the subunit oligomers (trimers, pentamers, hexamers, *etc.*) often exhibit pronounced asymmetry.

For X-ray crystal structures and cryo-reconstructions that have been solved to about 5 Å or better, the hand can be deduced directly from the appearance of secondary structural elements in the density map. For example, an α-helix-rich protein structure will exhibit helical features with a right-handed twist in the map of correct hand. Unfortunately, cryo-reconstructions at such high resolutions are still fairly difficult to achieve.

4 Image Reconstruction Example – PsV-F

A recent reconstruction of a fungal virus, PsV-F, serves to illustrate the image reconstruction process (Figure 6). A low-resolution (25–30 Å) starting model was obtained using the RMC (Section 3) with 150 particle images chosen from the micrographs (40 in total) with the highest defocus levels. AUTO3DEM was then run using the full set of 2605 images to generate a map with an estimated resolution of 8.1 Å in a completely automated manner. The hand of the resulting structure was chosen to be consistent with ScV-L-A.[40] The entire process required 70 min on a 16-processor Linux cluster, including the 8 min used to generate the starting model. Detailed statistics for the calculations (Table 1) show that, during the first three iterations, the value of the FSC never dropped below a value of 0.5 for the entire range of spatial frequencies over which the FSC was calculated. Therefore, the actual resolution should be higher than the value reported in Table 1 (highlighted with *). With image data that had been subjected to a 2×2 binning, an estimated resolution of 12.4 Å was reached in 13.8 min. Three more iterations of AUTO3DEM in search mode using unbinned image data led to a 3D reconstruction with an estimated

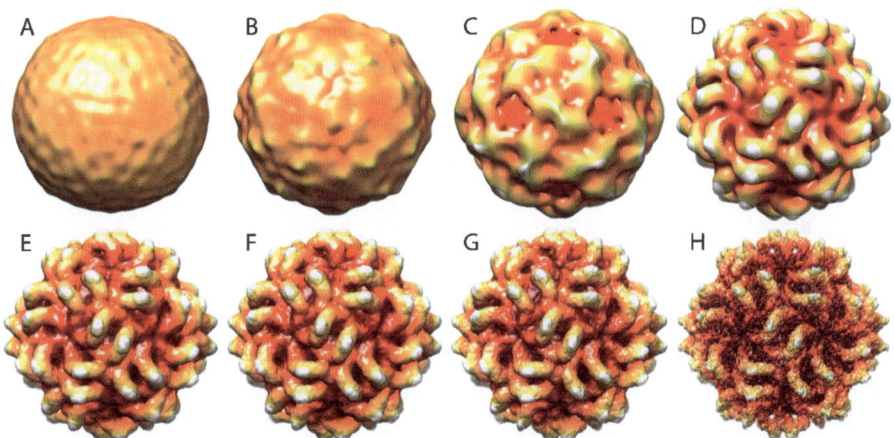

Figure 6   3D surface-shaded representations of PsV-F reconstruction viewed along a twofold axis. Radius in each 3D map is color coded from black (lowest radius) through dark red, orange, yellow, to white (highest radius). (A–D) and (E–H) represent models constructed from 150 and 2605 particle images, respectively. Timings and resolutions are given in Table 1. (A) Model constructed from images that had been assigned random orientations. (B–D) Models from iterations 1, 3 and 5 of the reconstruction process. (E–G) Iterations 9, 12 and 15 from an AUTO3DEM run that used (D) as the starting model. (E) and (F) correspond to the last iterations in search mode using binned and unbinned data, respectively. (G) Result after three further iterations in refine mode. (H) Model at estimated 6 Å resolution obtained using an 8 Å model as starting point and manually running P3DR, PO²R, PSF and PCUT as standalone programs. An inverse temperature factor of $1/300 \text{ Å}^{-2}$ was applied to enhance high-spatial frequency terms.[41]

Table 1 AUTO3DEM statistics for 3D reconstruction of PsV-F. Iteration 0
corresponds to the generation of the initial model from images that
had been assigned random orientations and hence the AUTO3DEM
mode is not applicable. Iterations 0–5 constitute the RMC, whereas
subsequent iterations were carried out with the full set of image data.
Results marked with * indicate that the value of the FSC never
dropped below 0.5 over the entire range of spatial frequencies for
which it was calculated. FSC data from iterations 0 to 4 are so noisy
that resolution estimates are considered to be unreliable, but suggest
that the maps are at resolutions worse than 30 Å.

Iteration	*Mode*	*Images*	*CPUs*	*Binning*	*t (min)*	*Total t (min)*	*Resolution (Å)*
0	–	150	4	2×2	0.9	0.9	>30.0
1	Search	150	4	2×2	1.3	2.2	>30.0
2	Search	150	4	2×2	1.3	3.5	>30.0
3	Search	150	4	2×2	1.5	5.0	>30.0
4	Search	150	4	2×2	1.5	6.5	>30.0
5	Search	150	4	2×2	1.5	8.0	30.0*
6	Search	2605	16	2×2	3.9	11.9	25.0*
7	Search	2605	16	2×2	3.3	15.2	16.7*
8	Search	2605	16	2×2	3.3	18.5	12.5*
9	Search	2605	16	2×2	3.3	21.8	12.4
10	Search	2605	16	1×1	12.5	34.3	10.7
11	Search	2605	16	1×1	12.4	46.7	10.2
12	Search	2605	16	1×1	12.4	59.1	10.2
13	Refine	2605	16	1×1	5.0	64.1	8.9
14	Refine	2605	16	1×1	3.2	67.3	8.5
15	Refine	2605	16	1×1	2.4	69.7	8.1

resolution of 10.2 Å in an additional 37 min of wall clock time. Three iterations
in refine mode further improved the resolution to 8.1 Å in just 11 min more.
Starting from this 8.1 Å map, we were able to increase the resolution further to
approximately 6 Å. This was accomplished by running P3DR and PO²R as
standalone programs outside the context of AUTO3DEM, manually adjusting
the input parameters (inverse temperature factors,[41] method of CTF correction[13] and angular step size used in orientation determination) and visually
inspecting the resultant maps. Changes to the input parameters that resulted in
'better' maps (*e.g.* secondary structural features became less ambiguous and
easier to interpret) were accepted and the corresponding map was used as the
starting point for the next round of origin and orientation refinement. Surface
renderings at various stages of the process illustrate how the quality of the
PsV-F reconstruction improved during refinement (Figure 6).

   Of course, although the results with PsV-F are not atypical, not all virus
image data will behave so well and structures refined to sub-nanometer resolutions may not be achieved so quickly. Nonetheless, the reconstruction of
PsV-F demonstrates that it is possible in some instances to generate models to
sub-nanometer resolution within 1 h after particle images have been boxed and

CTF parameters estimated. Indeed, solving larger viruses such as the 1850 Å diameter Chilo iridescent virus[42] to comparable resolution requires significantly more computation and the construction of a suitable starting model is much more difficult if the virus has a relatively smooth, featureless surface such as dengue virus.[20]

5 Summary

Recent advances in image reconstruction software have made it possible to solve the structures of icosahedral viruses to moderate resolutions in a completely automated manner. What had formerly been an extremely time-consuming step, often requiring many months of concerted effort, can now be completed in a matter of hours in the best-case scenarios once boxed image data are available.

Some of the roadblocks to structure determination have been eliminated but others remain. For example, sub-nanometer resolution cryo-reconstructions are now being achieved more frequently by users with little computational knowledge or experience with the image reconstruction process, but reaching the highest possible resolutions given the limitations of the data set has only been carried out by expert users. Our aim is to quantify and build this expert knowledge into the software so that higher resolutions can more easily and routinely be achieved. The image preprocessing steps, most notably particle boxing and defocus estimation, still require significant manual effort. Fully automating these steps would relieve the structural biologist of much tedium and pave the way to our eventual goal of performing image reconstructions directly at the microscope and providing the cryo-microscopist timely feedback on specimen quality.

Acknowledgments

This work was supported in part by grants R37 GM-033050 and R01 AI-055672 and shared instrumentation grant 1S10 RR020016-01 from the National Institutes of Health to T.S.B. The San Diego Supercomputer Center (SDSC) provided access to TeraGrid computing and support from the University of California, San Diego and the Agouron Foundation (to T.S.B.) were used to establish and to equip cryo-TEM facilities at the University of California, San Diego. R.S.S. received partial support from SDSC.

References

1. R. A. Crowther, L. A. Amos, J. T. Finch, D. J. DeRosier and A. Klug, *Nature*, 1970, **226**, 421.
2. S. C. Harrison, A. J. Olson, C. E. Schutt, F. K. Winkler and G. Bricogne, *Nature*, 1978, **276**, 368.

3. C. Abad-Zapatero, S. S. Abdel-Meguid, J. E. Johnson, A. G. W. Leslie, I. Rayment, M. G. Rossmann, D. Suck and T. Tsukihara, *Nature*, 1980, **286**, 33.

4. R. A. Crowther, D. J. DeRosier and A. Klug, *Proc. R. Soc. London, Ser. A*, 1970, **317**, 319.

5. M. Adrian, J. Dubochet, J. Lepault and A. W. McDowall, *Nature*, 1984, **308**, 32.

6. J. Dubochet, M. Adrian, J.-J. Chang, J. -C. Homo, J. Lepault, A. W. McDowall and P. Schultz, *Q. Rev. Biophys.*, 1988, **21**, 129.

7. T. S. Baker, N. H. Olson and S. D. Fuller, *Microbiol. Mol. Biol. Rev.*, 1999, **63**, 862.

8. D. L. D. Caspar and A. Klug, *Cold Spring Harbor Symp. Quant. Biol.*, 1962, **27**, 1.

9. X. Yu, L. Jin and Z. H. Zhou, *Nature*, 2008, **453**, 415.

10. X. Zhang, E. Settembre, C. Xu, P. R. Dormitzer, R. Bellamy, S. C. Harrison and N. Grigorieff, *Proc. Natl. Acad. Sci. USA*, 2008, **105**, 1867.

11. H. P. Erickson and A. Klug, *Philos. Trans. R. Soc. London*, 1971, **261**, 105.

12. T. S. Baker, J. Drak and M. Bina, *Proc. Natl. Acad. Sci. USA*, 1988, **85**, 422.

13. J. Frank, *Three-dimensional Electron Microscopy of Macromolecular Assemblies: Visualization of Biological Molecules in Their Native State*, Oxford University Press, Oxford, 2006.

14. R. M. Glaeser, K. Downing, D. DeRosier, W. Chiu and J. Frank, *Electron Crystallography of Biological Macromolecules*, Oxford University Press, Oxford, 2007.

15. S. P. Mallick, B. Carragher, C. S. Potter and D. J. Kriegman, *Ultramicroscopy*, 2005, **104**, 8.

16. Z. H. Zhou, S. Hardt, B. Wang, M. B. Sherman, J. Jakana and W. Chiu, *J. Struct. Biol.*, 1996, **116**, 216.

17. X. Yan, R. S. Sinkovits and T. S. Baker, *J. Struct. Biol.*, 2007, **157**, 73.

18. W. Jiang, Z. Li, Z. Zhang, C. R. Booth, M. L. Baker and W. Chiu, *J. Struct. Biol.*, 2001, **136**, 214.

19. G. Tang, L. Peng, P. R. Baldwin, D. S. Mann, W. Jiang, I. Rees and S. J. Ludtke, *J. Struct. Biol.*, 2007, **157**, 38.

20. X. Yan, K. A. Dryden, J. Tang and T. S. Baker, *J. Struct. Biol.*, 2007, **157**, 211.

21. T. S. Baker and R. H. Cheng, *J. Struct. Biol.*, 1996, **116**, 120.

22. Y. Ji, D. C. Marinescu, W. Zhang, X. Zhang, X. Yan and T. S. Baker, *J. Struct. Biol.*, 2006, **154**, 1.

23. V. Lucic, F. Forster and W. Baumeister, *Annu. Rev. Biochem.*, 2005, **74**, 833.

24. V. D. Bowman, E. S. Chase, A. W. Franz, P. R. Chipman, X. Zhang, K. L. Perry, T. S. Baker and T. J. Smith, *J. Virol.*, 2002, **76**, 12250.

25. G. Cardone, D. C. Winkler, B. L. Trus, N. Cheng, J. E. Heuser, W. W. Newcomb, J. C. Brown and A. C. Steven, *Virology*, 2006, **361**, 426.

26. J. T. Chang, M. F. Schmid, F. J. Rixon and W. Chiu, *J. Virol.*, 2006, **81**, 2065.
27. W. Jiang, Z. Li, Z. Zhang, M. L. Baker, P. E. Prevelige Jr and W. Chiu, *Nat. Struct. Biol.*, 2003, **10**, 131.
28. J. Chang, P. Weigele, J. King, W. Chiu and W. Jiang, *Structure*, 2006, **14**, 1073.
29. G. C. Lander, L. Tang, S. R. Casjens, E. B. Gilcrease, P. Prevelige, A. Poliakov, C. S. Potter, B. Carragher and J. E. Johnson, *Science*, 2006, **312**, 1791.
30. M. van Heel and M. Schatz, *J. Struct. Biol.*, 2005, **151**, 250.
31. P. B. Rosenthal and R. Henderson, *J. Mol. Biol.*, 2003, **333**, 721.
32. W. Jiang, M. L. Baker, J. Jakana, P. R. Weigele, J. King and W. Chiu, *Nature*, 2008, **451**, 1130.
33. S. J. Ludtke, M. L. Baker, D. H. Chen, J. L. Song, D. T. Chuang and W. Chiu, *Structure*, 2008, **16**, 441.
34. S. J. Ludtke, P. R. Baldwin and W. Chiu, *J. Struct. Biol.*, 1999, **128**, 82.
35. E. H. Egelman, *J. Struct. Biol.*, 2007, **157**, 83.
36. W. T. Baxter, A. Leith and J. Frank, *J. Struct. Biol.*, 2007, **157**, 56.
37. M. van Heel, G. Harauz and E. V. Orlova, *J. Struct. Biol.*, 1996, **116**, 17.
38. M. Hohn, G. Tang, G. Goodyear, P. R. Baldwin, Z. Huang, P. A. Penczek, C. Yang, R. M. Glaeser, P. D. Adams and S. J. Ludtke, *J. Struct. Biol.*, 2007, **157**, 47.
39. D. M. Belnap, N. H. Olson and T. S. Baker, *J. Struct. Biol.*, 1997, **120**, 44.
40. H. Naitow, J. Tang, M. Canady, R. B. Wickner and J. E. Johnson, *Nat. Struct. Biol.*, 2002, **9**, 725.
41. W. A. Havelka, R. Henderson and D. Oesterhelt, *J. Mol. Biol.*, 1995, **247**, 726.
42. X. Yan, Z. Yu, P. Zhang, A. J. Battisti, H. A. Holdaway, P. R. Chipman, C. Bajaj, M. Bergoin, M. G. Rossmann and T. S. Baker, *J. Mol. Biol.*, 2009, **385**, 1287.

CHAPTER 6

X-ray Crystallography of Virus Capsids

LAKSHMANAN GOVINDASAMY, MAVIS AGBANDJE-MCKENNA AND ROBERT MCKENNA

Department of Biochemistry and Molecular Biology, Center for Structural Biology, The McKnight Brain Institute, University of Florida, Gainesville, FL 32610, USA

1 Introduction

Two basic principles govern the assembly of spherical (icosahedral) viruses: (i) genetic economy – the encapsidated genome encodes a single or few capsid proteins (CPs) that assemble a protective shell (the viral capsid); and (ii) specificity – the CPs have to fold to recognize each other and form exact CP–CP interfacial interactions during the assembly pathway. The study of virus structure benefited greatly from both the technical and methodological advances in single-crystal X-ray crystallography (in the 1970s) that facilitated the determination of intact virus capsids structures,[1–3] which was pivotal to understanding the nature of the interactions between protein–protein subunits and protein–nucleic acid that facilitate viral capsid assembly.

The method of virus capsid crystallography involves the same essential steps as protein or small molecule crystallography: (1) the isolation and purification of the virus (see Chapter 1), (2) the crystallization of the virus, (3) the acquisition and processing of X-ray diffraction data, (4) the phasing of the measured data and (5) the building and refinement of the protein and nucleic acid structures into the experimentally determined electron density maps (Figure 1).

RSC Biomolecular Sciences No. 21
Structural Virology
Edited by Mavis Agbandje-McKenna and Robert McKenna
© Royal Society of Chemistry 2011
Published by the Royal Society of Chemistry, www.rsc.org

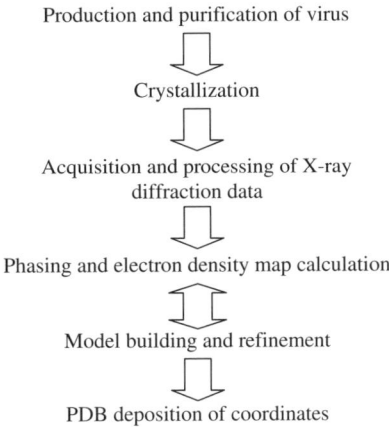

Figure 1 Steps involved in virus capsid crystallography.

This chapter will only be concerned with a non-mathematical discussion of the structure determination of intact virus capsids consisting of a protein shell with icosahedral (532 point) symmetry, commonly termed spherical viruses.

2 Experimental Procedures

Crystallization

The requirement for crystals stems from the criteria for the measurement of diffracted X-rays (a consequence of not being able to focus X-rays): a three-dimensional lattice of the virus is necessary to create a regularly ordered array of the virus, to create an amplified, measurable and interpretable X-ray diffraction pattern.

Crystallization conditions for viruses are not significantly different from those for proteins (see Chapter 8). Crystal growth is achieved by the slow removal of the water of solvation from the virus in a controlled manner that prevents precipitation and takes the virus out of solution and into the crystalline state. However, just like proteins, the success of this process is highly dependent on sample purity. The standard 'rule-of-thumb' is that the virus must be at least 95% pure, and preferably 99%, as the greater the percentage of impurities present, the less likely the sample is to crystallize. The purity of the virus should always be checked before starting a crystallization experiment or trial; hence it is good practice to check the virus sample by SDS-PAGE (ensuring that the virus has been boiled for long enough, 5–10 min, to ensure that the capsid is disassembled) developed by silver staining and by negative-stain electron microscopy, to verify that there are no contaminating entities (such as lipids) and capsid integrity.

The virus should be stable in the crystallization buffer of choice for at least the duration of the crystallization process (the optimum time required for virus crystal growth). Note that this can be on the order of a few days to a few months. The concentration of the sample should be at least $5\,mg\,mL^{-1}$, preferably $10\,mg\,mL^{-1}$ or higher. In some cases, high solubility may require the addition of detergents, such as *n*-octyl-β-D-glucoside, in the sample buffer. Finally, before setting up any crystallization experiment, standard spectrophotometric [optical density (OD)] measurements should be taken at 260, 280 and 310 nm, to check concentration and possible virus aggregation. If the A_{310} reading is significantly high (for example, $>10\%$ of the 280 nm OD), the sample should be microcentrifuged for 4–5 min to remove insoluble virus, as this will adversely affect the crystallization process. However, even if the solubility of the virus is inherently low, it is still worth attempting a crystallization trial at whatever concentration can be achieved.

The most common crystallization experiments are manually set up in plates (referred to as trays), which can contain 24 or 96 individual crystallization compartments or wells. A different condition can be tested in each well of the plate and the most widely used technique for virus capsid crystallization is the sitting or hanging drop vapor-diffusion method.[4] These approaches are easy to perform and require only small amounts of sample, and are therefore the methods of choice when trying to identify new crystallization conditions for a virus sample.

In the sitting drop method, a 2–40 µl droplet of the sample is mixed with an equal (or varied ratio) amount of the precipitant solution (also referred to as mother solution, mother liquor or reservoir solution) and placed on a bridge/post sitting inside a well, in vapor equilibration with the precipitant solution (500–1000 µl) at the bottom of the well. In the hanging drop method, the droplet of sample, also mixed with the precipitant, is suspended from a cover-slip over the top of the precipitant solution in the well. With both of these methods the sealing of the each crystallization well is essential to prevent air evaporation of the drop. The wells are sealed by creating an interface between the cover-slip and the rim of each well on the plate using vacuum grease, oil or sealing tape in the case of the sitting drops. Hence the crystallization experiment is 'set up' where the initial precipitant concentration in the droplet is less than that in the reservoir (generally at 50%), hence over time the reservoir solution will 'draw' water from the droplet in a vapor phase until equilibrium exists between the droplet and the reservoir. During this equilibration process [which can happen at room temperature, cold room (4 °C) or other desired temperature] the virus is also concentrated, increasing its relative supersaturation, thus slowly 'bringing it out' of solution and into a crystalline state.

The major advantages of the sitting and hanging drop techniques are speed and simplicity. The disadvantage of the sitting drop technique is that crystals can sometimes adhere to the sitting drop surface, making harvesting for data collection difficult. The hanging drop method avoids the problems of surface crystal adherence, but the droplet volume is limited compared with sitting drops. Despite these minor disadvantages, both the sitting drop and hanging drop methods are excellent for crystallization condition screening and optimization.

Screening for initial crystallizations for a new virus project often begins with the testing of conditions that have been used for the crystallization of homologous viruses or testing of conditions listed in the Virus Particle Explorer (VIPER) crystallization condition database webpage (http://viperdb.scripps.edu/crystals.php). Screening can also begin with commercially available kits (containing ~ 50 solutions of varying precipitant, buffer, salt and pH) which can be used in a sparse matrix 'trial and error procedure', although these tend to be more suitable for the crystallization of proteins since they generally contain high concentrations of salt or polyethylene glycol precipitants. Regardless of the route for selecting a screen, the goal is to find a condition which brings the virus out of solution in a controlled manner, *i.e.* crystallize it, which may take several months to years. Hence researchers routinely check crystal trays using a light microscope once per week or month to monitor for crystal growth during screening (Figure 2A). Once a condition has been found to produce crystals, it is normal practice to fine tune this condition to obtain optimal crystal growing parameters, which may include varying the temperature and the possible addition of an additive, for example detergents or salts.

The process of crystallization screening has now become less manual and more automated with the development of robotics, which can screen tens of thousands of crystallization conditions in a single experiment. However, robotics are mostly routinely used for protein crystallization, where sample quantity is not a bottleneck. Other than plant viruses, sample yield is often in

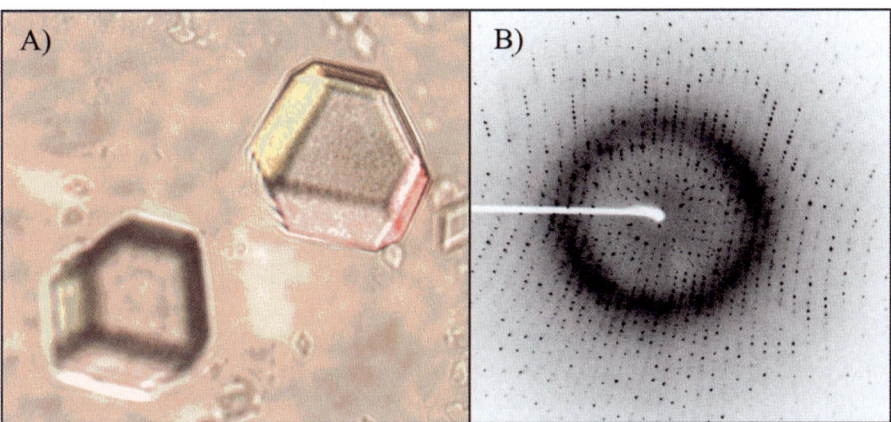

Figure 2   (A) An optical photograph of the AAV4 crystals in a hanging drop, taken with a Zeiss Axioplan 2 microscope. The approximate dimensions of the crystals are 0.2×0.2×0.2 mm. (B) A typical 0.25° oscillation photograph for an AAV4 crystal diffracting X-rays to 3.2 Å resolution. The image was collected on an ADSC Quantum4 CCD detector at the F1 beamline ($\lambda = 0.916$ Å) at the Cornell High Energy Synchrotron Source with an exposure time of 90 s and a crystal to detector distance of 300 mm.

the low microgram region and as such manual set-ups, as described above, are still common practice in most virus crystallography laboratories.

X-ray Data Collection and Processing

Even though virus capsid crystals are in the solid state, they still have an average solvent content of 30–70% and therefore are sensitive to dehydration. Hence when manipulating crystals, for example when transferring them for data collection, caution must always be taken to ensure that the crystal solution environment mimics the crystal. Hence a crystal stabilizing solution is often required which mimics the final concentration of the crystallization drop after vapor diffusion. This can be approximated by using the reservoir solution of crystallization.

Crystal Mounting

Crystals are prepared for X-ray data collection by one of two methods, commonly termed 'wet' and 'cryo' mounting.

The less common of these two methods is wet mounting, where the crystal is aspirated from the drop, still in solution, into a quartz capillary. Then the solution surrounding the crystal is slowly removed by pipette or some absorbent material (for example, a piece of filter-paper or paper wick) until most of it is removed. Care has to be taken to ensure that the crystal is not touched or dried out. This process allows the crystal to attach to the side of the capillary tube and become stationary. A solution plug is placed on either side of the crystal, but not in contact, and the capillary tube is then sealed with wax or oil at both ends. This method ensures that the crystal is maintained in the same environment as it was in the crystal drop. The major advantage of mounting the crystals using this method is that diffraction data can be collected at room temperature, but the major disadvantage is that the crystals will undergo radiation damage during the X-ray data collection. The cooling of the crystals to 4 °C will reduce the speed of radiation damage, by slowing both heating effects and the diffusion rate of the free radicals formed, although not all crystals tolerate the temperature transition if grown at a different temperature.

The more widely used method for crystal preparation prior to X-ray data collection is cryo-mounting, because this prevents (reduces) free radical radiation damage to the crystal, as cooling to 100 K prevents the diffusion of free radicals. In this method, the crystals are harvested from the drop with a small nylon loop[5] and flash cooled either directly in a flowing 100 K nitrogen gas stream on the X-ray data collection instrument itself or by plunging into liquid nitrogen. Sometimes a crystal may not freeze completely during the first exposure to the 100 K nitrogen stream and therefore a re-freezing is necessary for a better outcome, *i.e.* the nitrogen stream flow is interrupted for few seconds, using a thin card, which is then removed to allow the stream to flow on the crystal to re-cool it. The crystal can then be shipped at 100 K to a data collection facility, using dry shipping dewars, as shipping in liquid nitrogen is too hazardous. The disadvantage of this method is that the cooling of the crystal has to be very rapid and done in the presence of a cryo-protecting additive to prevent the formation of ice crystals within the solvent channels of the crystal lattice. The most commonly used of these cryo-protecting additives is glycerol (at ∼25%), which is added to the crystal stabilizing solution. Some other useful cryo-protectants are ethylene glycol, sucrose, PEG-400 and 2-methyl-2,4-pentanediol (MPD). Care has to be taken not to dilute the concentration of the stabilizing solution with the addition of the cryo-protectant. Caution is also necessary when introducing cryo-protectant solutions to the crystal stabilizing solution, since many crystals are extremely sensitive to environment change and may crack or dissolve. Therefore, 'crack' tests are performed in which small crystals are tested by addition of the cryo-protectant solution before larger crystals, more suitable for data collection, are exposed to the new environment. To circumvent these potential issues, screening conditions, including many versions of crystallization screening kits, are prepared with glycerol and are thus cryo-ready. As a final word of caution, the act of cryo-cooling can sometimes decrease the diffraction quality (resolution and order) of the crystal, hence it is always wise to test the diffraction quality at room temperature using the wet-mounting method as a bench mark prior to cryo-cooling.

Data Collection

X-ray diffraction data acquisition requires an X-ray source, optics and detector. These can be either 'in-house' laboratory instrumentation or nationally based synchrotron facilities. X-ray diffraction data collection from large-unit cells, such as is the case with whole virus capsids, can be challenging. Hence the use of in-house laboratory Cu Kα ($\lambda = 1.5418$ Å) X-rays, although feasible using good optics, has disadvantages because of the long exposure times (30–90 min) for a single useful image, which creates a large background signal and thereby reduces the quality of the data (low signal-to-noise ratio). Hence most virus capsid crystal diffraction X-ray data sets are routinely collected at a high-energy synchrotron radiation facility, which has many advantages compared with laboratory source X-ray generator data collection. The most obvious of these is that the synchrotron beams have a much greater flux of X-rays (1000-fold greater intensity) and optimal optics setup to deliver a high-quality stable beam, which in practical terms means that a virus crystal data set can be collected from a single crystal with image exposure times of seconds. Another advantage of synchrotron data collection is that many beamlines have optics systems that allow the selection and tuning of the wavelength (typically $\lambda = 2.0$–0.8 Å) used for data collection, other than the alternative low-flux fixed wavelength that is produced in the in-house laboratory setting. Having a high flux source also means that the beam size can be reduced using a collimator or slit to produce a beam size favorable for collecting large unit cell data for viruses without overlap between the reflections.

Before exposing a virus crystal to the X-ray beam (in-house or at a synchrotron), the optical bench parameters should be optimized for data acquisition. These should be optimized taking into consideration the crystal unit cell dimensions [the repeating motif that makes the crystal lattice (typically > 300 Å)] and the expected resolution of diffraction (not typically expected to be better than 2.0–3.0 Å resolution). These considerations will determine the ideal crystal-to-detector distance and the oscillation angle for each collected diffraction image. These values can vary from sample to sample and are affected by the wavelength of the X-rays used, but typically values are 300–600 mm for the crystal-to-detector distance and 0.1–0.3° for the oscillation angle. Usually, what is termed a 'snap-shot' is first taken to determine if these experimental parameters are suitable for data collection. After several images have been recorded, the intensities of the reflections on each image should be assessed to ensure that they have good signal-to-noise ratios [the average $I/I(\sigma)$ should be at least above 2.0]. The space group (the packing arrangement of the virus in the unit cell) can also be identified from a few initial images. This information is required to assess how many oscillation images should be collected for a complete diffraction data set.

The percentage data completeness is very important for successful virus structure determination. A theoretical expected number of collected diffracted waves can be determined from the known unit cell dimensions as a function of resolution. Complete coverage of this diffraction space is assured by collecting as large a volume (sweep) of reciprocal space as possible and avoiding reflection spot overlap. In some cases the crystals, even if cryo-cooled, are still susceptible to radiation damage and a complete data set has to be built from the merging of data collected from several crystals that were in a different orientation during the data collection process.

The orderedness of the crystal (regularity of the arrangement of the crystal unit cells) can be judge by the mosaicity of the crystal, which is defined as the average spread of the orientation of the building blocks in the whole lattice. Good quality crystals have mosaicity ranging from 0.1 to 0.2° and ideally should be no more than 1°.

Most often, virus capsid crystals that diffract X-rays between 6.0 and 2.5 Å resolution will provide reasonable biological information from the structures determined. Again, this is very dependent of the biological question being asked. At ∼5.0 Å resolution the main-chain fold of the virus can be determined and at ∼3.5 Å resolution side-chains can be assigned.

Data Processing

The main purpose of X-ray diffraction data processing is to derive the plane (Bragg plane), within the crystal lattice, from which each X-ray beam was diffracted and to measure the intensity of each diffracted wave (also called a reflection). This information is determined from the individually collected diffraction patterns, each depicting a slight, but precise, orientation change (due to oscillation) of the crystal from the previous image. There are several software packages that can perform these calculations, including MOSFLM, HKL2000 and D*TREK (Table 1).[6–8] Essentially the orientation of the unit cell is determined for the first image and used to predict the subsequent diffraction pattern that was recorded (Figure 2B). This process is termed 'indexing'. Typically, data processing requires the display of the diffraction frames, selection of reflections (peak picking), indexing (defining the unit cell parameters for the crystal and Bragg planes for each diffracted wave) and peak integration (measuring the intensities of each diffracted wave). This incorporates the experimental parameters, for example, crystal-to-detector distance, oscillation angle, beam position and detector type. Due to recent advances in the programs used and computational power, the processing of diffraction data has become almost automated. Once each oscillation image has been indexed and the intensity of each diffracted wave measured, all the individual images are scaled and/or merged together and reduced, based on the space group assignment (a definition of the internal symmetry which relates the virus capsids orientations and positions within the unit cell) to one data file. This generates the reflection 'observed data set', which can be saved as intensities or converted to the amplitudes of the waves (which is approximately equivalent to the square root of the measured intensity of each reflection). Each of the recorded intensity also has a measurable associated signal-to-noise error, which is also included in

Table 1 X-ray crystallography programs (indexing, phasing and refinement).

Program	Website	Ref.
Indexing		
MOSFLM	http://www.mrc-lmb.cam.ac.uk/harry/mosflm/	6
HKL2000	http://www.hkl-xray.com/	7
d*TREK	http://www.rigaku.com/software/dtrek.html	8
Structure determination (phasing and refinement)		
Envelope	http://bilbo.bio.purdue.edu/~viruswww/ Rossmann_home/softwares/other.php	11
AMORE	http://www.ccp4.ac.uk/autostruct/amore/	12
CNS	http://cns.csb.yale.edu/v1.2/	13
PHASER	http://www-structmed.cimr.cam.ac.uk/phaser/	14
CCP4	http://www.ccp4.ac.uk/index.php/	15
PHENIX	http://www.phenix-online.org/	16
SHELX	http://shelx.uni-ac.gwdg.de/SHELX/	17
COMO	http://como.bio.columbia.edu/tong/Public/Como/ como.html	18
EPMR	http://www.msg.ucsf.edu/local/programs/epmr/ epmr.html	19
MOLREP	http://www.ysbl.york.ac.uk/~alexei/molrep.html	20
SHARP	http://www.globalphasing.com/sharp/	21
SnB	http://www.hwi.buffalo.edu/SnB/	22
REFMAC	http://www.ysbl.york.ac.uk/~garib/refmac/	23
TNT	http://www.uoxray.uoregon.edu/tnt/welcome.html	24

the data set. Analysis of the scaling statistics of the entire data set allows for the removal of individual reflections or whole images to be rejected or reprocessed to improve the overall quality of data set. Information on the quality of the data is provided by the agreement of multiple recorded or symmetry-related data (based on the assigned space group of the crystal) in the merged/scaled diffraction images, termed the R_{sym}. The value of this term can range from 5 to 20% for virus data sets and is dependent on diffraction resolution and quality of the crystal.

3   Phase Determination

The method of X-ray crystallography allows the direct measurement of the intensity of the diffracted waves (as discussed above), but not the phases required to construct the electron density maps, the so-called phase problem. Hence several methods have been developed to obtain this phase information, including the methods of isomorphous replacement and molecular replacement. However, other methods to determine the phase for protein structures exist, such as multi-wavelength anomalous dispersion (MAD), but they have not been used successfully to solve virus capsid structures so far (Table 1).[11-29]

The icosahedral architectural nature of spherical viruses, with their 532 point group symmetry, directly implies that symmetry not consistent with crystal packing (a fivefold symmetry operator) and that can never occur as part of the crystal lattice will be observed. This virus symmetry is termed non-crystallographic symmetry (NCS) and was first identified by Crick and Watson in 1956. Later, in 1962, Rossmann and Blow demonstrated how NCS operators could be used to determine the orientation of macromolecules (later applied to virus capsids) within the crystal unit cell, termed the rotation function (Figure 3).[9,10] The use of NCS operators has also been shown to be critical for improving the initial phases by electron density map averaging and also helpful for phase extension from low resolution to high resolution of the X-ray diffraction data.

Isomorphous Replacement

This method of obtaining phase information is used when there is no homologous virus structure available. The method involves introducing one or two heavy (electron-rich) atoms into the native virus crystal by soaking them into the lattice without changing the conformation of virus or distorting the crystal lattice. Usually, two or more X-ray diffraction data sets are collected, one for the native virus capsid crystal and at least two derivative data sets from the virus crystal incorporated with different heavy atoms. Examples of heavy atoms used include the salts of gold, lead, mercury, platinum and uranium compounds. The mercury and platinum salts typically bind covalently to cysteine and methionine residue side-chains. Other heavy atoms are usually less specific and often bind through electrostatic interactions. The native and derivative heavy atom soaked crystals should be isomorphous with each other with the only difference between

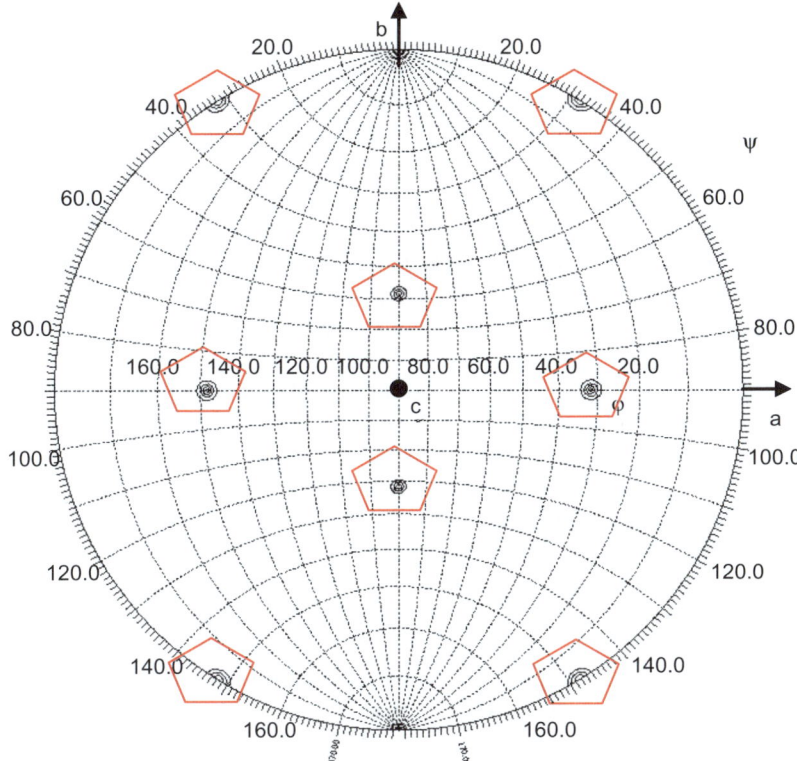

Figure 3 Stereographic projection showing the self-rotation function result for the AAV4 X-ray diffraction data for $\kappa = 72°$. Shown are the fivefold icosahedral symmetry elements, using the observed data set in the 10–3.5 Å resolution range, with a radius of integration of 120 Å. The map is contoured at 3σ. The virus fivefold symmetry axes peaks are highlighted as open red pentagons.

them being the additional heavy atom(s). This means that the crystal unit cell (and space group) and the contribution of the virus and solvent atoms to the diffraction are the same. Then, using a process of calculating difference Patterson maps that only requires the diffracted intensity data and no phase information, the heavy atom sites within the crystal lattice are determined and refined using the native and derivative data sets. Determining the known position of these heavy atoms allows the calculation of their contribution to the diffraction data, and this information can be used to obtain the initial phases for all the diffracted waves, derived from the unit cell including the virus capsid. Several virus capsid structures have been determined successfully using this method.

Molecular Replacement

The molecular replacement method is the most often approach applied to phase virus capsid diffraction data when determining crystal structures, provided that there is an available homologous structure from which initial phases can be obtained. It is a technique that matches homologous structural models (20–30% sequence identity or higher), once correctly oriented and positioned in the crystal unit cell, with the amplitudes of the diffracted waves for the structure being determined. This method is easy, straightforward and reduces the time and effort required for structure determination by isomorphous replacement, because there is no need to prepare heavy atom derivatives and collect the derivative data sets.

For this method to be successful, the structure of the known phasing model must be a good approximation (at the resolution of phasing) to that of the structure being determined. Several computer programs, including Envelope, AMoRe, CNS, PHASER, CCP4 and MOLREP, have been written for this purpose (Table 1).[11–24] The concept is simple: place the phasing model in the same orientation and position in the crystal unit cell of the unknown virus structure, through rotation and translation function searches, and use the calculated phases from the model as the initial values for the unknown structure. To avoid the possibility of model bias using this method, it is normal to remove the side-chains of the search model, *i.e.* use a polyalanine model, and if the method works the side-chains will show up in the calculated electron density maps because of the influence of the observed diffraction data amplitudes. A phase is assigned to each reflection in the data set, which will now contain the assigned Bragg plane, an amplitude value, a signal-to-noise ratio value and an initial phase angle for each diffracted wave.

Virus capsid crystallography has successfully used low-resolution single-particle cryo-EM reconstructed density maps or pseudo-atomic models built into reconstructed maps for initial phase determination and used phase extension methods (in which approximate phases are extended and improved by averaging, one reciprocal lattice point at a time, following improvement of the initial phases) to achieve higher resolution. There are a number of methods to improve the quality of the initial phases, including exploiting the idea of capsid *versus* bulk solvent boundaries (a virus capsid contains large amounts of solvent) in solvent flattening; and using the NCS operators to average the electron density equivalent points. Both of these methods improve the quality of the electron density maps. Recently, a few cryo-EM structures have pushed to near-atomic 3.8 Å resolution.[25–34]

4   Structure Refinement and Model Building

Refinement programs[11–24] and graphic user interfaces[35–39] aid in this process of improving the starting phasing model to match the observed diffraction data better (Tables 1 and 2).[11–24,35–39] This process of structure refinement and model building aims to match calculated structure factors (amplitude and phase information for the model) to those of the observed data (amplitude and improved phases) in an iterative manner until convergence. Phases are improved by the refinement process, which employs the strategies of rigid body,

Table 2 X-ray crystallography programs (model building and visualization).

Program	Website	Ref.
Model building		
O	http://xray.bmc.uu.se/alwyn/	35
COOT	http://www.ysbl.york.ac.uk/~emsley/coot/	36
MAID	http://www.msi.umn.edu/~levitt/	37
PRODRG	http://davapc1.bioch.dundee.ac.uk/prodrg/	38
ARP/wARP	http://www.embl-hamburg.de/ARP/	39
Structure validation and analysis		
Uppsala Electron Density Server	http://eds.bmc.uu.se/eds/	42
PROCHECK	http://www.csb.yale.edu/userguides/datamanip/ procheck/manual/index.html	43
HIC-UP	http://xray.bmc.uu.se/hicup/	44
Crystal Twinning	http://nihserver.mbi.ucla.edu/Twinning/	45
Molprobity	http://molprobity.biochem.duke.edu/	46
SSM	http://www.ebi.ac.uk/msd-srv/ssm/	47
Molecular graphics and visualization		
PYMOL	http://pymol.sourceforge.net/	48
UCSF-CHIMERA	http://www.cgl.ucsf.edu/chimera/	49
BOBSCRIPT	http://www.strubi.ox.ac.uk/bobscript/	50
GRASP	http://wiki.c2b2.columbia.edu/honiglab_public/ index.php/Software:GRASP	51
RASTER3D	http://skuld.bmsc.washington.edu/raster3d/	52
RIBBONS	http://www.cbse.uab.edu/ribbons/	53
LIGPLOT	http://www.msg.ucsf.edu/local/programs/ligplot/ index.html	54

simulated annealing, energy minimization (bond lengths, bond angles, torsion angles), atomic position and temperature factor (thermal motion of atoms) refinement of the model towards matching the observed structure factors[13,16,17] with the application of NCS operators. Averaged electron density Fourier maps [$2F_o - F_c$ (Figure 4A) and $F_o - F_c$, in which F_o represents the observed and F_c the calculated structure factors] are calculated using the experimentally determined amplitudes and the improved phases following structure refinement. This averaging procedure generally uses a molecular mask (which represents with a viral capsid protein monomer or a viral asymmetric unit) to allow the application of NCS operators. Model building into the density map is then carried out in steps of increasing finer detail. Several programs can be used for this process, with O[35] and Coot[36] being the most widely used for manual building. Initially, the polypeptide main-chain is traced, the backbone amino acids are built guided by the viruses' capsid protein sequence (into density consistent with their type and hence the side-chain density envelope shape) and the secondary structural elements, α-helices and β-strands, assigned (Figure 4B). The side-chains are initially built based on the most common rotamer conformation available in a database of structures within the model building program. Ordered nucleic acid (which is rare given that it does not

A)

B)

Surface loops

Interior β-barrel

C)

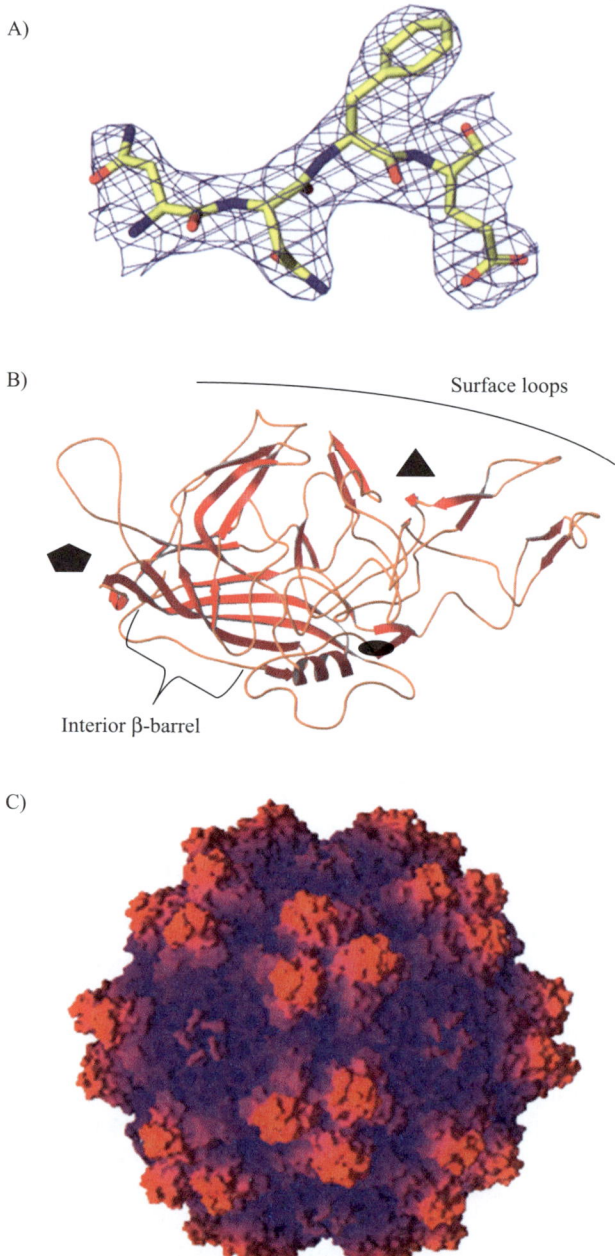

Figure 4   (A) A section of $2F_o - F_c$ electron density map of AAV4 (3.2 Å resolution) (blue mesh) contoured at 1.5σ. (B) A ribbon diagram of AAV4 VP3 showing the core eight-stranded strands, stretches of anti-parallel β-strands, loops and helical regions. Location of the five-, three- and twofold icosahedral axes of symmetry are depicted as geometric solid shapes. (C) Radially colored surface rendition of the virus capsid consisting of 60 VP3s.

obey the icosahedral NCS applied during the structure-averaging procedures described above) is built in the same way, based on nucleotide base type. Following model building, another round of refinement/improvement is carried out. At the end of each model improvement step, new phases are calculated and a new averaged electron density map is calculated (with observed amplitudes and improved phases) in preparation for the next round of model building. Sometimes the more flexible surface loop regions are difficult to interpret (disordered) and these should not be built until phases have been improved further, to avoid wrong interpretations. Calculation of 'omit' electron density maps, in which structural regions that are disordered are omitted from the model (and thus no phase information for the region), will sometimes result in a better density definition for these loops, due to the influence of the observed data. Final steps of model building involve the addition of solvent molecules, ligands and ions into the density map, to produce a model more closely matched to the observed data.

Important factors to consider during refinement and model building are the geometry restraints of bond lengths, bond angles and torsion angles for the amino acids within the polypeptide chain and for ordered nucleic acids that are energy minimized. These should be constrained to standard molecular geometry obtained from ultra-high-resolution peptide and protein structures. Many programs aid in the validation, visualization and analysis of the final structure (Table 2).[42–54]

The structure is said to be refined when the model cannot be adjusted any further to 'agree' better with the observed data. The agreement is given as a difference, fraction or percentage between the observed and calculated structure factors and is termed the R_{cryst} in X-ray crystallography. Due to the use of high-redundancy NCS operators during the refinement of virus capsid structures, the R_{free}, a refinement progress monitor used to avoid wrong density map interpretations, is very close to that of R_{cryst}.[40] The R_{cryst}/R_{free} values are generally in the range 20–35% for published virus structures.[69]

The final refined structure model is presented as a coordinate file, assigning six parameters to every atom; the atom type (C, N, O, *etc.*), the atomic positions (x, y, z) within the crystal unit cell, the occupancy n (the average fractional occurrence of an atom in the unit cell) and the thermal vibration parameter B (providing an indication of the motion of the atom). The final refined coordinate file and corresponding structure factor file are submitted to the Protein Data Bank (PDB),[41] an international repository for determined structures.

5   X-ray Structure Determination Example: AAV4

Adeno-associated viruses (AAVs), members of the dependovirus genus of the *Parvoviridae*, are helper dependent parvoviruses isolated from a number of different species, including humans.[55] Despite a requirement for co-infection with a helper virus such as adenovirus or herpes virus for productive replication, AAVs capsids are similar to those of the autonomous parvoviruses with $T = 1$ icosahedral symmetry and an overall diameter of ~ 260 Å that encapsulates an ssDNA genome size of ~ 5000 bases.[56-59] The capsid consists of three overlapping viral proteins (VPs), VP1 (90 kDa), VP2 (72 kDa) and VP3 (60 kDa), in the ratio 1:1:10, generated by alternative splicing and translational initiation during productive infection. VP3 constitutes 90% of the capsid, but all three proteins contain a common C-terminal domain of about 530 amino acids.

Tropism differences dictated by the capsid sequence and the need to improve the effectiveness of AAV gene delivery applications through capsid manipulation have generated a need to understand the basic biology of the different serotypes, particularly the mechanism(s) of cellular attachment and entry, antigenic reactivity and capsid structure. The homology between the capsid sequences of the AAV serotypes is high, although primary cell-surface receptor recognition properties are dramatically different. For example, AAV2 and AAV3, which are $\sim 87\%$ identical, utilize heparin sulfate as their primary receptor, but with different binding affinities.[60-62] Hence to delineate differences on the capsid surface with function (such as receptor binding sites), there is a need to determine the crystal structures of several AAVs.

Briefly, described here are the experimental steps that were taken to obtain AAV4's 3.2 Å resolution crystal structure using the method of molecular replacement described earlier in this chapter. This study also identified AAV capsid regions that can tolerate compensating structural loop insertions and deletions when compared with another serotype, AAV2, without detriment to capsid assembly. In addition, ordered nucleic acid density was observed inside the capsid.

Virus Production, Purification and Crystallization

Wild-type AAV4 particles were produced in Cos cells and purified using a CsCl density gradient and Centriprep filtration. Approximately 1×10^5 DNAse-resistant particles/cell were obtainable in each preparation. SDS-PAGE and negative stain electron microscopy confirmed the purity and quality of the virus particles, respectively. The crystallization conditions were initially screened, based on previously reported conditions used for autonomous parvoviruses,[63-65] using hanging drop vapor diffusion[4] in VDX 24-well plates and siliconized cover-slips (Hampton Research, Laguna Niguel, CA, USA). The crystallization drops contained 2 µl of virus mixed with 2 µl of precipitant solution consisting of 20 mM Tris–HCl, pH 7.5, with 2 mM $MgCl_2$, 150 mM NaCl and PEG-8000. The drops were equilibrated by vapor diffusion against 1 mL of precipitant solution at room temperature. Small diamond-like crystals, approximately $0.2 \times 0.2 \times 0.2$ mm in size, were obtained in 4–8 weeks (Figure 2A).

X-ray Diffraction Data Collection

X-ray diffraction data collection was performed at the F1 beamline at the Cornell High Energy Synchrotron Source (CHESS, Ithaca, NY, USA), using an ADSC Quantum4 CCD detector system. The X-ray diffraction data were collected at cryo-temperatures (~ 100 K), the crystals being transferred to a freshly prepared cryo-protectant solution consisting of 30% glycerol in the precipitant solution (20 mM Tris–HCl, pH 7.5, with 2 mM $MgCl_2$, 150 mM NaCl and PEG-8000). The data were collected at a wavelength of $\lambda = 0.916$ Å with a 0.2 mm collimator and a crystal-to-detector distance of 300 mm. All images were collected using a 0.25° oscillation angle with exposure times ranging from 60 to 120 s per image. A total of 276 images were collected from six crystals (average of ~ 45 images each) that all diffracted X-rays to at least 3.5 Å resolution (Figure 2B). The collected diffraction data were indexed using the software DENZO and scaled and reduced with SCALEPACK,[7] to the orthorhombic crystal system *I*222 with unit cell parameters $a = 339.6$, $b = 319.2$ and $c = 285.0$ Å. The data set was merged to 163 565 independent reflections (84.3% complete overall and 83.2% complete in the outermost resolution shell), resulting in an R_{sym} of 15.9% (26.9% in the outermost resolution shell).

Determination of Particle Orientation and Position

The orientations of the two AAV4 particles in the unit cell were determined with self-rotation functions.[9] The function was explored by searching for five-, three- and twofold icosahedral symmetry axes of the virus particles, with $\kappa = 72$, 120 and 180°, respectively. The rotation function for $\kappa = 72°$ (to search for the fivefold non-crystallographic symmetry axes directions) is shown in Figure 3. Packing of the AAV4 particles in the *I*222 unit cell, with their crystallographically constrained orientations required them to be situated at (0, 0, 0) and (1/2, 1/2, 1/2). This would allow the particles to have a maximum diameter of ~ 285 Å. This is consistent with the diameter of parvovirus particles that are approximately 260 Å.[66]

Phasing, Refinement and Model Building

The initial phase for the AAV4 X-ray diffraction data was calculated to 3.2 Å resolution using the CNS program[13] from a polyalanine model of feline panleukopenia virus (FPV)[64] directly oriented and positioned in the *I*222 unit cell of the AAV4 data.[67]

The refinement of the AAV4 capsid VP structure was performed by alternating cycles of refinement with the CNS program[13] and model building into averaged Fourier $2F_o - F_c$ and $F_o - F_c$, electron density maps using the O program.[35] A test data set of 5% was partitioned for monitoring the refinement process,[40] *i.e.* R_{free} calculations. The initial phases were improved using simulated annealing, energy minimization, conventional positional and temperature factor refinement followed by real-space electron density map averaging, using a molecular mask, while applying strict 15-fold NCS operators in the CNS program.[13] The final AAV4 capsid VP model, built into averaged maps, was then used to calculate the root mean square deviation from ideal bond lengths, bond angles and torsion angles using the PROCHECK program[43] to ensure that standard bond geometry had been maintained. The structure of AAV4 is shown in Figure 4.

The structural determination of AAV4 and its comparison with AAV2 showed the conservation of the core β-strands (βB–I) and helical (αA) secondary structure elements, which also exist in all other known parvovirus structures.[68] However, the structure also showed that surface loop variations (I–IX), some containing compensating structural insertions and deletions in adjacent regions, result in local topological differences on the capsid surface. These include AAV4 having a deeper twofold depression, wider and rounder protrusions surrounding the threefold axes and a different topology at the top of the fivefold channel compared with AAV2. Also, the previously observed 'valleys' between the threefold protrusions, containing AAV2's heparan sulfate binding residues, are narrower in AAV4. The observed differences in loop topologies at subunit interfaces were also consistent with the inability of AAV2 and AAV4 VPs to combine for mosaic capsid formation in an effort to engineer novel tropisms.[62] Interestingly, despite the application of NCS operators, density for a nucleotide was observed inside the AAV4 capsid, which was unexpected given that only one copy of the genome is packaged. This nucleotide binding site is also conserved in other AAV structures which have been subsequently determined (unpublished data). Hence the significance of this observation requires further study.

Of functional importance was the observation that all the surface loop variations were associated with amino acids reported to affect receptor recognition, transduction and anti-capsid antibody reactivity for AAV2. This observation suggested that these capsid regions may also play similar roles in the other AAV serotypes and has since been used in the development of new AAVs to be used in gene therapy.

6   Virus Database: VIPER

As mentioned previously, the PDB is an international repository site for the solved crystal structures of proteins and a subset of this database is virus structures.[41] As more virus structures are determined and more virologists have the need to use structural information for functional annotation, a database dedicated to the analysis of high-resolution virus structures, VIPERdb (http://viperdb.scripps.edu),[69] has been developed. This database is not only a source of the coordinates of known virus structures but also a one-stop site dedicated to helping virologists examine the many icosahedral virus structures contained within the PDB. It provides an easy-to-use database containing current data and a variety of analytical tools[69] in addition to crystallization conditions. The website describes various icosahedral virus capsid structures in terms of their complete capsid and allows computational analysis of the surface and capsid protein interfaces. The virus asymmetric unit (building block) coordinates are stored in a single icosahedral convention and are classified in terms of their quasi-symmetry (*e.g.* T = 3, T = 4, T = 7) and pseudo-symmetry (*e.g.* P = 3), and several tools are available to study the capsids in terms of structural, energetic and assembly aspects.

7 Summary

Advances in sample preparation, crystallization tools, X-ray diffraction data collection resources, computer algorithms and computational power have played a major role in our ability to determine entire spherical viral capsid structures using X-ray crystallography. In turn, these structures play a pivotal role in efforts to understand the basic functions of virus proteins and their interactions (with other proteins and nucleic acids), to annotate determinants of tissue tropism and pathogenicity and in the use of viruses in vaccine development, nano-materials and therapeutics. The next steps in the field will be the continued development of these approaches for obtaining the structure of viruses with more complicated capsid arrangements, including those with lipid envelopes and asymmetric morphologies, some of which were discussed in Chapter 5.

8 Acknowledgments

This work was supported in part by NSF grant MCB-0718948 and NIH R01 GM082946.

References

1. E. Arnold, G. Vriend, M. Luo, J. P. Griffith, G. Kamer, J. W. Erickson, J. E. Johnson and M. G. Rossmann, *Acta Crystallogr., Sect. A*, 1987, **43**, 346.
2. E. Fry, R. Acharya and D. Stuart, *Acta Crystallogr., Sect. A*, 1993, **49**, 45.
3. R. McKenna, D. Xia, P. Willingmann, L. L. Ilag and M. G. Rossmann, *Acta Crystallogr., Sect. B*, 1992, **48**, 499.
4. A. McPherson, *Preparation and Analysis of Protein Crystals*, 1st edn, Wiley, New York, 1992, p. 96.
5. T. Y. Teng, *J. Appl.Crystallogr.*, 1990, **23**, 387.
6. A. G. Leslie, *Acta Crystallogr., Sect. D*, 1999, **55**, 1696.
7. Z. Otwinowski and W. Minor, *Methods Enzymol.*, 1997, **276**, 307.
8. J. W. Pflugrath, *Acta Crystallogr., Sect. D*, 1999, **55**, 1718.
9. M. G. Rossmann and D. M. Blow, *Acta Crystallogr.*, 1962, **15**, 24.
10. L. Tong and M. G. Rossmann, *Acta Crystallogr.*, 1990, **46**, 783.
11. M. G. Rossmann, R. McKenna, L. Tong, D. Xia, J. Dai, H. Wu, H. Choi and R. E. Lynch, *J. Appl. Crystallogr.*, 1992, **25**, 166.
12. J. Navaza and P. Saludjian, *Methods Enzymol.*, 1997, **276**, 581.

13. A. T. Brunger, P. D. Adams, G. M. Clore, L. W. Delano, P. Gros, R. W. Grosse-Kunstleve, J.-S. Jiang, J. Kuszewski, M. Nilges, N. S. Pannu, R. J. Read, L. M. Rice, T. Simonson and G. L. Warren, *Acta Crystallogr., Sect. D*, 1998, **54**, 905.

14. A. J. McCoy, R. W. Grosse-Kunstleve, P. D. Adams, M. D. Winn, L. C. Storoni and R. J. Read, *J. Appl. Crystallogr.*, 2007, **40**, 658.

15. CCP4, *Acta Crystallogr., Sect. D*, 1994, **50**, 760.

16. P. D. Adams, R. W. Grosse-Kunstleve, L. -W. Hung, T. R. Ioerger, A. J. McCoy, N. W. Moriarty, R. J. Read, J. C. Sacchettini, N. K. Sauter and T. C. Terwilliger, *Acta Crystallogr., Sect. D*, 2002, **58**, 1948.

17. G. M. Sheldrick, *Acta Crystallogr., Sect. A*, 2008, **64**, 112.

18. G. Jogl, X. Tao, Y. Xu and L. Tong, *Acta Crystallogr., Sect. D*, 2001, **57**, 1127.

19. C. R. Kissinger, B. A. Smith, D. K. Gehlhaar and D. Bouzida, *Acta Crystallogr., Sect. D*, 2001, **57**, 1474.

20. A. Vagin and A. Teplyakov, *J. Appl. Crystallogr.*, 1997, **30**, 1022.

21. G. Bricogne, C. Vonrhein, C. Flensburg, M. Schiltz and W. Paciorek, *Acta Crystallogr., Sect. D*, 2003, **59**, 2023.

22. R. Miller, S. M. Gallo, H. G. Khalak and C. M. Weeks, *J. Appl. Crystallogr.*, 1994, **27**, 613.

23. G. N. Murshudov, A. A. Vagin and E. J. Dodson, *Acta Crystallogr., Sect. DA*, 1997, **53**, 240.

24. D. E. Tronrud, *Acta Crystallogr., Sect. A*, 1992, **48**, 912.

25. T. C. Terwilliger and J. Berendzen, *Acta Crystallogr., Sect. D*, 1999, **55**, 849.

26. T. C. Terwilliger, *Acta Crystallogr., Sect. D*, 2000, **56**, 965.

27. J. Navaza and P. Saludjian, *Methods Enzymol.*, 1997, **276**, 581.

28. R. McKenna, D. Xia, P. Willingmann, L. L. Ilag and M. G. Rossmann, *Acta Crystallogr., Sect. B*, 1992, **48**, 499.

29. P. Nissen, J. Hansen, N. Ban, P. B. Moore and T. A. Steitz, *Science*, 2000, **289**, 920.

30. E. J. Dodson, *Acta Crystallogr., Sect. D*, 2005, **57**, 1405.

31. F. Coulibaly, C. Chevalier, I. Gutsche, J. Pous, J. Navaza, S. Bressanelli, B. Delmas and F. A. Rey, *Cell*, 2005, **120**, 761.

32. M. Selmer, C. M. Dunham, F. V. Murphy, A. Weixlbaumer, S. Petry, A. C. Kelley, J. R. Weir and V. Ramakrishnan, *Science*, 2006, **313**, 1935.

33. A. Korostelev, S. Trakhanov, M. Laurberg and H. F. Noller, *Cell*, 2006, **126**, 1065.

34. Z. H. Zhou, *Curr. Opin. Struct. Biol.*, 2008, **18**, 218.

35. T. A. Jones and M. Kjeldgaard, *Methods Enzymol.*, 1997, **277**, 173.

36. P. Emsley and K. Cowtan, *Acta Crystallogr., Sect. D*, 2004, **60**, 2126.

37. D. G. Levitt, *Acta Crystallogr., Sect. D*, 2001, **57**, 1013.

38. A. W. Schuettelkopf and D. M. F. van Aalten, *Acta Crystallogr., Sect. D*, 2004, **60**, 1355.

39. G. Langer, S. X. Cohen, V. S. Lamzin and A. Perrakis, *Nat. Protocols*, 2008, **3**, 1171.

40. A. T. Brunger, *Nature*, 1992, **355**, 472.
41. H. M. Berman, J. Westwood, Z. Feng, G. Gilliland, T. N. Bhat, H. Weissig, I. N. Shindyalov and P. E. Bourne, *Nucleic Acids Res.*, 2000, **28**, 235.
42. G. J. Kleywegt, M. R. Harris, J. Y. Zou, T. C. Taylor, A. Wählby and T. A. Jones, *Acta Crystallogr., Sect. D*, 2004, **60**, 2240.
43. R. A. Laskowski, M. W. MacArthur, D. S. Moss and J. M. Thornton, *J. Appl. Crystallogr.*, 1993, **26**, 283.
44. G. J. Kleywegt, *Acta Crystallogr., Sect. D*, 2007, **63**, 94.
45. T. O. Yeates, *Methods Enzymol.*, 1997, **276**, 344.
46. I. W. Davis, A. Leaver-Fay, V. B. Chen, J. N. Block, G. J. Kapral, X. Wang, L. W. Murray, W. B. Arendall, J. Snoeyink, J. S. Richardson and D. C. Richardson, *Nucleic Acids Res.*, 2007, **35** (Web Server Issue), W375.
47. E. Krissinel and K. Henrick, *Acta Crystallogr., Sect. D*, 2004, **60**, 2256.
48. W. L. DeLano, The PyMOL Molecular Graphics System, 2002, DeLano Scientific, San Carlos, CA, USA.
49. E. F. Pettersen, T. D. Goddard, C. C. Huang, S. G. Couch, D. M. Greenblatt, E. C. Meng and T. E. Ferrin, *J. Comput. Chem.*, 2004, **25**, 1605.
50. R. M. Esnouf, *Acta Crystallogr., Sect. D*, 1999, **55**, 938.
51. A. Nicholls, K. A. Sharp and B. Honig, *Proteins Struct. Funct. Genet.*, 1991, **11**, 281.
52. E. A. Merritt and D. J. Bacon, *Methods Enzymol.*, 1997, **277**, 505.
53. M. Carson, *Methods Enzymol.*, 1997, **277**, 493.
54. A. C. Wallace, R. A. Laskowski and J. M. Thornton, *Protein Eng.*, 1995, **8**, 127.
55. N. Muzyczka and K. I. Berns, Parvoviridae: the viruses and their replication in *Fields Virology*, 4th edn, ed. D. M. Knipe and P. M. Howley, Lippincott Williams and Wilkins, New York, 2001, pp. 2327.
56. M. S. Chapman and M. Agbandje-McKenna, Atomic structure of viral particles, in *Parvoviruses*, ed. J. R. Kerr, S. F. Cotmore, M. E. Bloom, R. M. Linden and C. R. Parrish, Edward Arnold, New York, 2006, pp. 107.
57. E. Padron, V. Bowman, N. Kaludov, L. Govindasamy, H. Levy, P. Nick, R. McKenna, N. Muzyczka, J. A. Chiorini, T. S. Baker and M. Agbandje-McKenna, *J. Virol.*, 2005, **79**, 5047.
58. R. W. Walters, M. Agbandje-McKenna, V. D. Bowman, T. O. Moninger, N. H. Olson, M. Seiler, J. A. Chiorini, T. S. Baker and J. Zabner, *J. Virol.*, 2004, **78**, 3361.
59. Q. Xie, W. Bu, S. Bhatia, J. Hare, T. Somasundaram, A. Azzi and M. S. Chapman, *Proc. Natl. Acad. Sci. USA*, 2002, **99**, 10405.
60. C. Summerford and R. J. Samulski, *J. Virol.*, 1998, **72**, 1438.
61. A. S. Handa, S. Muramatsu, J. Qiu, H. Mizukami and K. E. Brown, *J. Gen. Virol.*, 2000, **81**, 2077.
62. J. E. Rabinowitz, F. Rolling, C. Li, H. Conrath, W. Xiao, X. Xiao and R. J. Samulski, *J. Virol.*, 2002, **76**, 791.

63. J. Tsao, M. S. Chapman, H. Wu, M. Agbandje-Mckenna, W. Keller and M. G. Rossmann, *Acta Crystallogr., Sect. B*, 1992, **48**, 75.
64. M. Agbandje, R. McKenna, M. G. Rossmann, M. L. Strassheim and C. R. Parrish, *Proteins*, 1993, **16**, 155.
65. A. L. Llamas-Saiz, M. Agbandje-McKenna, W. R. Wikoff, J. Bratton, P. Tattersall and M. G. Rossmann, *Acta Crystallogr., Sect. D*, 1997, **53**, 93.
66. M. Agbandje, C. R. Parrish and M. G. Rossmann, *Semin. Virol.*, 1995, **6**, 299.
67. N. Kaludov, E. Padron, L. Govindasamy, R. McKenna, J. A. Chiorini and M. Agbandje-McKenna, *Virology*, 2003, **306**, 1.
68. L. Govindasamy, E. Padron, R. McKenna, N. Muzyczka, N. Kaludov, J. A. Chiorini and M. Agbandje-McKenna, *J. Virol.*, 2006, **80**, 11556.
69. M. Carrillo-Tripp, C. M. Shepherd, I. A. Borelli, S. Venkataraman, G. Lander, P. Natarajan, J. E. Johnson, C. L. Brooks and V. S. Reddy, *Nucleic Acids Res.*, 2009, **37**, 436.

CHAPTER 7

Structural Studies of Viral Proteins – X-ray Crystallography

JOHN DOMSIC AND ROBERT MCKENNA

Department of Biochemistry and Molecular Biology, Center for Structural Biology, The McKnight Brain Institute, University of Florida, Gainesville, FL 32610, USA

1 Introduction

For the last 50 years, macromolecular X-ray crystallography has dominated structural biology, from the study of single globular proteins to large multiple complexes involving proteins, RNA, DNA and various other ligands. The method allows the precise mapping of macromolecular surfaces, enzymatic active sites and interactions at interfaces. Hence the knowledge it provides, the atomic details of interactions, has proven to be invaluable in the understanding of the host–virus immune response, host–receptor and viral release from the host interactions. It has also provided insight into intra-viral protein and nucleic acid interactions in the assembly and maturation of the virus. This detailed structural knowledge has led to advances in the development of small-molecule drugs that are designed to disrupt the lifecycle of the virus infection.

The fundamental requirement of this technique is the formation of a crystal, a solid regular array of identical molecules of the sample under investigation. This is often the most difficult and limiting step in the technique, but even with the great advances made in cryo-EM, cryo-ET (see Chapters 4 and 5) and NMR spectroscopy (see Chapter 8), this is still the method of choice for high-resolution structure determinations, if crystals can be obtained. The crystal is

RSC Biomolecular Sciences No. 21
Structural Virology
Edited by Mavis Agbandje-McKenna and Robert McKenna
© Royal Society of Chemistry 2011
Published by the Royal Society of Chemistry, www.rsc.org

required to permit the formation of lattice planes to provide the geometry for diffraction to occur and to create many repeating units of the crystal, which provide for the amplification of the diffraction event. X-rays are required for the method to work, as they have a wavelength short enough (~ 1 Å), to permit the creation of diffracted waves that allow the observed separation (resolution) to identify the position of individual atoms. The method exploits the interaction of electrons surrounding each atom in the crystal with the directed X-ray beam to obtain data that can be manipulated to produce the three-dimensional structure of the sample under consideration.

Generally, the method can be divided up into several steps: (1) the expression, purification and crystallization of the molecules of study, (2) the collection of the diffraction data and (3) the phasing and determination of the three-dimensional structure (Figure 1). This chapter discusses general methods for expressing and purifying proteins and then presents several case studies of viral proteins with emphasis on the means by which the structure was obtained and the key mechanistic discoveries made. The data collection and phasing methods have been previously discussed in Chapter 6, hence these will be only briefly described in this chapter.

2   Sample Preparation

Protein Expression

The first step requires the expression of sufficient quantities of the protein under study (usually milligram amounts) to allow the feasibility of trying to obtain crystals. The choice of expression system depends on the complexity of the desired protein. The simplest system in terms of materials and time are the bacterial platforms, with *Escherichia coli* being the dominant bacteria used. Bacterial cultures can be grown to very high densities in well-defined and readily available media, thus yielding a large amount of protein with minimal cost. However, these systems are not able to handle either large proteins or those with any sort of post-translational modification. Other expression systems may be utilized in such cases, including yeast, insect cells and mammalian cells. Yeast systems have the benefit of assistance in proper protein folding and also glycosylation. Although several species tend to hyperglycosylate proteins, several modifications have been made in a few species to yield strains that perform human-like glycosylation.[1] Baculovirus expression systems can lead to proteins with more mammalian-like post-translational modifications.[2] Mammalian cells, insect larvae and, more commonly, insect cell cultures are used as the protein production factories *via* infection by baculovirus containing the appropriate expression vector(s). The benefits of baculovirus expression are dependent on the infected cell, but generally this system is used when proteins require chaperone-assisted folding and appropriate post-translational modifications. It should be noted, however, that insect cells will not produce glycans with terminal galactose or sialic acid moieties. More details on expression systems can be found in Chapter 1.

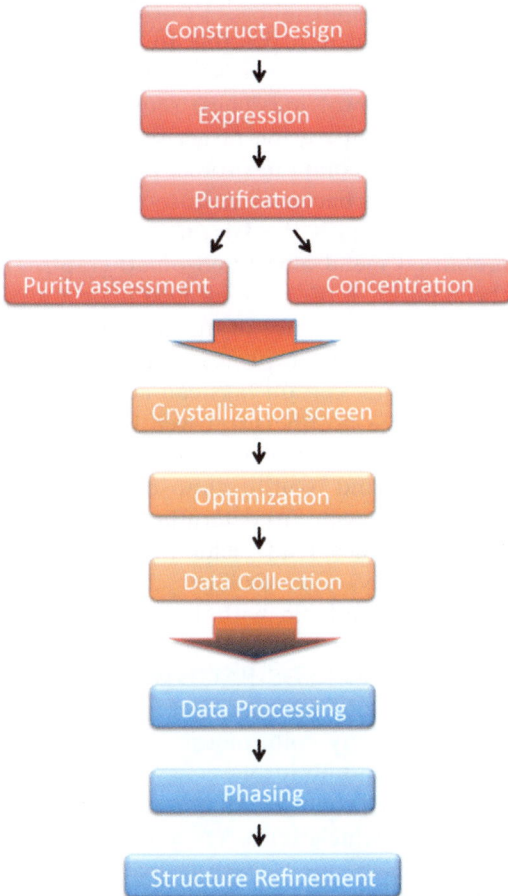

Figure 1   Flow chart of the major steps in protein crystallography. The chart is
divided into three phases. Phase I (red) involves the placement of the target
gene into an appropriate vector and then expression, purification and yield
assessment. Phase II (orange) is the growth of protein crystals starting with a
screen and optimizing any hits, followed by data collection. Phase III (blue)
is the determination of the structure, which first requires data reduction,
then initial phasing and finally structure refinement. Ideally, each of these
phases is a distinct step and if at any point there is an unacceptable result the
first step would be to go back and begin at the beginning of that phase. Only
if several rounds of repeated failure should one have to jump backwards to a
previous phase.

Protein Purification

Many of the standard techniques of protein purification are applicable to
viral proteins. Gel filtration, ion exchange and hydrophobic interaction
chromatography can play on the various physical properties of the protein.
These techniques, especially when used in combination, offer a wide range of

selectivities and can be used at any stage of purification. However, by manipulating the expression plasmid it is possible to improve the speed and yield
of purification. Several proteins and peptide sequences have been used to aid
in the purification of many proteins. These sequences are attached either at
the amino- or carboxy-terminus of the target sequence depending on the tag
being used. Among the more common protein affinity tags are glutathione *S*-
transferase (GST) and various bacterial maltose-binding proteins (MBPs). The
most common peptide attached is the His_6 tag which binds to nickel affinity
resins used in immobilized metal affinity chromatography (IMAC). Once the
target protein has been purified, the tag is cleaved off at the peptide that links it
to the protein. Generally, the sequence of this peptide is designed so as to be a
high-specificity substrate for a given protease. There is another advantage to
tagging a protein, namely that certain tags can add to its solubility, although
the mechanism by which this occurs is not well known and may in fact vary
from protein to protein. Examples of well-studied tags that have solubility-
enhancing effects are MBP[3] and N-utilization substance A (NusA).[4]

Upon purification, it is necessary to determine the purity and concentration
of the protein. The most widespread technique used for assessing purity is
sodium dodecyl sulfate polyacrylamide gel electrophoresis (SDS-PAGE).[5]
There are a variety of stains available, with the most common being Coomassie
Brilliant Blue and silver stains. Also, a few water-soluble metal salts allow for
rapid (1 min) negative staining of SDS-PAGE gels with greater sensitivity than
Coomassie staining.[6] Assessing concentration is a more difficult task and is
prone to error. There are many colorimetric assays available, the two most
common being the Lowry and Bradford assays.[7] Probably the most widely
used, but significantly error-prone, technique is the UV spectrophotometric
absorbance at $\lambda = 280$ nm, which relies on a predetermined (or calculated)
extinction coefficient that is unique to each protein. Each of these techniques,
however, is very sensitive to the presence of contaminating proteins.

Once a protein has been purified (usually deemed to be useful if <95% pure)
and is determined to be homogeneous, one can proceed with structural techniques. The techniques that result in atom coordinate models of proteins
require the largest amount of sample. NMR and X-ray crystallography each
require high sample concentrations (> 10 mg mL^{-1}) that may not be readily
achievable if a sample is prone to aggregation. This problem can, in some cases,
be alleviated by the use of low salt concentrations, detergents and/or glycerol.
In addition, glycosylated proteins may need to be deglycosylated.

3   Single-crystal Protein Crystallography

Crystallization

X-ray diffraction studies additionally require the growth of protein crystals.
Crystallization occurs when a protein is slowly concentrated to the point where
a few protein molecules form the beginnings of a crystal lattice, known as a

nucleation event. Once nucleation occurs, more and more protein molecules are "added" on to the crystal lattice until the protein concentration is too low to allow any further growth, the metastable zone. There are many different techniques available for growing crystals, the most common being vapor diffusion.[8] In vapor diffusion, a reservoir solution is made that contains a defined concentration of precipitant. Above this is the drop of protein solution mixed with a, typically, equal volume of precipitant suspended on a cover-slip. The cover-slip is then placed over the reservoir so as to create a completely sealed environment. Slowly, *via* vapor diffusion, the water in the protein drop (50% precipitant) will diffuse into the reservoir solution (100% precipitant). During this time, the protein in the drop will hopefully reach the nucleation zone and pass into the metastable zone, leading to the formation of crystals.

Sometimes the growth of crystals can be a severe bottleneck in the determination of a structure. The crystallization of viral proteins can be challenging for several reasons. In order to achieve diffraction quality crystals, it is necessary to have a protein that is capable of forming a tightly packed lattice. Therefore, the same problems arise as in obtaining high concentrations of protein. To determine initially if a protein is crystallizable in its current state, one of the best options is to use a sparse matrix screen that contains a wide variety of precipitants, salts and pHs. With the advent of crystallization robots, it has become possible to screen thousands of crystallization conditions using a small amount of sample. If a crystallization condition 'hit' is obtained, it then needs to be optimized, and this is usually achieved by screening around the initial condition (*i.e.* slight adjustments of pH, salt/precipitant concentrations and possibly additive screening). If no initial hits are found, one must go back to molecular biological techniques. One option is to perform limited proteolysis to remove surface features such as highly flexible loop regions (*e.g.* hemagglutinin; see Section 4, Influenza Hemagglutinin). Another, more complicated, option is an overhaul of the expression vector with the goal of removing flexible loops (*e.g.* HIV ENV, gp120; see Section 4, HIV Envelope Glycoprotein), posttranslational modification sites and mutation of surface residues that may aid in crystallization (*e.g.* removal of surface cysteines).

X-ray Data Collection

The next step in determining the crystal structure is the acquisition of diffraction data. The details behind diffraction are outside the scope of this chapter, but there are numerous texts with very detailed explanations of the mathematics and physics of X-ray crystallography. Several decisions must be made along the course of the diffraction experiment.[9] An initial test of the diffraction quality of the crystal is most easily done at room temperature with the crystal sealed in a quartz or polymer capillary. If the crystals are deemed of acceptable quality (for a protein, this typically is a crystal displaying diffraction to $<2.5\,\text{Å}$ resolution), then a complete data set can be collected. Additionally, and in the case of poorly diffracting crystals, the crystal can be cryo-cooled in a nitrogen gas cryo-stream after being soaked in an appropriate cryo-protectant. One of the most widely used cryo-protectants is a low percentage (10–25%) of glycerol mixed with reservoir solution. The length of time that the crystal spends in the cryo-protectant generally has to be optimized for each protein. Other options for improving diffraction quality include slow, controlled dehydration of the crystal, chemical cross-linking of proteins in the crystal lattice and the growth of new crystals in the initial condition but with a variety of salts, solvents, detergents or organics added as an additive screen, amongst many others.

Ideally, the diffraction tests mentioned previously would be carried out on an in-house diffractometer to save valuable national synchrotron beam time. Typically, an in-house source uses a rotating copper anode supplying X-rays with a wavelength of 1.5418 Å. The use of synchrotron radiation allows for higher resolution data collection (if the crystal is capable, that is, the crystal lattice is ordered enough to permit the collection of higher resolution data) due to the shorter wavelengths achievable at these sources (<1 Å). Additionally, synchrotron radiation also allows for much faster data acquisition times due to the higher flux of these beamlines. The raw diffraction data are collected by rotating the orientation of the crystal and collecting slices (at 0.5–1° intervals) of diffraction data. Usually, $\sim 180°$ or less, of data is required to complete a data set; this is dependent on the symmetrical arrangement of the proteins in the crystal lattice.

Once diffraction data have been obtained, usually on an image plate or CCD detector, they must be processed using any of the numerous programs available. One of the most widely used program suites available is HKL2000.[10] Each individual diffraction image is read, as raw diffraction images, the crystal lattice is indexed (meaning the location of the crystal lattice plane is determined) and the intensities of the individual diffracted waves are integrated and the complete data set is scaled. Because of the internal symmetry of a crystal lattice, the data can be scaled internally to itself to provide a 'reliability', R_{symm}, value for the data set, this error is typically between 5 and 10%. In addition, it is expected that about 90–95% of the theoretically possible diffraction data, 'reflections', at any given resolution limit are measured.

Structure Determination

The intensities of these collected reflections are then converted to wave amplitudes, which can then be used in any of the myriad software packages available to determined the structure (Chapter 6, Tables 1 and 2). The so-called phase problem in crystallography is the determination of the phase angle of each of these measured wave amplitudes. With the correct phase assignment to each wave, the diffracted wave can be summed together to calculated the resultant electron density at each point in the crystal building block, the unit cell. Then a set of atomic coordinates can be built into the electron density; these coordinates are often referred to as the structure. It should always be made clear these coordinates are fitted to the electron density map and therefore the quality of this fit is governed by many factors, including the quality of the resolution of the data. Among the more common program suites available for protein crystallography to determine these phases and refine the atomic model built are Crystallography and NMR Systems (CNS),[11] CCP4[12] and Phenix.[13] These packages are capable of structure solution using any of the currently utilized phasing techniques. Some of these methods have already been discussed in Chapter 6. Also, model refinement can then be performed within the same package. A typical 'solved' protein crystal structure will have a refined R_{factor} of between 15 and 25%; this number provides an agreement value between the built atomic model and the observed diffraction data. This value varies with resolution and orderedness of the crystal and the quality of the collected diffraction data. A more extensive listing of current crystallographic methods can be found in Chapter 6.

4 Case Examples

Four examples of virus protein crystallography are presented here.

Influenza Hemagglutinin

The recognition of host cells by a virus occurs *via* receptor proteins located on the surface of the capsid (*e.g.* adenovirus fiber protein) or anchored in the outer membrane (*e.g.* influenza hemagglutinin). The action of these proteins, that is, the recognition of specific receptors, is better understood through detailed structural analysis of protein–receptor complexes. However, there are also many challenges in the analysis of the proteins, particularly in acquiring soluble sample. This is especially true for the integral membrane receptors of membranous viruses, as these proteins generally contain very hydrophobic domains. To resolve this problem, researchers usually will express only that part of the receptor that is on the one face of the membrane, thus alleviating the solubility issues while still examining an interaction-competent protein.

The earliest example of X-ray crystallographic analysis of a viral membrane-bound receptor is that of influenza hemagglutinin (Figure 2A).[14] Hemagglutinin is a highly glycosylated (seven sites), homo-trimeric, membrane-bound protein containing a membrane-spanning domain that is connected *via* a coiled coil to an antigenic region and sialic acid-binding domain. Overall this results in a very large complex that is approximately 135 Å in total length. Initially, hemagglutinin exists as a complex of three independently continuous peptide chains (HA0). Each chain is subsequently proteolytically processed *via* the removal of Arg328 (resulting in HA1 and HA2), but remains linked *via* a disulfide bond. To overcome the challenge of solubilizing the full-length protein, Wilson *et al.* cleaved off the membrane-spanning domain using the protease bromelain.[14]

Hemagglutinin exhibits two functions that are vital for influenza viral infection: (1) sialic acid binding and (2) viral–host membrane fusion. To begin,

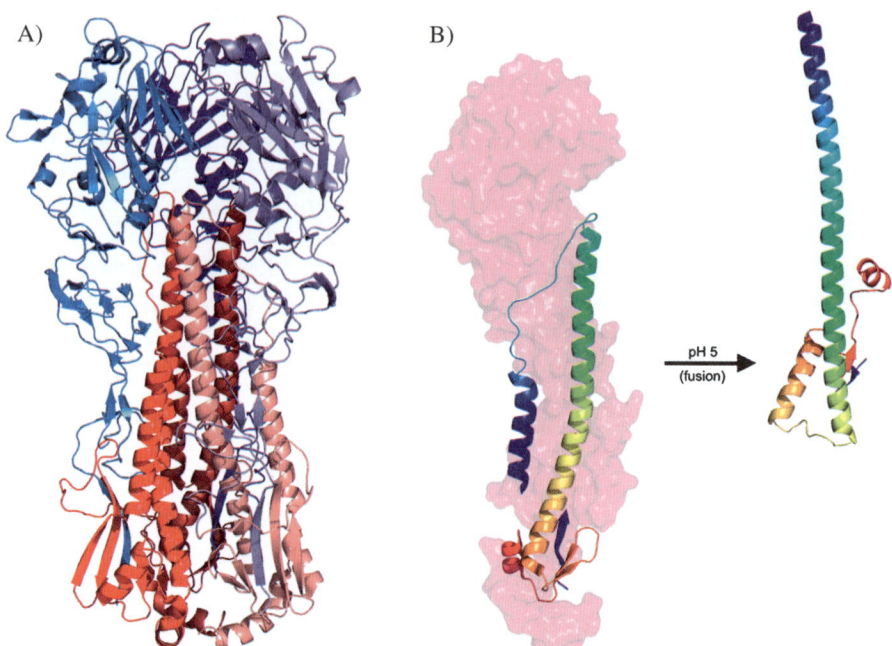

Figure 2   Influenza hemagglutinin. (A) The overall structure of influenza hemagglutinin with the viral membrane-binding HA2 domains shaded red and the
host cell membrane-binding HA1 domains shaded blue (PDB ID 1RD8).[30]
(B) In order for host–viral membrane fusion to occur, the N-terminal fusion
peptide (circled) must translocate by approximately 100 Å as indicated by
the structure of the low-pH form of the proteolyzed HA2 domain, which is
suggested to mimic the structure of HA in the endosome (PDB ID 1HTM).[17]
It can also be seen here that an extended helix forms by altering the entire
C-terminus.

the membrane distal portion of the protein interacts *via* a well-characterized
sialic acid recognition motif with glycosylated proteins on the host cell's surface.[15] Weis *et al.* solved the structure of hemagglutinin bound to sialyllactose
by soaking crystals of hemagglutinin with the glycan.[16] This study provided a
detailed look at the interactions that are necessary for host cell recognition and
revealed that there apparently are no major conformational changes upon
receptor binding. However, the sialyllactose occupies the entire highly conserved receptor binding domain, which suggests that it is the major host cell
recognition factor for influenza virus.

Once the virus has selected a host, it must be internalized *via* receptor-
mediated endocytosis. During the internalization process, the viral and host cell
membranes fuse in a mechanism driven by hemagglutinin. The low-pH environment is thought to cause a large conformational change in hemagglutinin.
The structural effects of this change were revealed by the structure of a partially
digested, mature form of hemagglutinin that consisted of residues 38–175 of

HA2 and residues 1–27 of HA1.[17] This change exposes the N-terminal fusion peptide by causing a 38-residue extension of the central helix of HA2, thus translating the peptide by approximately 100 Å (Figure 2B). This would allow for the fusion peptide to interact with the host cell membrane, thus increasing the interaction between the viral and host cell membranes. The mechanism by which the two membranes fuse is still unknown, but is likely to involve structural rearrangements that ultimately relocate the fusion peptide nearer to its original position.

HIV Envelope Glycoprotein

Another example of a membrane protein involved in recognition and membrane fusion is the human immunodeficiency virus (HIV) envelope glycoprotein (ENV). ENV exists as a 'spike' on the surface of the HIV membrane and is divided into two parts surface protein SU (gp120) and transmembrane protein TM (gp41). This division occurs after Env is expressed as a 160 kDa precursor that is subsequently proteolytically processed in the Golgi apparatus into the two domains that associate non-covalently to form a trimeric complex [(TM-SU)$_3$]. A complete review of the HIV lifecycle is given in Chapter 15.

TM, being an integral membrane protein, has not lent itself to any complete structural analyses. However, there are a few structures of TM peptides, which would be located on the surface of the membrane, in complex with their respective antibodies,[18,19] and also that of the core N- and C-terminal peptides in a trimeric complex.[20] On the other hand, several structures of the core region of the SU have been solved both alone and in complex with a Fab and CD4 and the cell surface receptor CD4 (Figure 3).[21–23] These structures were achieved through two means: an elegant re-engineering of the protein and, in the case of the uncomplexed protein, the use of the nearly identical simian immunodeficiency virus (SIV) SU. The re-engineering was accomplished, as was done with hemagglutinin, *via* removal of the highly flexible regions: N- and C-termini, glycosylation sites and the V1/V2 and V3 loops.[22] The loops that were removed were replaced with a Gly–Ala–Gly tripeptide, which left the protein as a single chain, allowing the authors to avoid the use of proteases that may result in cleavage at undesirable locations.

A comparison of the complexed and uncomplexed structures of SU indicates that a major conformational shift occurs in order for membrane fusion to occur. The shift results in 40 Å domain movements that are probably necessary for the cell membrane, which interacts *via* the Env fusion peptide, to fuse with the viral membrane. These domain shifts have been further supported by small-angle X-ray scattering (SAXS) studies that confirm CD4 binding-induced reorganizations of full-length gp120.[24] Further evidence of massive structural rearrangement comes from the previously mentioned structure of the core of TM. This structure suggests that the C-terminal peptides move from an extended (pre-fusion) conformation to a conformation in which they run anti-parallel to the central N-terminal peptides.

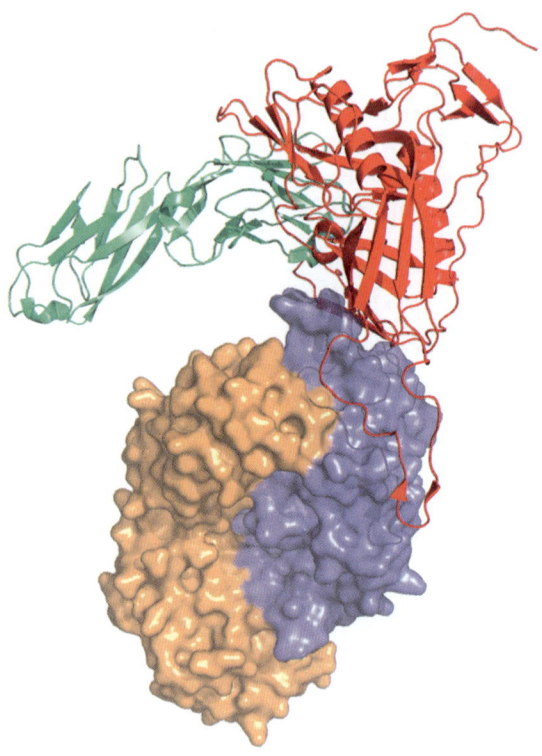

Figure 3   HIV-1 gp120. The structure of the HIV-1 gp120 core (red) in complex with
the CD4 receptor (green) and the heavy (blue) and light chain (orange) of
antibody X5 (PDB ID 2B4C).[31] One can distinctly see how the Fab makes
extensive contacts with the gp120/CD4 complex, especially between the
heavy chain and the V3 loop, thus allowing for stabilization and
crystallization.

Due to the heterogeneity of the ENV spike, it remains doubtful whether a
crystal structure of the full complex will ever be acquired. Therefore, it is
necessary to turn to other techniques, in this case cryo-electron tomography.[25]
Due to the low expression that was seen in wild-type HIV virions, the authors
turned to a mutant of SIV that has increased accumulation of Env spikes. The
results suggest that Env exists as a globular receptor that is attached to the
membrane by three 'feet', forming a tripod with a hollow area directly below
the receptor. This indicates that the external domains of the TM trimer do not
associate to form a stalk as was previously suggested.

HIV Reverse Transcriptase

Another class of proteins that provide a great challenge to structural analysis,
due in part to their complex mode of action and also their great size, are the

viral polymerases. There are three classes of viral polymerases based on the substrate and product: DNA-dependent DNA polymerases (DdDp), RNA-dependent DNA polymerases (RdDp), DNA-dependent RNA polymerases (DdRp) and RNA-dependent RNA polymerases (RdRp). Indeed, the variety and complexity lend themselves to nearly a full textbook explaining the biophysical details of these enzymes. As an example of one of the most studied viral polymerases, we will briefly examine the structure of reverse transcriptase (RT) (an RdDp) from HIV-1.

Reverse transcriptase is responsible for the conversion of the single-stranded RNA genome of HIV into a double-stranded DNA 'copy'. The structure reveals an asymmetric dimer of two proteins, p66 and p51.[26] The full-length p66 contains two domains, a ribonuclease H domain (p66$_{RNaseH}$) and a polymerase domain (p66$_{Pol}$). The p51 protein is simply a copy of p66 with a proteolytic removal of the RNase H domain. The polymerase domain can further be divided into four subdomains: palm, fingers, thumb and connection. The palm, fingers and thumb domains are named based on the original structure of the Klenow fragment of *E. coli* DNA polymerase. The connection subdomain is a structural motif that is responsible for allowing the association of p66 and p51 (Figure 4A).

Perhaps the most interesting aspect of this structure is in the remarkably high degree of asymmetry seen between p66$_{Pol}$ and p51. The connection and thumb subdomains virtually swap positions, with the connection subdomain now making contacts with the fingers, thumb and palm subdomains. Overall it undergoes a rotation of approximately 155° and about a 17 Å translation. It is believed that the alteration in conformation of p51 provides a function that differs from that of p66$_{Pol}$. The binding of dsDNA by p66$_{Pol}$ has been described in detail by Jacobo-Molina *et al. via* a structure of HIV RT complexed with a dsDNA molecule.[27] This structure mimics the active formation of the dsDNA version of the HIV genome. The activity of p51, on the other hand, is thought to be a tRNA binding function. This allows for initiation of the pol reaction *via* priming of synthesis with the 3′ end of the tRNA. To grasp clearly the peculiarity of the differences in function between p66 and p51, one must remember that these two proteins share nearly exactly the same parental nucleic acid sequence (save only the removal of the RNase H domain). This makes HIV RT one of the best examples of the degree to which viruses must evolve to package a diverse set of viral machinery into the capsid-limited size of the viral genome.

Structural studies of proteins have also been utilized to help elucidate the function of proteins for which one was undetermined. This utilization of functional assignment is especially prevalent in the structural genomics initiatives where the goal is to obtain structures for as many novel proteins as possible, generally focusing on one viral pathogen or other microorganism at a time. One of the best examples of this is the determination of the structures of several SARS viral proteins (Figure 4B).[28] SARS is attractive to structural proteomics because of its relatively small genome size and due to its recent emergence as a lethal disease-causing virus. The majority of studies focus on the so-called non-structural proteins (nsps), which are responsible for initiating viral replication.

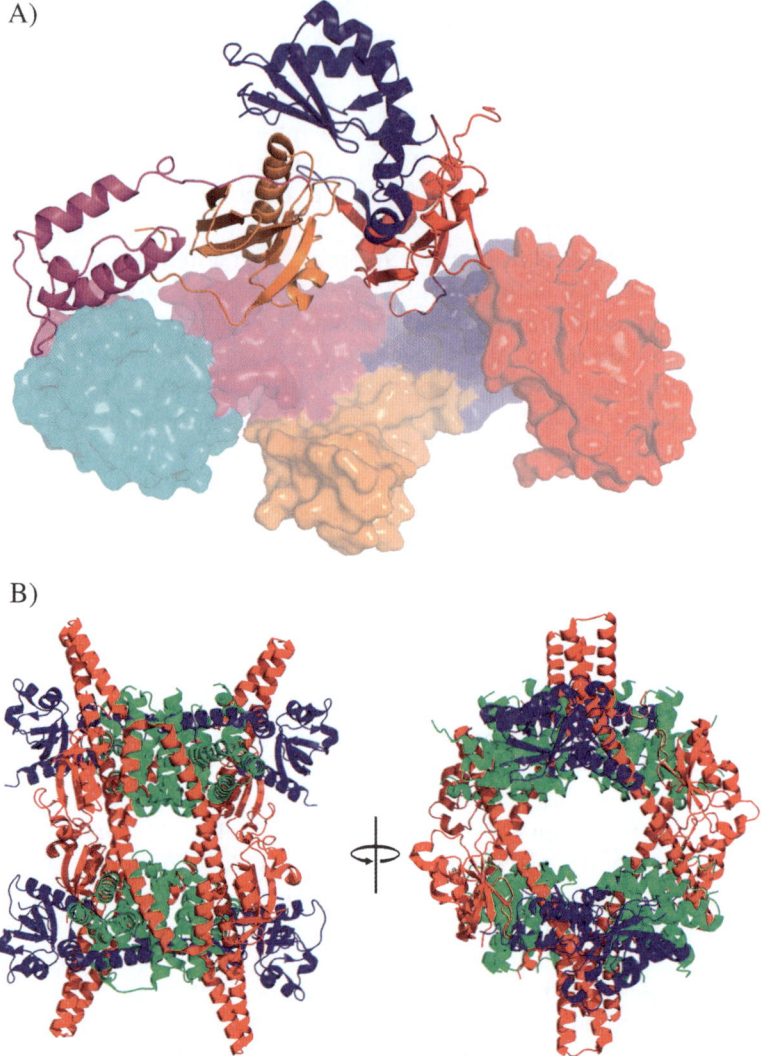

Figure 4   (A) HIV-IRT. There is a great deal of asymmetry between the PSI (ribbon-top) and p66 (bottom-surface) domains of HIV-IRT. The Finger (red), palm (blue), thumb (orange), and connection (magenta) domains differ in position and orientation with the connection domain of p51 interacting with RNase H domain (cyan) of p66. (B) The structure of the SARS nsp7–nsp8 complex provides insights into its function (PDB ID 2AHM).[29] The nsp8 protein exists in two distinct conformations: nsp8I (red) and nsp8II (blue). Together with nsp7 (green), an apparent dsRNA interacting pore (right side).

One of the more interesting examples of a viral protein whose function was hypothesized from its structure is the hexadecameric complex of nsp7 and nsp8.[29] The overall structure of this supercomplex consists of eight copies of both nsp7 and nsp8. The structure of the 78 residue nsp7 reveals a helical bundle consisting of a three-helix core with an additional short C-terminal helix. The four monomers of nsp7 present in the supercomplex are nearly identical in structure, with the most variation occurring in the C-terminal helix. There are two different forms of nsp8 present, nsp8I and nsp8II. The fold of nsp8I is termed the 'golf-club' fold and consists of two distinct domains. The N-terminal shaft domain consists of three α-helices with the majority of the domain forming a nearly 70 residue long helix. The C-terminal head domain is made up of seven β-strands, five of which form an anti-parallel sheet, and three α-helices. The head domain of nsp8II is nearly identical with that of nsp8I. The shaft domain, however, is markedly different from that of nsp8I, with a nearly 90° bend approximately at its midpoint (Figure 4B).

The overall structure of the supercomplex forms a nearly 30 Å cylindrical hole running directly down its center. Additionally, the N-terminal end of the shaft domains of the nsp8I monomers forms four 'handles' on the complex's surface. Because of this hole, one immediately gets the impression that this protein has some function in interacting with nucleic acids. This is further supported by surface charge analysis, which reveals that the cylindrical hole has a high level of positive potential and the outer surface is mostly negative. This suggests that the electrostatic potential is tuned so as to direct nucleic acid into the hole. SARS requires a double-stranded (ds) RNA duplex to form during genome replication, so the authors modeled in dsRNA and discovered that it is a nearly perfect fit. Based on the solution of the structure of the nsp7–nsp8 hexadecameric supercomplex, it is now thought that this protein complex acts to stabilize the dsRNA intermediate that is formed during SARS genome replication, thereby improving replication efficiency.

5 Summary

The insights into protein function provided by structural analyses afford researchers the means by which to design inhibitory compounds to block the progression of the viral lifecycle. As an example, the influenza medications oseltamivir (Tamiflu) and zanamivir (Relenza) were based on structural studies of neuraminidase. Refer to Chapters 15 and 16 to see how structural studies have aided in drug design.

References

1. E. Böer, G. Steinborn, G. Kunze and G. Gellissen, *Appl. Microbiol. Biotechnol.*, 2007, **77**, 513.
2. T. A. Kost, J. P. Condreay and D. L. Jarvis, *Nat. Biotechnol.*, 2005, **23**, 567.
3. J. D. Fox, K. M. Routzahn, M. H. Bucher and D. S. Waugh, *FEBS Lett.*, 2003, **537**, 53.
4. V. De Marco, G. Stier, S. Blandin and A. de Marco, *Biochem. Biophys. Res. Commun.*, 2004, **322**, 766.
5. U. K. Laemmli, *Nature*, 1970, **227**, 680.

6. J. K. Dzandu, J. F. Johnson and G. E. Wise, *Anal. Biochem.*, 1988, **174**, 157.
7. B. J. Olson and J. Markwell, *Curr. Protoc. Protein Sci.* 2007, Ch 3, Unit 3.4.
8. N. E. Chayen and E. Saridakis, *Nat. Methods*, 2008, **5**, 147.
9. B. Heras and J. L. Martin, *Acta Crystallogr., Sect. D*, 2005, **61**, 1173.
10. Z. Otwinowski and W. Minor, *Methods Enzymol.*, 1997, **276**, 307.
11. A. T. Brunger, P. D. Adams, G. M. Clore, L. W. Delano, P. Gros, R. W. Grosse-Kunstleve, J. -S. Jiang, J. Kuszewski, M. Nilges, N. S. Pannu, R. J. Read, L. M. Rice, T. Simonson and G. L. Warren, *Acta Crystallogr., Sect. D*, 1998, **54**, 905.
12. CCP4, *Acta Crystallogr.*, 1994, **50**, 760.
13. P. D. Adams, R. W. Grosse-Kunstleve, L. -W. Hung, T. R. Ioerger, A. J. McCoy, N. W. Moriarty, R. J. Read, J. C. Sacchettini, N. K. Sauter and T. C. Terwilliger, *Acta Crystallogr., Sect. D*, 2002, **58**, 1948.
14. I. A. Wilson, J. J. Skehel and D. C. Wiley, *Nature*, 1981, **289**, 366.
15. J. J. Skehel and D. C. Wiley, *Annu. Rev. Biochem.*, 2000, **69**, 531.
16. W. Weis, J. H. Brown, S. Cusack, J. C. Paulson, J. J. Skehel and D. C. Wiley, *Nature*, 1988, **333**, 426.
17. P. A. Bullough, F. M. Hughson, J. J. Skehel and D. C. Wiley, *Nature*, 1994, **371**, 37.
18. G. Ofek, M. Tang, A. Sambor, H. Katinger, J. R. Mascola, R. Wyatt and P. D. Kwong, *J. Virol.*, 2004, **78**, 10724.
19. R. M. Cardoso, M. B. Zwick, R. L. Stanfield, R. Kunert, J. M. Binley, H. Katinger, D. R. Burton and I. A. Wilson, *Immunity*, 2005, **22**, 163.
20. D. C. Chan, D. Fass, J. M. Berger and P. S. Kim, *Cell*, 1997, **89**, 263.
21. B. Chen, E. M. Vogan, H. Gong, J. J. Skehel, D. C. Wiley and S. C. Harrison, *Nature*, 2005, **433**, 834.
22. P. D. Kwong, R. Wyatt, J. Robinson, R. W. Sweet, J. Sodroski and W. A. Hendrickson, *Nature*, 1998, **393**, 648.
23. C. C. Huang, M. Tang, M. Y. Zhang, S. Majeed, E. Montabana, R. L. Stanfield, D. S. Dimitrov, B. Korber, J. Sodroski and I. A. Wilson, *et al.*, *Science*, 2005, **310**, 1025.
24. R. G. Ashish, J. Anguita and J. K. Krueger, *Biophys. J.*, 2006, **91**, L69.
25. P. Zhu, J. Liu, J. Bess, E. Chertova, J. D. Lifson, H. Grise, G. A. Ofek, K. A. Taylor and K. H. Roux, *Nature*, 2006, **441**, 847.
26. L. A. Kohlstaedt, J. Wang, J. M. Friedman, P. A. Rice and T. A. Steitz, *Science*, 1992, **256**, 1783.
27. A. Jacobo-Molina, J. Ding, R. G. Nanni, A. D. Clark, X. Lu, C. Tantillo, R. L. Williams, G. Kamer, A. L. Ferris and P. Clark, *et al.*, *Proc. Natl. Acad. Sci. USA*, 1993, **90**, 6320.
28. M. Bartlam, Y. Xu and Z. Rao, *J. Struct. Funct. Genomics.*, 2007, **8**, 85.
29. Y. Zhai, F. Sun, X. Li, H. Pang, X. Xu, M. Bartlam and Z. Rao, *Nat. Struct. Mol. Biol.*, 2007, **12**, 980.
30. J. Stevens, A. L. Corper, C. F. Basler, J. K. Taubenberger, P. Palese and I. A. Wilson, *Science*, 2004, **303**, 1866.
31. L. Fan, S. Kim, C. L. Farr, K. T. Schaefer, K. M. Randolph, J. A. Tainer and L. S. Kaguni, *J. Mol. Biol.*, 2006, **358**, 1229.

CHAPTER 8

Solution NMR Spectroscopy in Characterizing Structure, Dynamics and Intermolecular Interactions of Retroviral Structural Proteins

KANG CHEN AND NICO TJANDRA

Laboratory of Molecular Biophysics, National Heart, Lung, and Blood Institute, National Institutes of Health, Bethesda, MD 20892, USA

1   Introduction

Solution nuclear magnetic resonance (NMR) spectroscopy has been an important tool in modern structural biology owing to the development of isotope labeling methods, triple-resonance pulse sequence techniques, specialized cryogenic probes and higher magnetic field strengths. NMR spectroscopy allows the characterization of the solution structure and dynamics of biomolecules at atomic resolution. Both the precision and accuracy of the structures determined by NMR are continually improving, due to advances in instrumentation and also newly developed methodologies that provide additional new and independent classes of structure information. Often, structure information alone does not reveal the functional mechanism of a biomolecule; dynamic information at many time-scales is also needed. The dynamics of biomolecules, ranging from picosecond (ps) to millisecond (ms) time-scales, can be studied with NMR spin relaxation measurements. The slower micro- to millisecond (μs–ms) dynamic motions are frequently

RSC Biomolecular Sciences No. 21
Structural Virology
Edited by Mavis Agbandje-McKenna and Robert McKenna
© Royal Society of Chemistry 2011
Published by the Royal Society of Chemistry, www.rsc.org

correlated with the function of the biomolecules.[1] A wider range of structural flexibility, sometimes involving minor conformers, can now be probed by NMR using residual dipolar coupling and paramagnetic relaxation enhancement,[2] providing new insights into conformational sampling in molecular recognition. Potential inter-biomolecule or ligand interactions can readily be tested and structurally characterized in aqueous samples with NMR. The improvement in efficiency of these types of titration experiments impacts efforts in general large-scale screening, yielding information at atom-specific resolution.[3]

Structures from X-ray crystallography and NMR spectroscopy have been obtained for many retroviruses and, interestingly, the sizes and tertiary folds of common scaffold proteins from different species, capsid proteins for example, are similar to each other despite their primary sequence differences. Some structural concepts are general and interchangeable among them.[4,5] Three-dimensional structures for nearly all HIV-1 proteins were thoroughly reviewed in 1999 by Turner and Summers.[6] Another chapter in this book focused on X-ray crystallography methodology in solving viral protein structures. Here we review some recent progress in the application of NMR spectroscopy to dynamics and interactions of the retroviral structural proteins matrix (MA), capsid (CA) and nucleocapsid (NC), which are proteolytic products of their precursor protein Gag. Gag of HIV-1 is a 55 kDa multi-domain protein, necessary and sufficient for the formation of the immature viral particle.[7] Prior to virus budding, Gag is targeted to the cytosolic side of cell membrane through its N-terminal MA domain. Gag oligomerization leads to virus budding. Meanwhile, two copies of the viral genome RNA strands, carried by the NC domain at the C-terminus of Gag, are packed into the virions. The subsequent proteolysis of Gag completes virus maturation. For HIV-1, the release of three major Gag components results in a significant virion morphological transition from a spherical- to conical-shaped inner shell. Gag itself and a number of cellular factors regulate the budding and maturation processes.[8–14]

Although the precise order and structural mechanisms of interactions among the components of Gag are not fully understood, recent NMR data elucidate some of their basic features. These include the structural mechanisms of Gag membrane targeting[15] and genome RNA encapsulation,[16] the fast dynamics of Gag components[17,18] and the slow dynamics of the CA proline loop caused by cyclophilin A (CypA) catalysis,[19] to name just a few. We start with a brief description of general NMR measurements for protein structure and dynamics, followed by examples of NMR studies of viral proteins and RNA. NMR spectroscopy for RNA molecules is a fast-moving field and although many technical details will be omitted here, this field will be of increasing importance to the study of retroviruses.[20–22] As more cellular partners are discovered that regulate viral maturation, NMR spectroscopy will play a crucial role in illustrating mechanisms and dynamics of those important biomolecular interactions.

2 Experimental Methods

NMR spectroscopy measures interactions between magnetically active nuclei in a strong external magnetic field. In biological applications, one typically prefers, for favorable relaxation properties, nuclei with spin quantum number $\frac{1}{2}$. For instance, for proteins and nucleic acids this includes ^{1}H, ^{15}N, ^{13}C and ^{31}P. For ^{15}N and ^{13}C this requires isotope labeling of the biomolecules. Structure and dynamics studies of biomolecules by NMR start by the identification of each observed resonance line with a particular nucleus, thus an atom in the molecule. This is followed by probing certain types of interaction between these nuclei in the magnetic field to provide internuclear distance, dihedral or orientation information for structural study and relaxation rates that are relevant for dynamic information. The bulk of the data analysis is concentrated on these first two steps. Typically, this entails analyzing data sets containing thousands of site-specific interactions. In the final phase of the analysis, a group of structures is calculated that best satisfy all of the measured restraints and the relaxation data are then interpreted in terms of various motional models to describe the dynamics of the solved structure. The flowchart diagram in Figure 1 illustrates the necessary steps, introduced below, toward structure and dynamics studies using NMR spectroscopy.

Sample Preparation

Most NMR studies of protein require labeling of the protein with ^{15}N and ^{13}C stable isotopes. This can be achieved by expressing the protein in *Escherichia coli* grown in a minimal medium supplemented with appropriate labeled nutrients. For the nitrogen source one typically uses ^{15}NH$_4$Cl, whereas the carbon source can vary depending on the type of labeling required. For instance, for uniform labeling one would use [^{13}C]glucose, whereas for side-chain-specific methyl labeling one would add monomethyl[^{13}CH$_3$]ketoisovalerate or -ketobutyrate to the medium, along with non-labeled glucose.[23] The site-specific labeling is crucial for the study of much larger proteins ($>$ 50 kDa), which also typically requires deuteration of the protein.[24] A high level of deuteration is needed to reduce the numbers of protons, which interact with other nuclei and cause fast relaxation of their magnetization and is achieved by replacing the glucose with [^{2}H]glucose and growing the bacteria in a minimal medium in D$_2$O. For RNA molecules, many labeling schemes are possible, such as uniform, base-specific or segmental, by adding various isotope-enriched nucleotide triphosphates (NTPs) at different steps during *in vitro* transcription.[20]

Until recently, the protein concentration for NMR study needed to be 0.5–1 mM in a minimal sample volume of 250 µL. With the availability of cryogenic probes, this concentration can now typically be lowered by a factor of two or more. Buffer that minimizes proton exchange is preferred, and also low ionic strength. High salt concentration degrades the performance of the radio-frequency circuitry inside the NMR probe and therefore should be avoided.

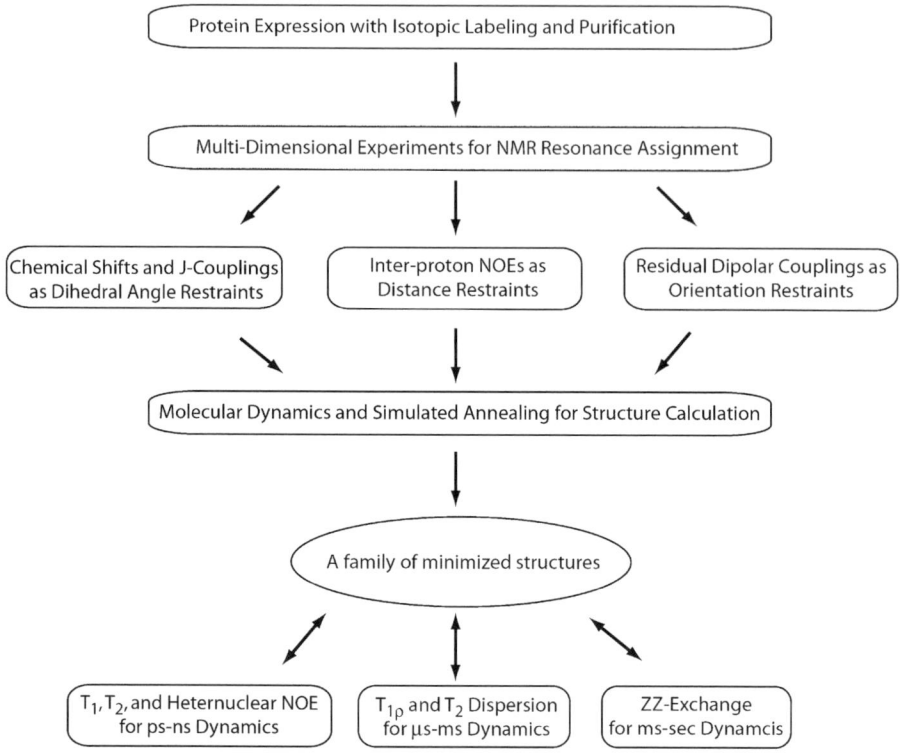

Figure 1   Flowchart for protein NMR structure and dynamics characterizations. The dual-arrow lines indicate that the relaxation data can also provide structural information.

Addition of other factors, such as detergent or lipid micelles, preservatives (protease inhibitor, EDTA, *etc.*), reducing agents and other surfactants, can readily be checked by NMR to ensure that the sample of interest is not altered.[25]

Resonance Assignments

Utilizing only ^{1}H in studying biomolecules by NMR rapidly becomes impractical due to the large number of resonances (in the hundreds) that appear in a relatively small spectral range. Resolving the overlapping resonances requires extension to multi-dimensional NMR experiments in which interactions of the proton with its attached nitrogen or carbon nucleus are utilized. Typically it is the through-bond interaction, known as scalar or *J*-coupling, which is selected. This can differentiate various bonding networks in the bio-molecules by the unique strength of the through-bond interaction ($^1J_{\mathrm{NH}} \approx 93\,\mathrm{Hz}$, $^1J_{\mathrm{CH}} \approx 140\,\mathrm{Hz}$, $^1J_{\mathrm{CC}} \approx 55\,\mathrm{Hz}$ and $^1J_{\mathrm{NC}} \approx 15\,\mathrm{Hz}$).[26] There are many experiments designed to link specific sets of atoms in a protein backbone, in addition to side-chains, through the pathways of scalar coupled nuclei.[27] For instance, the 3D-HN(CO)CA experiment correlates resonance frequencies, also called chemical shifts, of a backbone HN pair with the C^α of the preceding residue, whereas the 3D-HNCA experiment correlates chemical shifts of the HN pair with the C^α both in the same residue and in the preceding residue. This pair of experiments provides sequential connection through C^α chemical shifts and thus stepwise identification of the full backbone, a process called resonance assignment. These experiments can be expanded to include the C^β chemical shift to alleviate any remaining ambiguities due to resonance overlap. The backbone assignment can be extended to include the side-chain. For example the CC(CO)NH and HCC(CO)NH experiments correlate chemical shifts of a backbone HN pair with the side-chain carbons and protons of the previous residue, respectively. Sensitivity of these types of experiments depend on the efficiency of magnetization transfer from one nucleus to the next, which is governed by the transverse relaxation times (T_2) of the nuclei involved. The relaxation times depend on how fast the molecule is tumbling in solution, and thus the size of the molecule. Above a certain limit ($> 30\,kDa$), the molecule tumbles so slowly that T_2 becomes too short to allow efficient use of the above experiments to establish resonance assignments. An alternative approach, along with deuteration of the protein, can overcome this limitation. Transverse optimized relaxation spectroscopy (TROSY)[28] is becoming a standard method to study larger molecules by NMR. It offers better sensitivity and increased resolution by resonance narrowing. It relies on the ability to select the slower decaying components of the resonance signals, resulting from the cancellation of T_2 relaxation contributions from two different sources. In the case of the HN pair, their internuclear dipolar relaxation contribution cancels the contribution from the chemical shift anisotropy. The chemical shift anisotropy arises from the electron density that surrounds the nitrogen nucleus, whereas in a CH_3 group different dipolar interaction contributions cancel each other.[29]

Solution NMR Structure Determination

NMR structures are typically given as an ensemble of individual structures where each is consistent with the experimental structural restraints. The restraints can be in the form of distances, dihedral angles and bond orientations. These are all derived experimentally from nuclear dipole interactions or empirically from the NMR resonance frequencies. More experimental and empirical methods to obtain new restraints are continuously being developed. We will restrict this review to the most commonly used restraints.

Distance Restraints

Homonuclear (^1H–^1H) dipolar interaction depends on the distance between the two nuclei. These interactions can be measured as nuclear Overhauser effects (NOEs) and are the most common distance restraints used in NMR.[30]

The NOE is measured as the intensity of the signal correlating the two interacting ^{1}H nuclei. NOEs are generally weak because of their $1/r^6$ dependence, where r is the distance between the two nuclei. As a result, NOEs are only commonly observed when the two protons are relatively close, less than 5–6 Å apart. Homonuclear NOEs can also be measured using the frequencies of the attached ^{13}C or/and ^{15}N nuclei, in addition to the protons, to provide three- and four-dimensional experiments that can resolve ambiguities arising from ^{1}H resonance overlap. For example, ^{15}N-edited 3D-NOESY only measures NOEs between ^{1}H^N attached to ^{15}N and other protons. For structure determination, the long-range NOE restraints, which correlate ^{1}H–^{1}H pairs that are close in space but far apart in primary sequence, are essential to derive the correct protein fold.

Intermolecular NOEs between protein and ligand, protein and DNA/RNA and protein and protein are observable if the binding is sufficiently strong. Methods used to measure intramolecular NOEs can be used to obtain intermolecular NOEs when both protein and ligand are isotopically labeled. Alternatively, just one partner can be labeled and an NMR experiment performed to detect only the interactions from a ^{1}H attached to a magnetic nucleus to another ^{1}H attached to a non-magnetic nucleus, thus yielding only intermolecular distance information. For instance, a ^{13}C-edited/^{12}C-filtered 3D NOESY specifically measures NOEs between ^{13}C-attached ^{1}H nuclei of the protein and ^{12}C-attached ^{1}H nuclei of the ligand, in a sample of labeled ^{13}C protein and unlabeled ligand. Complementary information can be obtained by reverse labeling of the two components and performing the same NMR experiment.[31]

Dihedral Angle Restraints

When backbone resonance assignments are available, the protein backbone dihedral angles (Φ and Ψ) can be predicted empirically using the program TALOS.[32] TALOS compares the measured backbone chemical shifts (H$^\alpha$, C$'$, C$^\alpha$, C$^\beta$ and N) of the protein against the chemical shifts for 20 high-resolution X-ray structures. It then generates predicted Φ and Ψ angular restraints, along with error ranges, based on the level of agreement with the values in the database.

The dihedral angles can also be derived experimentally through homo- and heteronuclear J-coupling constants. These constants depend on the dihedral angles through relations known as Karplus equations. A typical example is the intra-residue three-bond J-coupling between amide- and α-proton $^3J_{\alpha N}$, which, because the J-coupling occurs through H^N–N–C$^\alpha$–H$^\alpha$, is sensitive to the backbone angle Φ.[30] This approach can be extended to define the side-chain conformations also, through the χ_1, χ_2 and χ_3 dihedral angles, by proper choice of the J-coupling to be measured.[33]

Orientation Restraints

In the last 10 years, much effort has been expended to develop ways of introducing a bias in the alignment of biomolecules in a magnetic field. With weak alignment (order $\sim 10^{-3}$–10^{-4}), nuclear interactions with the magnetic field are reintroduced that otherwise would have been averaged to zero by the isotropic rotational diffusion.[34,35] Weak alignment can be achieved by adding a co-solvent to the sample. Some of these co-solvents are Pf1 phage,[36] bicelles,[37] mixtures of polymer and alcohol[38,39] and strained polyacrylamide gels,[40,41] to name just a few. One interaction that is reintroduced is dipolar coupling between two nuclei. The residual dipolar coupling (RDC) values (D) are deduced from the difference between two apparent J-coupling values measured on similarly prepared samples with one in an isotropic medium (J) and the other in an anisotropic medium ($J + D$). RDC depends on the projection of the interaction vectors, which for directly bonded nuclei are equivalent to the bond vectors, in the alignment frame. This frame is a common frame for all bonds within the biomolecule. The most frequently measured RDCs involve the backbone chemical bonds, such as $^{1}D_{HN}$ of the backbone amide bond. When $^{1}D_{HN}$ at multiple sites along protein backbone are measured, the N–H bond orientations and the common alignment frame can be determined simultaneously. This turns out to be an extremely effective long-range restraint for structure determination, especially for nucleotides where other long-range restraints are often lacking.[42–44]

Structure Calculation

Combination of the above restraints allows protein structure to be determined by various computational methods. A common method uses a molecular dynamics (MD), simulated annealing protocol.[45,46] Every restraint is expressed as a potential energy and summed to yield a target potential function. The overall goal is to minimize the total target plus MD potential energy. A random starting structure is chosen and an MD/simulated annealing run is performed. The calculation is carried out initially at high temperature with weak potential forces. The system is gradually cooled while increasing the force constants. This calculation is repeated many times with different starting structures. This is crucial to overcome the possibility of the system being trapped in a local energy minimum. If all of the experimentally measured structural restraints are consistent, the calculations will converge to a common structure, which most likely is the global minimum. Typically, a set of structures is reported that satisfies all of the structural restraints, the standard deviation of which is taken to be the precision of the resulting structure.[47]

Spin Relaxation and Dynamics

Nuclear spin relaxation occurs due to the fluctuations in the local nuclear interactions in the magnetic field. In solution, the primary source of fluctuations comes from the reorientation of the internuclear vectors with respect to the magnetic field due to the rotational diffusion, and thus the hydrodynamics of the molecule. For proteins, the data most frequently reported are for the spins of backbone ^{15}N bonded to amide ^{1}H. The measured relaxation parameters, longitudinal relaxation T_1, transverse relaxation T_2 and heteronuclear NOE, are sensitive to the dynamics of the N–H bond in time-scales of ps to ns. Furthermore, the apparent T_2 (T_2 dispersion) and T_1 (ZZ exchange) can be modulated by radiofrequency pulses to probe slower dynamics in the μs to ms time-scale.

Spectral Density Function

The molecular rotational correlation time (τ_m) for a protein tumbling in solution, which drives the relaxation of nuclear spin, is typically a few nanoseconds in water at room temperature. This is the time constant for the correlation function describing the motion of the interaction vector (in the case of ^{15}N relaxation, this is effectively the N–H bond vector). The quantitative description of the probability of motional fluctuations at a particular frequency (ω) is best calculated using spectral density function $J(\omega)$ (SDF) [Equation (1)], which corresponds to the Fourier transform of the correlation function.[48] The SDF is essential in understanding NMR spin relaxation results. Commonly, the Model Free form of the SDF [Equations (2) and (3)] is used to analyze relaxation data, which also accounts for local ps time-scale motion (τ_e) of the HN bond with an amplitude S^2, the square of the generalized order parameter, without any assumption of a specific type of motion.[49] For a well-folded protein, the average S^2 value is about 0.85 for structured regions and can be significantly smaller for flexible loops and termini, and the local motion τ_e is typically less than a few hundred ps. Still the Model Free formulae in Equations (2) and (3) are a relatively simple treatment as it assumes isotropic tumbling for the overall motion τ_m. When the measured T_1 and T_2 values deviate significantly from their expected behavior in isotropic systems, methods accounting for anisotropic motion are required.[50,51]

$$J(\omega) = \frac{\tau_m}{1 + \omega^2 \tau_m^2} \tag{1}$$

$$J(\omega) = S^2 \frac{\tau_m}{1 + \omega^2 \tau_m^2} + \left(1 - S^2\right) \frac{\tau'}{1 + \omega^2 \tau'^2} \tag{2}$$

$$\tau'^{-1} = \tau_m^{-1} + \tau_e^{-1} \tag{3}$$

T_1, T_2 and Heteronuclear NOE

The process by which magnetization of a nuclear spin is restored to its Boltzmann population along the axis of the external field is called longitudinal T_1 relaxation. T_1 is measured by inverting the population with a radio frequency pulse and measuring the magnetization as it recovers at time points δ

(~ 0.1–1 s). Curve fitting of δ *versus* resonance intensities with an exponential yields the T_1 value. The relaxation is caused by the local field fluctuations described by the SDF. Similarly to other types of spectroscopy, T_1 relaxation is most effective when the frequency of the fluctuations matches the resonance frequency, the spin Larmor frequency ω. The probability of molecular motion at the ^{15}N Larmor frequency ω_N, SDF $J(\omega_N)$, is the dominant term in ^{15}N T_1 relaxation, hence its T_1 relaxation is sensitive to motions on the time-scale of $1/\omega_N$.

The decay of magnetization or dephasing of coherence in the transverse xy-plane perpendicular to the external field is the transverse T_2 relaxation. T_2 can be measured with experiments employing a spin-lock (SL) scheme or a spin–echo scheme, which measure the transverse magnetization at time points δ (~ 1–100 ms). Curve fitting of δ *versus* resonance intensities with an exponential yields the T_2 value. Unlike T_1, T_2 relaxation is dominated by zero frequency motion, SDF $J(0)$. Since $J(0)$ is equal to τ_m [Equation (1)], T_2 is a direct reporter of molecular correlation time and decreases monotonically with increasing protein size. T_2 is inversely related to the NMR resonance linewidth. Therefore, T_2 underlies the challenges encountered by NMR spectroscopy in studying large proteins.

Heteronuclear NOE between different types of nuclei is typically very weak, but between bonded ^{1}H and ^{15}N spins is observable as the N–H bond length (1.02 Å) is short. It is measured as a steady-state NOE by collecting two spectra, one with and the other without a low-power saturation on the amide ^{1}H. The difference in intensities of the ^{1}H–^{15}N signals is the heteronuclear NOE. The saturation on ^{1}H drives its population difference to zero, which, through the NOE interaction, alters the population difference of the ^{15}N spin as the system seeks to regain thermal equilibrium. The spectral densities $J(\omega_N - \omega_H)$ and $J(\omega_N + \omega_H)$, where ω_H and ω_N are the ^{1}H and ^{15}N Larmor frequencies, are dominant in heteronuclear NOE. The SDF $J(\omega_N \pm \omega_H)$ values are numerically close to $J(\omega_H)$ because ω_N is one-tenth of ω_H, therefore the ^{1}H–^{15}N NOE is sensitive to fast sub-ns dynamics on the time-scale of $1/\omega_H$.

T_2 Dispersion

The ps–ns dynamics described above are present in the thermal motion experienced by every spin. More interesting are dynamics involving the chemical or conformational exchange between states A and B, with corresponding distinct Larmor frequencies for a particular nucleus, ω_A and ω_B. Supposing the first-order rate constants from A to B and B to A are k_{fwr} and k_{rev}, respectively, in the fast exchange limit where $\Delta\omega$ ($= |\omega_A - \omega_B|$) is slower than the exchange rate k_{ex} [$= (k_{fwr} + k_{rev})$], a single coalesced resonance peak at the averaged frequency, $p_A\omega_A + p_B\omega_B$, is observed, where p_A and p_B are population fractions of the spin in state A and B, respectively. The two frequency components can be treated as an effective dephasing mechanism acting in a manner similar to other fluctuations contributing to T_2 relaxation. The relaxation rate term due to conformational exchange R_{ex} is equal to $p_A p_B \Delta\omega^2/k_{ex}$. This term can be appreciable and causes resonance line broadening proportional to R_{ex}/π.[1]

The experimental scheme for T_2 dispersion measurement is similar to the T_2 measurements. By measuring T_2 values at varying spin-lock field strengths or durations of the echo radiofrequency pulses (τ_{cp}), R_{ex} can be extracted. With sufficient T_2 measurements, both kinetic (k_{ex}) and thermodynamic (p_A, p_B and $\Delta\omega$) quantities can be probed.

ZZ Exchange

In slow exchange limit where k_{ex} is slower than $\Delta\omega$, two distinct peaks are observed at frequencies of ω_A and ω_B. The forward and reverse kinetics can be measured using the ZZ exchange pulse scheme, identical with the T_1 measurement. In a 2D spectrum, the signal correlating the two frequencies ω_A and ω_B is observed at time points δ (~ 0.1–1 s, $\delta < T_1$) on the order of $1/k_{ex}$. The signal intensity will build up with increasing δ; in contrast, the diagonal peak intensity, the source of exchange, will drop faster than the rate $1/T_1$ because of kinetic exchange in addition to the natural T_1 relaxation. Fitting the curve dependence on δ *versus* both peak intensities to theoretical equations yields T_1 and additional kinetic information.

In describing NMR relaxation both times (T_1, T_2, T_{ex}) and rates (R_1, R_2, R_{ex} which are inverse of the relaxation times) are sometimes used in the literature.

3   Solution Structure and Fast Dynamics

The crystal and NMR solution structures of individual Gag components from many retroviruses have been solved. Generally MA is a 15–17 kDa globular protein composed of five α-helices and one or two 3_{10}-helices. CA comprises two domains: its 14–15 kDa N-terminal domain (CA^N), relatively flat-shaped, consisting of seven α-helices and two short β-strands, and its 8–9 kDa C-terminal domain (CA^C) consisting of four α-helices. The 6 kDa NC protein lacks a global fold and contains one or two zinc-knuckles, each of which has turn and short α-helix structures. These domains represent tertiary structures of mature viral components. Knowledge of the quaternary packing of the precursor Gag and individual matured domains is directly related to the viral morphology transition during maturation. Data from cryo-EM and two-dimensional crystals have been informative regarding this subject.[52–55] NMR spectroscopy has also played a role. NMR spin relaxation studies on HIV-1 Gag N-terminal fragment composed of MA and CA^N demonstrated the flexibility on the ns time-scale between the two domains.[18] Similar analysis yields the partial flexibility between CA^N and CA^C of Rous sarcoma virus (RSV).[17] It may be reasonable to suggest the existence of a medium to high degree of mobility between Gag domains before proteolytic maturation. For isolated equine infectious anemia virus (EIAV) MA domain, the concentration-dependent NMR spin relaxation and chemical shift data reveal its monomer–trimer equilibrium and the trimeric MA packing may be conserved in mature stage.[56]

HIV-1 Matrix–Capsid^N Fragment

Solution Structure

The HIV-1 Gag MA–CA^N fragment, composed of N-terminal 283 residues, is the longest Gag sequence being studied using NMR spectroscopy.[18] Tang *et al.*[18] expressed and purified the ^{15}N/^{13}C and ^{2}H/^{15}N double-labeled 32 kDa MA–CA^N fragment for chemical shift assignments. The NMR sample contains 1 mM protein buffered at pH 5.0 and 100 mM NaCl. A total of 2046 distance restrains and 332 dihedral angle restrains, corresponding to 17.7 restrains per residue, were used for the structure calculation. No inter-domain NOEs were observed. The 20 structures with lowest energy were collected to yield an r.m.s.d. of 0.41 Å for MA (V7–T122) (Figure 2a) and 0.72 Å for CA^N (H144–S278) (Figure 2b), indicative of good convergence. For MA, the pairwise r.m.s.d to its X-ray structure is 0.9 Å. For CA^N, the global fold is consistent with both previous NMR and X-ray structures except for the missing anti-parallel β-sheet in the region of P133 to Q145, at the N-terminus of CA^N. The absence of β-sheet is fully consistent with the concept that proteolytic cleavage between MA and CA^N results in the β-sheet formation essential in forming cone-shaped capsid.[57] The β-sheet structure is also present in the solution structure of isolated CA^N domain, *e.g.* HTLV-1.[58,59]

Hydrodynamics

The backbone ^{15}N spin relaxation time constants T_1 and T_2 are sensitive to ps–ns fast dynamics, characterizing the overall and internal motion of proteins. The ratio T_1/T_2 ($= R_2/R_1$) is less dependent on internal ps motion and is generally used as a ruler for protein hydrodynamics on the ns time-scale. If a protein is a rigid spherical body, T_1/T_2 values for all well-structured ^{15}N spins should be the same. The discontinuity in T_1/T_2 profile demonstrated that in this Gag fragment the two domains tumble independently (Figure 2c). Numerical calculations taking into account the N–H bond orientations in an axial symmetric diffusion model yield effective correlation times of 10.0 ± 0.1 and 13.2 ± 0.2 ns for MA and CA^N, respectively. The significant slower tumbling of CA^N domain, similar in size to MA, may originate from two sources: first the partial dimerization of CA^N increasing its effective size, and second the presence of motional anisotropy, due to the non-spherical shape of CA^N. The anisotropic diffusion of CA^N is easily identified from a larger distribution in T_1/T_2 profile and elevated T_1/T_2 values in helices 1, 2 and 7 parallel to the long axis of its diffusion tensor. In addition, the heteronuclear NOE data, primarily sensitive to the sub-ns dynamics, are significantly below 0.5 in the regions of 105–150 and 220–230 in addition to the N- and C-termini (Figure 2d). Those

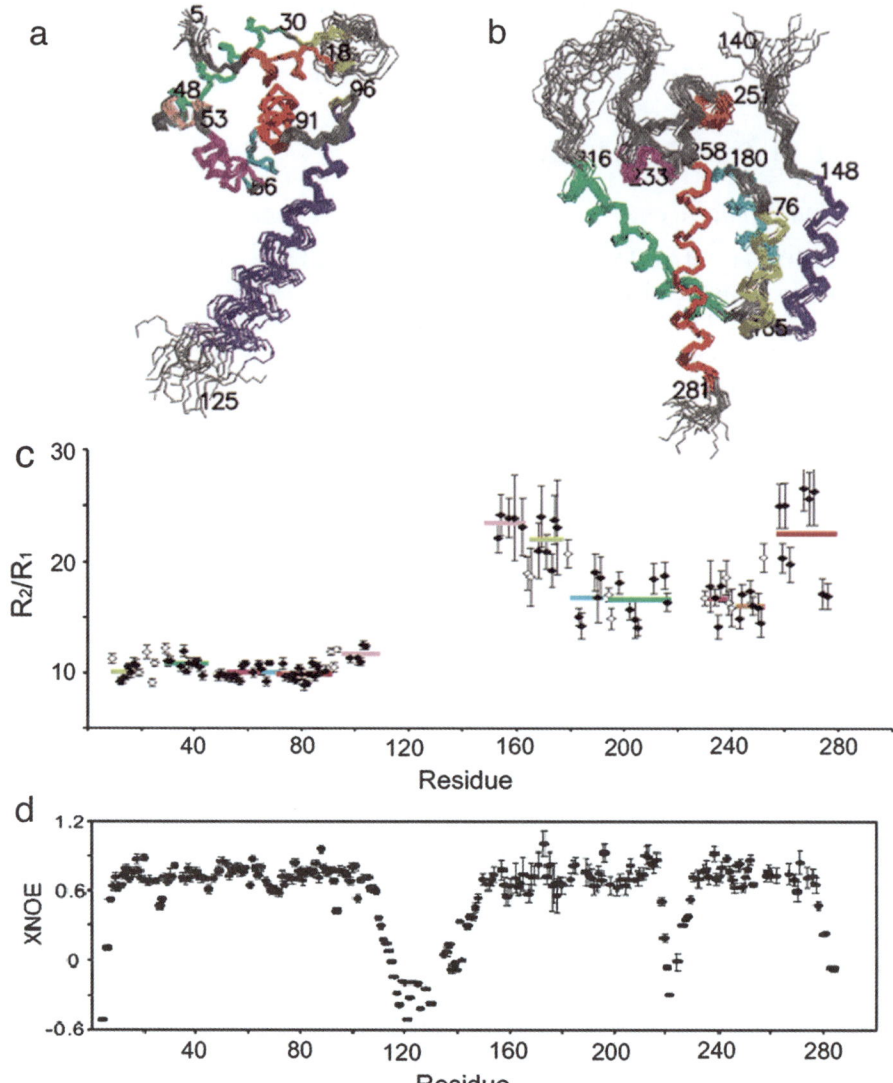

Figure 2   Solution structure and dynamics of HIV-1 MA–CA^N fragment. Superposition of 20 NMR structures of MA (a) and CA^N (b). Relaxation rate
ratio R_2/R_1 (c) and heteronuclear NOE (d) profiles for MA–CA^N. Parts (a)
and (b) are adopted from Figure 3, (c) from Figure 4a and (d) from Figure
2b in reference 18, with permission.

areas are the link regions between two domains and the loop between helix 4
and helix 5, the CypA binding site, experience large-amplitude fast motion.
Thus the heteronuclear NOE data further support the independent ps–ns
dynamics for the two domains and the mobility of the CypA loop.

RSV Capsid

Campos-Olivas *et al.*[17] carried out a similar structure and dynamics study on RSV CA protein comprised of both CA^N (14.3 kDa) and CA^C (8.4 kDa) domains. A short flexible linker (4–5 residues) between the two domains was identified from ^{15}N T_2 and heteronuclear NOE profiles. T_1/T_2 analysis yielded effective correlation times of 16.6 and 12.6 ns for CA^N and CA^C, respectively (Figure 3). It is reasonable to assume some degree of independent diffusion of the two domains because of the different tumbling rates. However, the correlation time obtained is significantly higher than those expected for individual domains. Also, the 16.6 ns tumbling for RSV CA^N is longer than the 13.2 ns correlation time of HIV-1 CA^N within its MA–CA^N fragment. The interpretation would be that the domain motion is correlated to a certain extent, but neither completely rigid nor completely independent. The partial correlation originates from the shorter linker between the domains, which does not allow large amplitude of the domain motion. Perhaps the dynamics of this partially

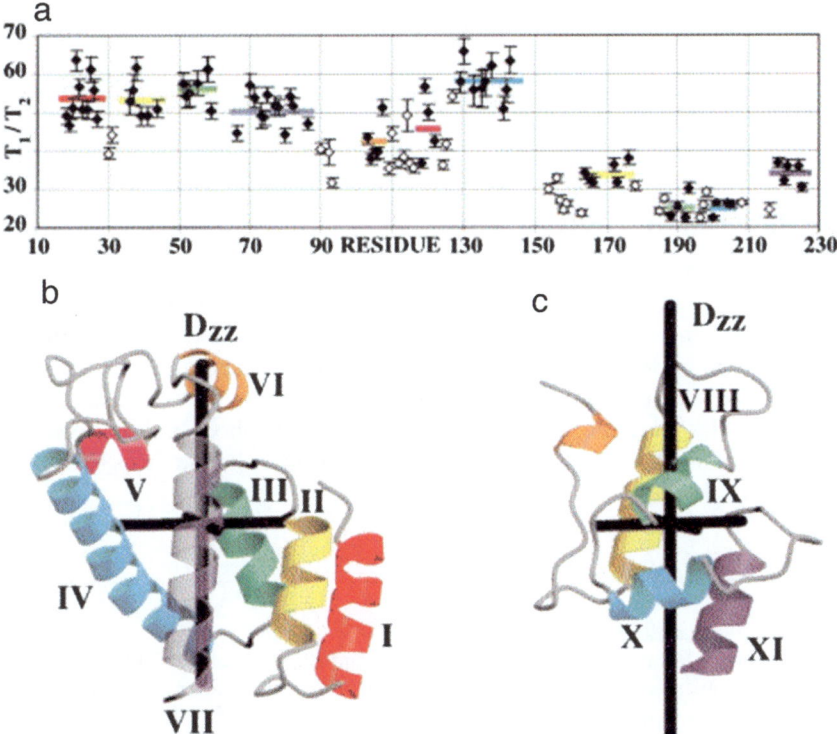

Figure 3 Solution structure and dynamics of RSV Capsid. (a) Relaxation rate ratio T_1/T_2 profile for RSV CA. Solution structures of CA^N (b) and CA^C (c) with the orientations of the calculated rotational diffusion tensors plotted as black rods. Adopted from Figure 6 in reference 17, with permission.

correlated domain motion requires a more complicated diffusion analysis to be meaningful.

EIAV Matrix

Chen *et al.*[56] carried out concentration-dependent ^{15}N spin relaxation measurements and chemical shift mapping for 15 kDa EIAV MA protein. A significant T_1/T_2 difference exists for two samples with a fivefold difference in concentration (Figure 4a). The average T_1/T_2 was 22.5 ± 3.6 for a 52.4 μM sample and 26.5 ± 3.3 for a 262 μM sample. The difference indicates a more populated slower tumbling species at higher concentration. The chemical shift difference identifies the interface residues (open boxes in Figure 4a) that are experiencing conformational exchange from monomer to oligomer (Figure 4b). Interestingly, those interface residues are not located at the dimeric interface of the crystal structure of EIAV MA.[60] Instead, they correspond to the trimeric interface of HIV-1 MA,[61] which suggests that the tertiary packing of HIV-1 MA is also present in EIAV MA (Figure 4c).

An interesting aspect of the EIAV MA study is that T_1/T_2 drops quickly beyond residue 109, indicative of lower order and fast tumbling on a sub-ns time-scale at the C-terminus, consistent with missing electron density beyond residue 109 in the crystal structures (Figure 4a).

4   Ligand Interaction and Complex Structure

Viral budding is a cooperative process involving many *trans* elements acting on Gag.[13] Cellular and biochemical experiments provide clues about the potential participants. Under optimized conditions, *e.g.* shifting the equilibrium more towards the intended complex without modifying the genuine interaction site, solution NMR can reveal detailed interaction mechanisms. Recent examples are HIV-1 MA targeting the lipid bilayer through phosphatidylinositol 4,5-bisphosphate [PI(4,5)P2] (**1**)[15] and Moloney murine leukemia virus (MoMuLV) NC packing the diploid genome RNA.[16]

Figure 4   Concentration-dependent change of oligomerization and dynamics of EIAV MA. Relaxation rate ratio R_2/R_1 (a) and the H–N chemical shift change (b) profiles for EIAV MA at a concentration difference of fivefold. Trimeric interface and PI(4,5)P2–C4 binding regions are depicted as empty and filled boxes, respectively, in (a). (c) Residues exhibiting concentration-dependent chemical shifts mapped on to the trimer model of EIAV MA. Residues H41 and D42 are colored dark green, residues V63, T64, T66, L67, S68 and E71 gold, residues F45 and D50 orange–red and residues T27 and S100 black. Parts (a) and (b) are adopted from Figure 3 and (c) from Figure 4 in reference 56, with permission.

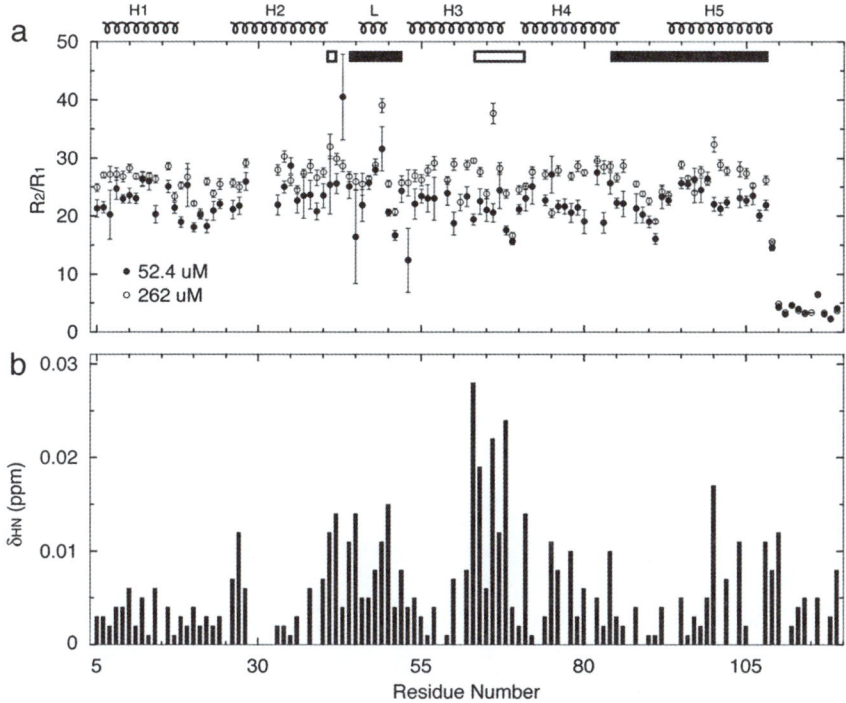

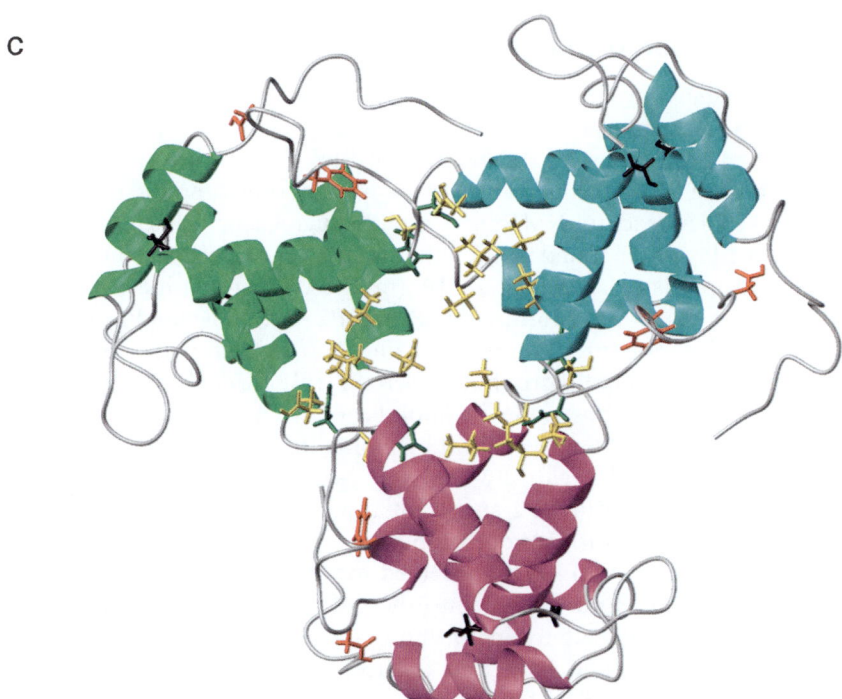

(*1*)

HIV-1 MA Coordinating PI(4,5)P2

PI(4,5)P2, a signaling lipid molecule in the inner leaflet of the plasma membrane (PM), is responsible for recruiting many cytosolic proteins. Cellular experiments also established the key role of PI(4,5)P2 in directing Gag localization.[62] Native PI(4,5)P2 molecules easily form micelles and their interactions with HIV-1 MA lead to NMR line broadening. Using truncated PI(4,5)P2 molecules with shorter 1′ and 2′ aliphatic chains, Saad *et al.* solved the complex structures of myristoylated HIV-1 MA (myrMA) coordinating PI(4,5)P2–C4 and unmyristoylated MA coordinating PI(4,5)P2–C8.[15] The binding clefts from the two complex structures are the same and were defined with intermolecular NOEs (Figure 5a). The hydrophobic cleft β–II–V formed among β-hairpin, helix 2 and helix 5, hosts the inositol head group and the 2′ fatty acid chain (Figure 5b). Other electrostatic interactions are established between phosphate groups and negatively charged residues. The interaction network allows selectivity for PI(4,5)P2 such that other PIP molecules with different phosphate sites such as PI(3)P do not bind.

No intermolecular NOEs were observed for the 1′ fatty acid chain of PIP2, which may be used as the anchor into the bilayer for the Gag complex. This observation leads to the interesting proposal that the micro domain of lipid raft containing only saturated lipid and the free 1′ fatty acid chain allows the accumulation of PIP2–Gag complex (Figure 5c).

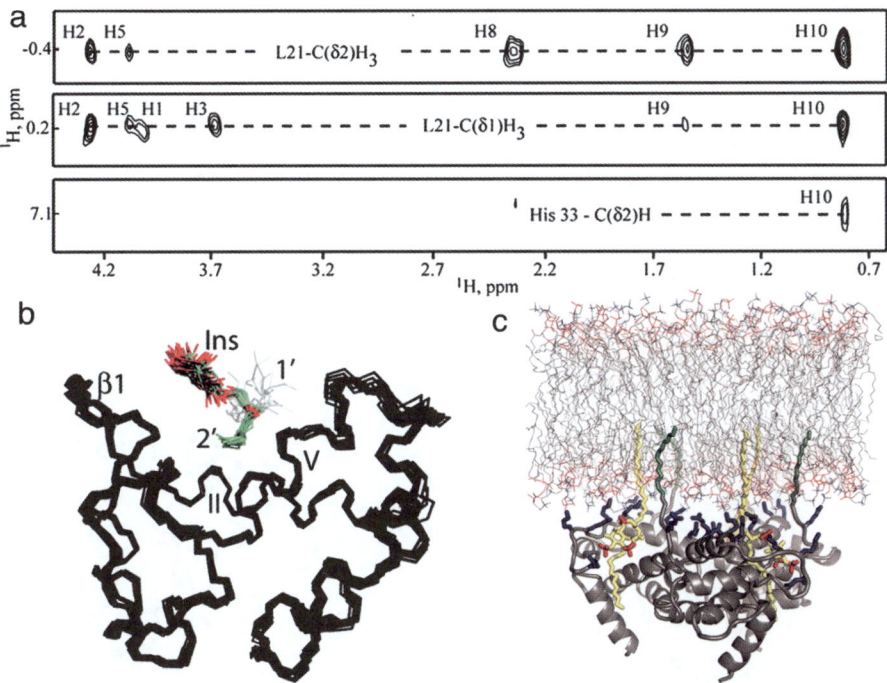

Figure 5 Interaction between HIV-1 MA and PI(4,5)P2. (a) Representative ^{13}C-edited/^{12}C-double-half-filtered NOE data showing unambiguously assigned intermolecular NOEs. (b) Superposition of 20 refined complex structures of MA and PI(4,5)P2–C4. (c) Membrane-binding model predicted from the structural studies. The myristyl groups are colored green. PI(4,5)P2 are colored yellow with red phosphates. The exposed 1′ fatty acids and myristyl groups project from a highly basic surface (Arg and Lys side-chains are shown in blue). Parts (a) and (b) are adopted from Figure 1c and d, respectively, and (c) from Figure 5b in reference 15, with permission.

MoMuLV Nucleocapsid Recognizing Duplex RNA

Retroviruses always pack two copies of genome RNA into a virion. The 350-nucleotide (nt) RNA fragment 5′ to the Gag's structural gene, named the Ψ site, is responsible for RNA duplex formation and NC recognition, which occur sequentially. There are four stem–loop structures in the Ψ site, DIS-1 and -2 and SL-C and -D. Palindromic sequences of DIS-1 (A204–G229) and -2 (C278–G309) provide a basis for equilibrium between mono- and dimeric RNA, *e.g.* when DIS-2 stem–loops dimerize to form intermolecular base pairs, the resulting register shifts leave a six-nucleotide unpaired linker (U304–G309) between DIS-2 and SL-C (Figure 6a). The 6-nt linker is base-paired in the monomeric state of DIS-2 and may be exposed for NC recognition only after dimerization. D'Souza and Summers carefully engineered a 101-nt RNA molecule composed of stem–loops DIS-2, SL-C and SL-D, termed the mutant

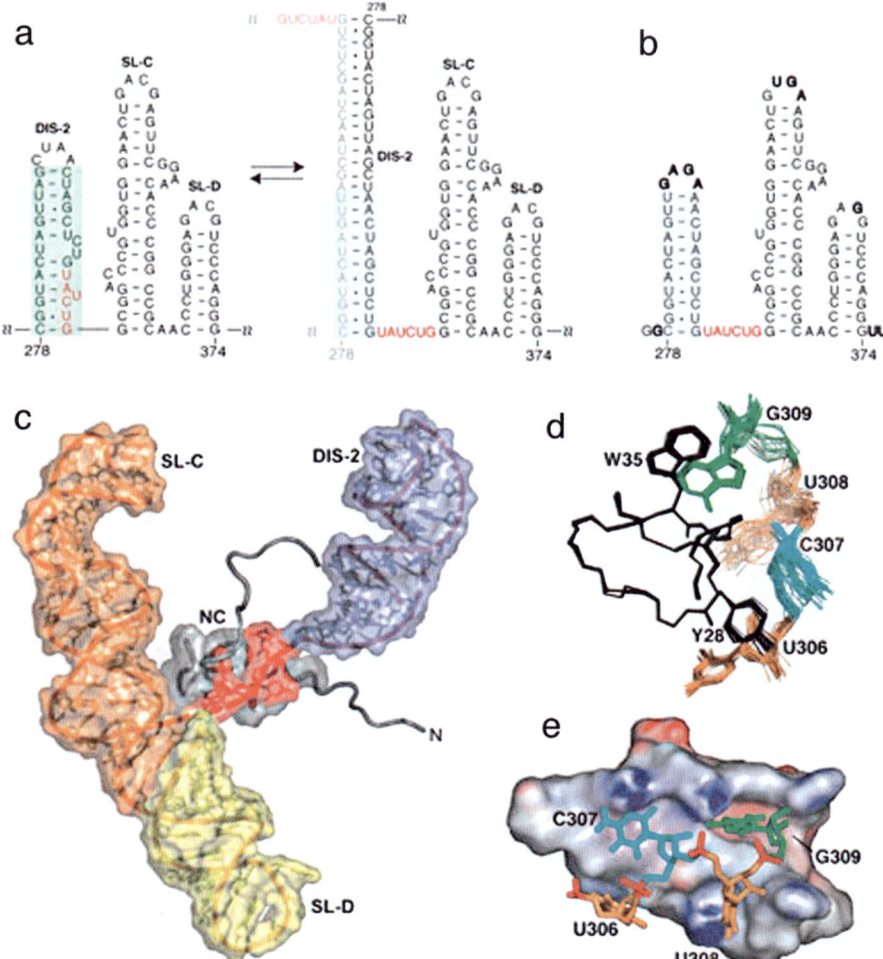

Figure 6   Interaction between MoMuLV NC and its mΨ^{ces} RNA fragment. (a) RNA
secondary structure of the core encapsidation signal (CES). DIS-2 exists in
two alternative monomeric conformations (shaded red and green) and
undergoes a frame shift upon dimerization (shaded blue) that exposes a
UAUCUG element (red). (b) RNA secondary structure of mΨ^{ces} with non-
native nucleotides shown in bold. DIS-2 base pairings match those of the
dimeric form of native CES. (c) Representative complex structure of
NC–mΨ^{ces} including DIS-2 (blue), SL–C (orange) and SL–D (yellow), the
UCUG segment (red) and NC (gray). (d) Superposition of 20 best-fit
backbone structures the CCHC zinc-knuckle coordinating the exposed
U306CUG element. (e) Interactions between the zinc knuckle (colored
according to electrostatic surface potential) and the U306CUG element.
Adopted from Figure 1 in reference 16, with permission.

Ψ 'core encapsidation site' (mΨ^{ces}) (Figure 6b).[16] The DIS-2 within the mΨ^{ces} segment has a dimeric base-pairing pattern and the same register shifts. Importantly the mΨ^{ces} segment exists as a monomer, favorable for NMR as T_2 relaxation in the RNA–protein complex is dominated by the size of the RNA. The 7 kDa NC binds this mΨ^{ces} segment with an affinity on the order of 100 nM, similar to the wild-type fragment.

NC binding induces chemical shift changes and intermolecular NOEs for four 3′ nucleotides U306–G309 of the 6-nt linker. All other resonances of mΨ^{ces} are unaffected. On the other hand, both N- and C-terminal residues of NC, A1–R17 and R44–L56 are disordered before and after binding and only the central zinc-knuckle region undergoes a slight conformational change to bind RNA (Figure 6c). A network of hydrophobic (Figure 6d) and electrostatic interactions was revealed between UCUG bases and phosphodiesters and side-chains of NC zinc-knuckle, *e.g.* the G309 base fits into the specific hydrophobic pocket formed by the side-chains of L21, A27, W35 and A36 (Figure 6e). The complex was also stabilized by complementary charges with the basic NC zinc-knuckle carrying positive charge and RNA exhibiting negative surfaces.

The detailed NC–RNA structure study revealed the essential role of the unpaired RNA fragment. The RNA duplex formation initiates the efficient Gag–NC recognition. It seems MoMuLV primarily uses a conformational switch mechanism of the RNA for its diploid genome packing. Similarly in HIV-1 the NC binding conformation is nearly pre-formed for the RNA.[63,64]

5 Slow and Functional Dynamics

The ps–ns dynamics of biomolecules are mostly due to thermal motion. Biologically interesting kinetics occur at a much slower time-scale because of the energy barriers separating the states. Although enzyme catalysis may reduce this barrier, the time-scale is still slower than ns hydrodynamics. NMR is capable of probing μs–ms dynamics provided that the chemical shift or Larmor frequency difference between the states is comparable to the rate of conversion. For viral structural proteins, slow dynamics were observed for EIAV MA during PI(4,5)P2–C4 binding and for HIV-1 CA^N when CypA was catalyzing the proline isomerization in the loop region.

PI(4,5)P2 Induced Sub-ms Dynamics on EIAV MA

The binding region of PI(4,5)P2–C4 on EIAV MA was established with chemical shift mapping to be the loop between helix 2 and helix 3 and the linker connecting helix 4 and helix 5 (filled boxes in Figure 7a). Because the NMR sample concentration was close to the K_d on the order of 100–200 μM, on–off hopping kinetics of PI(4,5)P2–C4 on EIAV MA should exist. The T_2 dispersion results on a PI(4,5)P2-MA complex sample confirmed the coupled dynamics in the binding region. The ratio of R_2 measured with $\tau_{cp} = 0.5$ ms to those measured at $\tau_{cp} = 0.1$ ms is significantly above 1 for residues T43, V46, E48, G88,

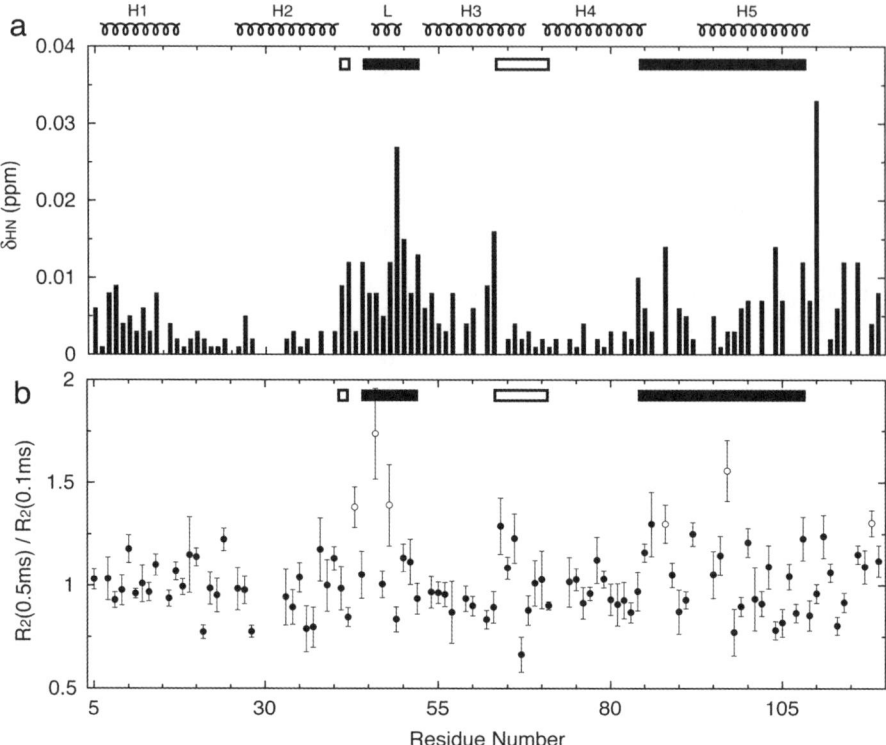

Figure 7   PI(4,5)P2–C4 binding on EIAV MA induced conformational change and sub-ms dynamics. (a) H–N chemical shift change profile of EIAV MA upon PI(4,5)P2–C4 binding. The EIAV MA concentration was 195 μM and the lipid:protein molar ratio is 1:1. (b) Relaxation rate ratio profile of R_2 measured at $\tau_{cp} = 0.5$ ms to R_2 measured at $\tau_{cp} = 0.1$ ms. Residues with significant R_2 enhancements are shown as empty circles. The protein concentration was 52.4 μM and the lipid concentration was 105 μM. Trimeric interface and PI(4,5)P2–C4 binding regions are depicted as empty and filled boxes, respectively. Parts (a) and (b) are adopted from Figures 5A and 9A, respectively, in reference 56, with permission.

G97 and S118 (Figure 7b), which are within the PI(4,5)P2–C4 binding region. The enhanced R_2 due to the exchange contribution (R_{ex}) indicates a motion on a time-scale close to sub-ms. The extraction of quantitative kinetics needs more measurements at various τ_{cp}. A control measurement on EIAV MA protein alone did not detect any R_{ex} contribution.

The sub-ms dynamics of PI(4,5)P2 suggest a weak energy barrier for Gag membrane targeting that maybe essential for Gag oligomerization. If the binding is too fast on the ns time-scale, there is no time left for Gag to adjust its rearrangement on the bilayer; and if the kinetics are too slow, ms–s, enzymatic catalysis might be needed for efficient budding. In fact, Gag alone is sufficient for the budding in a model cellular environment.

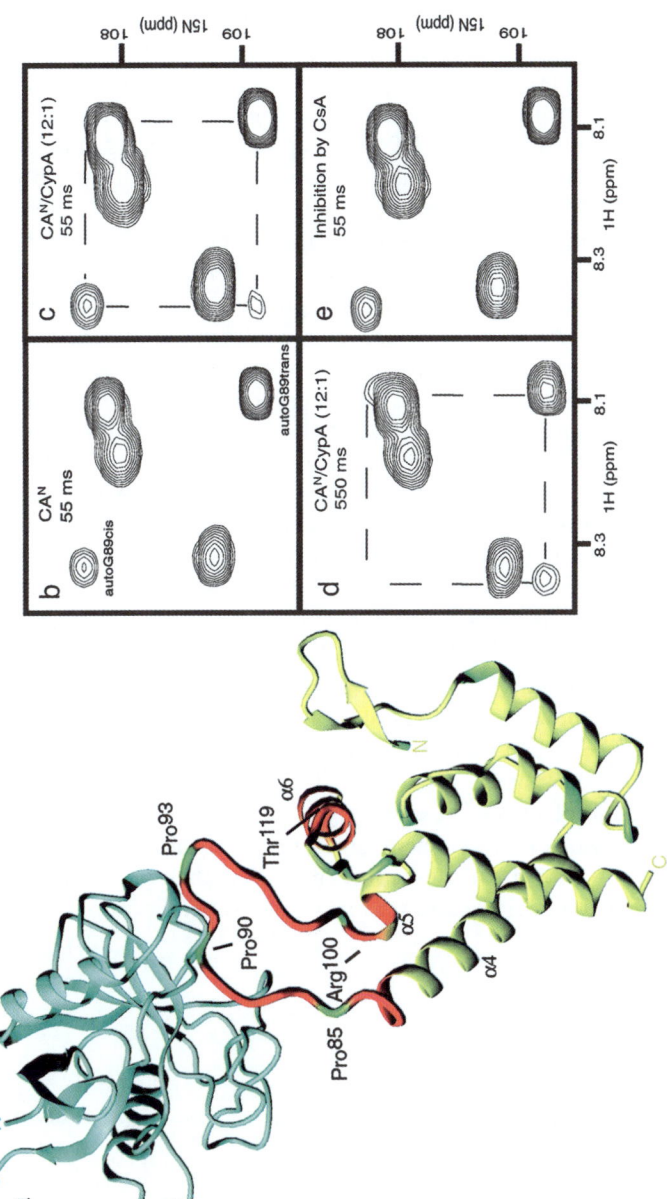

Figure 8 Slow dynamics of HIV-1 CA^N induced by CypA catalysis. (a) Complex structure of HIV-1 CA^N (yellow) and CypA (cyan). Backbone amides of CA^N that shift on CypA binding by more than 14 Hz are highlighted in red. These residues are not only located within the flexible loop (85–100), but also in α4, α5 and α6. Prolines within CA^N are shown in green. (b)–(e) Expansion of ^{1}H–^{15}N heteronuclear exchange spectra showing the amide signal of G89 for the *trans*- and *cis*-isomers. Chemical exchange between the *cis*- and *trans*-conformations of G89–P90 is slow in the absence of CypA (b) and is accelerated in the presence of catalytic amounts of CypA, as indicated by the appearance of exchange peaks between the *cis* and the *trans* auto peaks [(c) and (d)]. The intensity of the exchange peak increases with longer mixing time and concurrently the less abundant *cis* auto peak decreases because of additional loss in magnetization from chemical exchange and longitudinal relaxation (d). Inhibition of isomerase activity by CsA results in a loss of the exchange peaks (e). The mixing times used in NMR experiments are indicated. Part (a) is adopted from Figure 4 in reference 19 and (b)–(e) from Figure 1 in reference 19, with permission.

CypA Catalysis on HIV-1 CA^N

The cellular enzyme CypA is a peptidylprolyl isomerase and has been shown to be essential for HIV-1 infectivity.[65] CypA is packed into HIV-1 virion during the budding process. It specifically binds the *trans* peptide bond of G89–P90 within the loop between helix 4 and helix 5 of HIV-1 CA^N, as revealed by the crystal structure of the complex (Figure 8a).[66] However, its functional role is not clear in terms of viral capsid structure and infection. Structural studies of CA^N alone with NMR revealed that the G89–P90 peptide bond exists in both a *cis* (14%) and a *trans* (86%) conformation.[67] This provides a clue to the possible catalytic role of CypA. Using NMR ZZ exchange spectroscopy, Bosco *et al.* detected the slow kinetics of prolyl isomerization of the G89–P90 bond at a physiological molar ratio of CA^N to CypA of 12:1.[19] The ^{1}H–^{15}N NMR resonance peak of G89 of CA^N was used as a reporter of catalytic kinetics because P90 lacks amide proton resonance. The conformational exchange allows the initial magnetization from one conformer (*e.g. trans*) labeled with the ^{15}N chemical shift being transferred to the other conformer (*e.g. cis*) during the mixing time and subsequently be observed at the ^{1}H chemical shift of the other conformer (*e.g. cis*). As a result both *trans* to *cis* and *cis* to *trans* exchanges would yield two exchange peaks, each of which should correlate with two different resonances from two conformers. With a moderate ZZ mixing time of 55 ms, only the presence of CypA introduced new exchange peaks, in addition to the *cis* and *trans* auto peaks of G89 that were observable without CypA (Figure 8b and c). A longer mixing time builds up a stronger exchange peak (Figure 8d). The addition of CypA inhibitor effectively slowed the exchange process and eliminated the exchange peaks (Figure 8e).

The observation of the kinetics indicated the catalytic role of CypA in HIV-1 infection. CypA is less likely to participate in the thermodynamics of capsid. The calculated isomerization kinetics are about $10 \pm 5 \, \mathrm{s}^{-1}$ and this time-scale seems reasonable for virial infection.[19] CypA may catalyze the conformational fluctuation of capsid and leads to an efficient breaking of capsid. Actually, the binding of CypA on CA^N also introduced conformational change on residues away from the binding loop, indicating a possible allosteric mechanism (colored red in Figure 8a).

6 Summary and Discussion

We have discussed the applications of solution NMR spectroscopy in studying the structure and dynamics of retroviral Gag structural proteins. The structure coordinates obtained are complementary to and consistent with crystal structures. Under favorable situations where the intermolecular interactions are relatively strong and intermolecular NOEs are observable, the detailed interaction mechanisms between Gag components and cellular factors can be revealed. This effort is now being extended to study weak interactions using NMR spectroscopy. The current development of paramagnetic relaxation enhancement and related computational approaches are helpful in probing transient intermolecular interactions.[2]

In addition to structure, dynamics, another physical measure of biomolecules, are playing a significant role in defining biological functions. The ps–ns dynamics can be readily obtained from NMR spin relaxation measurements and are indicative of protein thermal motions. The spin relaxation data can be directly used to estimate the mobility of domains within a protein or the oligomerization of proteins. We have found that the linker between Gag MA and CA^N is more flexible than the linker between CA^N and CA^C within the capsid. The interdomain flexibility within the Gag allows the correct packing of protein shells before and after maturation is reached. The slower μs–ms dynamics correspond to the time-scale for biological events, *i.e.* binding, allosteric and enzymatic reactions, the knowledge of which furthers our understanding of viral functions.

Although obviously the continuing development of solution NMR methods has allowed it to tackle larger proteins and more interactions, the correct interpretations and cross-validations among different biophysical approaches are also desirable. For instance, the solution light scattering results on a mutant full-length Gag suggested a compact globular structure with the MA domain contacting the Gag C-terminus,[68] although there no NMR or other structural data have been collected on native full-length Gag in solution. This nevertheless seems contradictory to the flexibility between Gag domains. Clearly, more efforts are needed to define the conformation of immature viral structural proteins in solution.

References

1. A. G. Palmer, C. D. Kroenke and J. P. Loria, *Methods Enzymol.*, 2001, **339**, 204.
2. G. M. Clore, C. Tang and J. Iwahara, *Curr. Opin. Struct. Biol.*, 2007, **17**, 603.
3. M. Betz, K. Saxena and H. Schwalbe, *Curr. Opin. Chem. Biol.*, 2006, **10**, 219.
4. T. Wilk and S. D. Fuller, *Curr. Opin. Struct. Biol.*, 1999, **9**, 231.
5. M. S. Chapman and L. Liljas, *Adv. Protein Chem.*, 2003, **64**, 125.
6. B. G. Turner and M. F. Summers, *J. Mol. Biol.*, 1999, **285**, 1.
7. S. Scarlata and C. Carter, *Biochim. Biophys. Acta Biomembr.*, 2003, **1614**, 62.
8. H. G. Gottlinger, *Aids*, 2001, **15**, S13.
9. Y. Morikawa, *Curr. HIV Res.*, 2003, **1**, 1.
10. A. G. Bukrinskaya, *Arch. Virol.*, 2004, **149**, 1067.
11. C. S. Adamson and I. M. Jones, *Rev. Med. Virol.*, 2004, **14**, 107.
12. A. C. Steven, J. B. Heymann, N. Q. Cheng, B. L. Trus and J. F. Conway, *Curr. Opin. Struct. Biol.*, 2005, **15**, 227.
13. M. D. Resh, *Aids Rev.*, 2005, **7**, 84.
14. B. K. Ganser-Pornillos, M. Yeager and W. I. Sundquist, *Curr. Opin. Struct. Biol.*, 2008, **18**, 203.

15. J. S. Saad, J. Miller, J. Tai, A. Kim, R. H. Ghanam and M. F. Summers, *Proc. Natl. Acad. Sci. USA*, 2006, **103**, 11364.
16. V. D'Souza and M. F. Summers, *Nature*, 2004, **431**, 586.
17. R. Campos-Olivas, J. L. Newman and M. F. Summers, *J. Mol. Biol.*, 2000, **296**, 633.
18. C. Tang, Y. Ndassa and M. F. Summers, *Nat. Struct. Biol.*, 2002, **9**, 537.
19. D. A. Bosco, E. Z. Eisenmesser, S. Pochapsky, W. I. Sundquist and D. Kern, *Proc. Natl. Acad. Sci. USA*, 2002, **99**, 5247.
20. B. Furtig, C. Richter, J. Wohnert and H. Schwalbe, *ChemBiochem.*, 2003, **4**, 936.
21. X. Y. Sun, Q. Zhang and H. M. Al-Hashimi, *Nucleic Acids Res.*, 2007, **35**, 1698.
22. H. M. Al-Hashimi and N. G. Walter, *Curr. Opin. Struct. Biol.*, 2008, **18**, 321.
23. N. K. Goto and L. E. Kay, *Curr. Opin. Struct. Biol.*, 2000, **10**, 585.
24. K. H. Gardner and L. E. Kay, *Annu. Rev. Biophys. Biomol. Struct.*, 1998, **27**, 357.
25. S. Bagby, K. I. Tong and M. Ikura, *Methods Enzymol.*, 2001, **339**, 20.
26. G. M. Clore and A. M. Gronenborn, *Methods Enzymol.*, 1994, **239**, 349.
27. A. S. Edison, F. Abildgaard, W. M. Westler, E. S. Mooberry and J. L. Markley, *Methods Enzymol.*, 1994, **239**, 3.
28. K. Pervushin, R. Riek, G. Wider and K. Wuthrich, *Proc. Natl. Acad. Sci. USA*, 1997, **94**, 12366.
29. V. Tugarinov and L. E. Kay, *ChemBiochem.*, 2005, **6**, 1567.
30. K. Wüthrich, *NMR of Proteins and Nucleic Acids*, Wiley, New York, 1986.
31. G. M. Clore and A. M. Gronenborn, *Protein Sci.*, 1994, **3**, 372.
32. G. Cornilescu, F. Delaglio and A. Bax, *J. Biomol. NMR*, 1999, **13**, 289.
33. A. Bax, G. W. Vuister, S. Grzesiek, F. Delaglio, A. C. Wang, R. Tschudin and G. Zhu, *Methods Enzymol.*, 1994, **239**, 79.
34. N. Tjandra and A. Bax, *Science*, 1997, **278**, 1111.
35. J. R. Tolman, J. M. Flanagan, M. A. Kennedy and J. H. Prestegard, *Proc. Natl. Acad. Sci. USA*, 1995, **92**, 9279.
36. M. R. Hansen, L. Mueller and A. Pardi, *Nat. Struct. Biol.*, 1998, **5**, 1065.
37. M. Ottiger and A. Bax, *J. Biomol. NMR*, 1999, **13**, 187.
38. M. Ruckert and G. Otting, *J. Am. Chem. Soc.*, 2000, **122**, 7793.
39. L. G. Barrientos, C. Dolan and A. M. Gronenborn, *J. Biomol. NMR*, 2000, **16**, 329.
40. R. Tycko, F. J. Blanco and Y. Ishii, *J. Am. Chem. Soc.*, 2000, **122**, 9340.
41. H. J. Sass, G. Musco, S. J. Stahl, P. T. Wingfield and S. Grzesiek, *J. Biomol. NMR*, 2000, **18**, 303.
42. N. Tjandra, S. Tate, A. Ono, M. Kainosho and A. Bax, *J. Am. Chem. Soc.*, 2000, **122**, 6190.
43. A. Bax, G. Kontaxis and N. Tjandra, *Methods Enzymol.*, 2001, **339**, 127.
44. J. H. Prestegard, C. M. Bougault and A. I. Kishore, *Chem. Rev.*, 2004, **104**, 3519.

45. M. Nilges, G. M. Clore and A. M. Gronenborn, *FEBS Lett.*, 1988, **229**, 317.
46. G. M. Clore and A. M. Gronenborn, *NMR of Proteins*, CRC Press, Boca Raton, FL, 1993.
47. G. P. Gippert, P. F. Yip, P. E. Wright and D. A. Case, *Biochem. Pharmacol.*, 1990, **40**, 15.
48. P. J. Hore, J. A. Jones and S. Wimperis, *NMR, the Toolkit*, Oxford University Press, Oxford, 2000.
49. G. Lipari and A. Szabo, *J. Am. Chem. Soc.*, 1982, **104**, 4546.
50. D. E. Woessner, *J. Chem. Phys.*, 1962, **36**, 1.
51. L. K. Lee, M. Rance, W. J. Chazin and A. G. Palmer, *J. Biomol. NMR*, 1997, **9**, 287.
52. S. D. Fuller, T. Wilk, B. E. Gowen, H. G. Krausslich and V. M. Vogt, *Curr. Biol.*, 1997, **7**, 729.
53. J. F. Conway, W. R. Wikoff, N. Cheng, R. L. Duda, R. W. Hendrix, J. E. Johnson and A. C. Steven, *Science*, 2001, **292**, 744.
54. B. K. Ganser-Pornillos, A. Cheng and M. Yeager, *Cell*, 2007, **131**, 70.
55. K. Mayo, D. Huseby, J. McDermott, B. Arvidson, L. Finlay and E. Barklis, *J. Mol. Biol.*, 2003, **325**, 225.
56. K. Chen, I. Bachtiar, G. Piszczek, F. Bouamr, C. Carter and N. Tjandra, *Biochemistry*, 2008, **47**, 1928.
57. I. Gross, H. Hohenberg, C. Huckhagel and H. G. Krausslich, *J. Virol.*, 1998, **72**, 4798.
58. S. Khorasanizadeh, R. Campos-Olivas and M. F. Summers, *J. Mol. Biol.*, 1999, **291**, 491.
59. C. C. Cornilescu, F. Bouamr, X. Yao, C. Carter and N. Tjandra, *J. Mol. Biol.*, 2001, **306**, 783.
60. H. Hatanaka, O. Iourin, Z. H. Rao, E. Fry, A. Kingsman and D. I. Stuart, *J. Virol.*, 2002, **76**, 1876.
61. C. P. Hill, D. Worthylake, D. P. Bancroft, A. M. Christensen and W. I. Sundquist, *Proc. Natl. Acad. Sci. USA*, 1996, **93**, 3099.
62. A. Ono, S. D. Ablan, S. J. Lockett, K. Nagashima and E. O. Freed, *Proc. Natl. Acad. Sci. USA*, 2004, **101**, 14889.
63. R. N. De Guzman, Z. R. Wu, C. C. Stalling, L. Pappalardo, P. N. Borer and M. F. Summers, *Science*, 1998, **279**, 384.
64. B. M. Lee, R. N. De Guzman, B. G. Turner, N. Tjandra and M. F. Summers, *J. Mol. Biol.*, 1998, **279**, 633.
65. D. Braaten and J. Luban, *EMBO J.*, 2001, **20**, 1300.
66. T. R. Gamble, F. F. Vajdos, S. H. Yoo, D. K. Worthylake, M. Houseweart, W. I. Sundquist and C. P. Hill, *Cell*, 1996, **87**, 1285.
67. R. K. Gitti, B. M. Lee, J. Walker, M. F. Summers, S. Yoo and W. I. Sundquist, *Science*, 1996, **273**, 231.
68. S. A. K. Datta, J. E. Curtis, W. Ratcliff, P. K. Clark, R. M. Crist, J. Lebowitz, S. Krueger and A. Rein, *J. Mol. Biol.*, 2007, **365**, 812.

Section 2

CHAPTER 9

Evolution of Viral Capsid Structures – the Three Domains of Life

REZA KHAYAT AND JOHN E. JOHNSON

Department of Molecular Biology, The Scripps Research Institute, La Jolla, CA 92037, USA

1 Introduction

Viruses are pathogens of every cellular life form on the planet and these agents actively move between diverse ranges of hosts. The promiscuous nature of viruses is contributed by their mastery of horizontal gene transfer (HGT) and gene recombination. HGT and gene recombination allow viruses to evolve at an unprecedented rate compared with cellular life. A recurring theme among viruses believed to share a common ancestor is that while they share little to no sequence similarity between proteins that serve homologous functions, they do share structural homology between such proteins.

In this chapter, we take a historical approach and focus on two examples where the use of structural information, from electron microscopy and X-ray crystallography, have been crucial in identifying evolutionary relationships between viruses that infect members from all domains of life (archaea, prokarya and eukarya). The use of structural homology to infer common ancestry is based on the logic that viruses sharing structurally homologous capsid proteins and common capsid architectures are more likely to have arisen through divergent evolution from a common ancestor, rather than to have arisen from distinct ancestors and come to share such common traits through

RSC Biomolecular Sciences No. 21
Structural Virology
Edited by Mavis Agbandje-McKenna and Robert McKenna
© Royal Society of Chemistry 2011
Published by the Royal Society of Chemistry, www.rsc.org

convergent evolution. In other words, the probability of generating homologous coat proteins and viral architectures from morphologically distinct ancestors is lower than having such traits come from a common ancestor, *i.e.* Occam's razor. This is particularly true when considering the morphological diversity of viruses (*e.g.* the Virosphere).

The first example of a lineage encompassing the three domains of life is the adenovirus lineage, where a common ancestor is shared between the mammalian adenovirus, the bacteriophage PRD1, the algae virus PBCV-1 and the archaeal *Sulfolobus* turreted icosahedral virus (STIV).[1–4] The second example involves the bacteriophage HK97 lineage, where a common ancestor is shared between bacteriophage HK97, the animal herpesvirus and the archaeal *Pyrococcus furiosus* virus (PfV).[5–7]

2   The Adenovirus Lineage

The role of structural biology in viral phylogeny was initially appreciated when an unexpected relationship between RNA plant viruses (tomato bushy stunt virus and southern bean mosaic virus) and animal picornaviruses (poliovirus and rhinovirus) was recognized from their homologous capsid protein structures.[8–11] These proteins exhibited identical topological organizations that subsequently became known as the viral jellyroll.

Structural analysis of the protein database (PDB) indicates that there are more than 140 entries for virus capsid protein structures that incorporate the viral jellyroll. Of these entries, more than half contain less than 30% sequence identity with one another (R. Khayat and J. E. Johnson, unpublished results). Identifying phylogenetic relationships between proteins with less than 30% sequence identity remains a difficult task, hence structural information is crucial for identifying such relationships.

The Adenovirus

Adenoviruses (Ads) infect a variety of mammals in the animal kingdom (domain eukarya). These viruses were first isolated and cultured by Rowe *et al.* from adenoidal tissue removed from children.[12] Similar agents were isolated from military personnel with respiratory illnesses.[13,14] Human adenoviruses cause respiratory illnesses and conjunctivitis and have been associated with infantile gastroenteritis.[15–17]

The adenovirus was one of the earliest biological specimens to be imaged using electron microscopy.[18] These images identified an icosahedral particle with a proteinaceous capsid consisting of surface-protruding molecules. The molecules were termed hexons and pentons –which described the number of their capsid neighbors.[19]

Adenovirus type 2 (Ad2) has a mass of 150 MDa, measures $\sim 900 \text{Å}^2$ from opposite vertices and has fiber-like appendages decorating its vertices.[20–22] The predominant proteins in adenovirus include the hexon, the penton and the fiber protein arranged on a pseudo $T = 25$ surface lattice.[20,23,24] Supporting the capsid shell are the four minor capsid proteins IIIa, IX, VIII and VI. These proteins are also referred to as cementing proteins, for their particular role in helping stabilize the capsid structure. Proteins IIIa and IX attach to the towers of specific hexons. Protein VIII is situated below the peripentonal hexons and the hexons surrounding the icosahedral threefold axis (Figure 1).[25] The exact location of protein VI remains to be determined. At the core of each virion, and interacting with the linear dsDNA genome ($\sim 36\,000\,bp$), are the four non-structural proteins V, VII, terminal protein and μ.

The Major Capsid Protein (Hexon)

Hexon is a homo-trimeric assembly of a single gene product.[23] Each poly-peptide chain of the Ad2 hexon has a molecular weight of 109 kDa and contains 967 amino acid residues.[26] From negative stained electron microscopy images of disassembled virions, it was evident that whereas the hexons possessed a triangular top, the base possessed pseudo-hexagonal symmetry.[27,28] Hexon is extremely stable and requires extensive boiling in sodium dodecyl sulfate (SDS) to dissociate and denature. The samples used for structural studies were pur-ified by boiling the virion in SDS followed by centrifugation and chromato-graphy.[29] The crystal structure of Ad2 hexon, solved to 2.9 Å resolution, revealed that each chain contained two structurally homologous viral jellyroll domains (V1 and V2) connected to one another by a helix (αI) (Figure 1).[1,30] The six viral jellyrolls in each hexon, two from each chain, are arranged such that pseudo-hexagonal symmetry dominates at the base of the molecule. Extensive decorations in the loops of each viral jellyroll give rise to the trian-gular appearance of the hexons observed on the outer surface of the virus in the electron microscope images.

Each jellyroll can also be described as two β-sheets forming a β-sandwich. Each β-sheet is composed of four anti-parallel β-strands (labeled B-I for V1 and B'-I' for V2). The sheets are slightly tilted with respect to one another such that the sandwich appears to follow a right-handed helical twist. A hydrogen-bonding network between strands B–I–D–G form one of the sheets, and a similar network between strands C–H–E–F form the second sheet (Figure 1). There are no hydrogen bonds between strands from opposing sheets. The β-strands run parallel to the hexon's threefold symmetry axis and are normal to the surface of the icosahedral capsid.[22]

A hexon chain can be decomposed into eight sections (Figure 1). These include an N-terminal region (NT), the first jellyroll (V1), large loop insertions within V1 (DE1 and FG1), the jellyroll connector (VC), the second viral jel-lyroll (V2) and large loop insertions in V2 (DE2 and FG2). The extensive loop insertions in each jellyroll occur between homologous strands (DE1, FG1, DE2 and FG2) and create a 'tower'-like structure that extends 64 Å above the pseudo-hexagonal base.[21] The first jellyroll contains roughly twice the number of insertions as the second, but insertions in the second extend 15 Å further, creating the trimeric appearance on the outer surface. A 10-residue helix (αF) follows the F and F′ strand of each jellyroll and is almost orthogonal to the jellyroll. Within the hexon trimer, each αF helix is wedged between neighboring jellyrolls and points towards the axis of symmetry (Figure 1).

The Bacteriophage PRD1

Bacteriophage PRD1 was discovered in the sewers of Kalamazoo, MI, USA, and infects a broad range of Gram-negative prokaryotes (domain prokarya).[31] PRD1 has a mass of 66 MDa, measures $\sim 750 \text{Å}^2$ from opposite vertices and has fiber-like appendages decorating its vertices.[32]

The capsid shell is composed of the major capsid protein P3, the vertex protein P31, the fiber-like spike protein P5 and the receptor binding protein P2 arranged on a pseudo T = 25 surface lattice. Within the capsid shell are the tape measure protein P30, a host-derived viral membrane and a transmembrane-associated protein P16. Within the capsid are four non-structural proteins responsible for genome delivery (P11, P14, P18 and P32), a protein with muralytic activity (P15), a minor capsid protein (P6), a DNA packaging ATPase (P9), two DNA packaging proteins (P20 and P22) and a transglyco-sylase (P7). Associated with the linear dsDNA genome are the ssDNA-binding proteins P12 and P19 and the genome terminal protein P8.[33]

The Major Capsid Protein (P3)

The 43 kDa major capsid protein P3 is organized as homo-trimers in the cap-sid.[32] The trimer is extremely stable and requires extensive boiling in SDS to dissociate.[34] The crystal structure of P3, solved to 2.0 Å resolution, identified a topology and fold that are remarkably similar to those of the hexon from adenovirus.[2] Each P3 polypeptide chain is composed of two eight-stranded viral jellyrolls connected to one another *via* a helix (αI). The β-strands are

---

Figure 1   Structure and architecture comparison of the adenovirus lineage. (A) Radial coloring of the cryo-EM image reconstruction densities. Adenovirus type 5 at 10.3 Å, PRD1 at 15.3 Å, PBCV-1 at 25 Å and STIV at 27 Å. (B) Top view cartoon representation of the hexon, P3, Vp54 and B345 looking down the threefold axis (PDB entries 1P2Z, 1CJD, 1J5Q and 2BBD, respectively). The P3 and B342 l_{fg} loops are outlined in black. (C) Side-view cartoon repre-sentation of the isolated subunits. The 'towers' are indicated by brackets. (D) Topology figures of the subunits. Helices are shown as circles and strands as triangles. Triangles pointing up indicate strands pointing towards the reader and triangles pointing down indicate strands pointing away from the reader. The colors used are red for the terminal region, yellow for the first jellyroll, cyan for loop insertion DE1, green for loop insertion FG1, purple for the jellyroll connector, blue for the second jellyroll, pink for loop insertion DE2 and orange for loop insertion FG2. (E) Cartoon diagram representing the positions of the minor capsid proteins. Only the known components are indicated. The Ad5 and PRD1 figures have been modified from their original publications.[25,37]

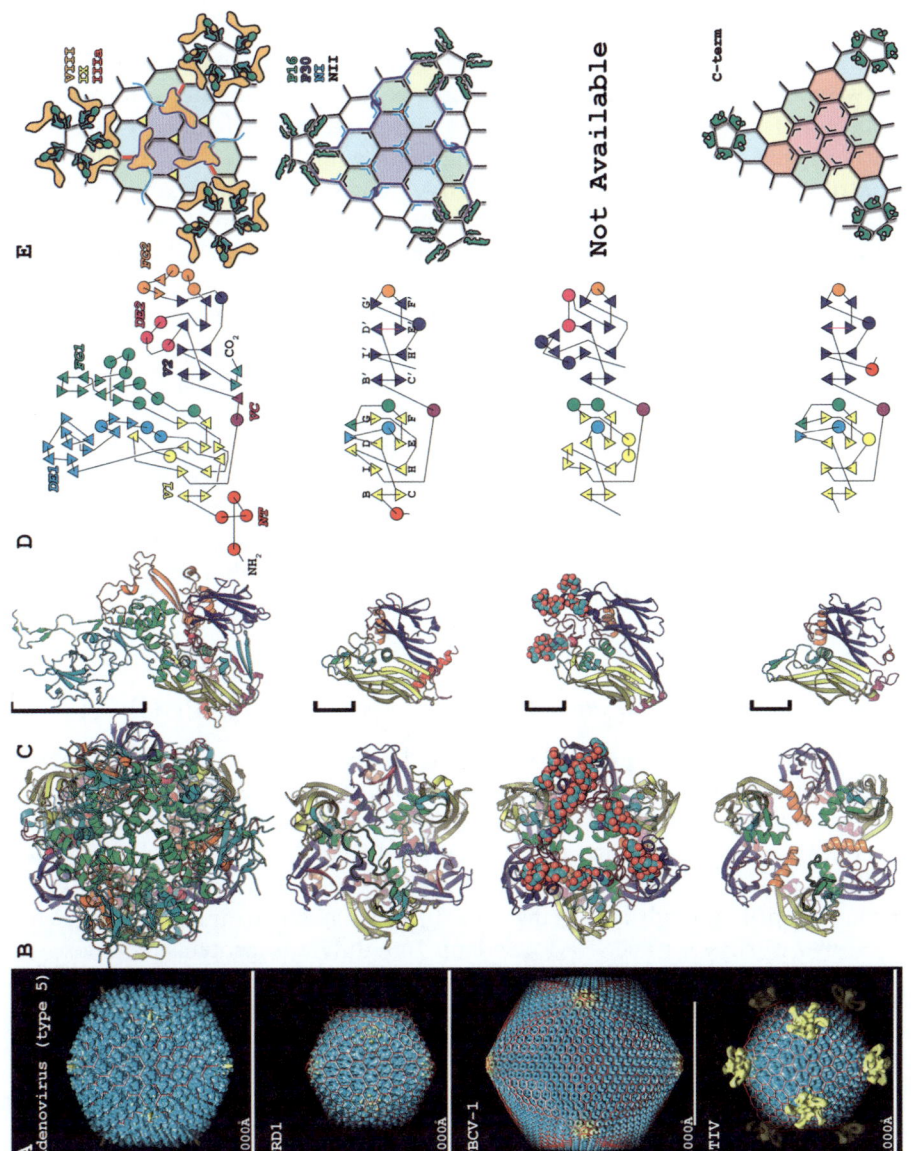

parallel to the trimer's axis of symmetry. There are large insertions in the DE and FG loops of both jellyrolls. The insertions in the first jellyroll are more extensive and define the 'tower' region of P3, which extends 22 Å above the second jellyroll. The FG loop insertions of both jellyrolls include a helix (αF and αF') that points to and is normal to the trimer symmetry axis and is wedged between neighboring jellyrolls (Figure 1).

The stability of the P3 trimer can be attributed to the extensive interaction between neighboring P3 molecules. Stable protein–protein interactions involve $1600 \pm 200 \,\text{Å}^2$ of buried surface area at the interface.[35] There is $\sim 3360 \,\text{Å}^2$ of surface area buried between two neighboring molecules. The predominant contribution to this interaction, $\sim 2220 \,\text{Å}^2$, comes from the lasso-like l_{fg} loop making extensive interactions with the neighboring subunits. The remaining $1140 \,\text{Å}^2$ of buried surface area comes from β-strand F' and helix αF' packing against the neighboring B–I–D–G β-sheet.

Structural studies of the intact virion show the strands in each jellyroll, and also the P3 symmetry axis, to be normal to the capsid surface.[36,37] The N-terminal helix is a molecular switch that adopts two distinct conformations in the icosahedral asymmetric unit (iASU). In one conformation, three basic residues at the N-terminus of the helix form an extended chain and interact electrostatically with the negatively charged viral membrane along the entire edge of the facet, essentially anchoring the membrane to the capsid shell.[36,37] In the second conformation, the N-terminal helix sweeps away from the membrane and interacts with the C-terminus of the neighboring subunit at the icosahedral and quasi-threefold axis (Figure 1).

The *Paramecium bursaria* Chlorella Virus Type 1 (PBCV-1)

Paramecium bursaria chlorella virus type 1 (PBCV-1) was isolated from zoo-chlorellae isolates of *P. bursaria* (kingdom plantae, domain eukarya).[38] PBCV-1 has a molecular mass of 1 GDa, measures $1900 \,\text{Å}^2$ from opposite vertices, encodes for ~ 375 proteins and encloses a linear 330 kbp protein-associated dsDNA genome.[39,40] The capsid shell is composed of the major capsid glyco-protein Vp54 and glycoproteins Vp280 and Vp260 arranged on a pseudo $T = 169d$ surface lattice.[41] More than 100 different proteins compose the infectious virion. Within the capsid shell is a host-derived lipid bilayer membrane that is sandwiched between the capsid shell and the viral genome.

The Major Capsid Protein (Vp54)

The major capsid protein Vp54 forms extremely stable homo-trimeric assemblies that require extensive boiling in SDS to denature. Each polypeptide chain consists of 437 residues (48.1 kDa), of which four are N- and two are O-linked glycosylated (6 kDa). The crystal structure of Vp54, solved to 2.0 Å resolution, is trimeric and shows each chain to be composed of two consecutive viral jellyroll domains, once again connected *via* a helix (αI). There are large insertions in the loops connecting strands DE and FG of both jellyrolls. The lengths of these insertions are nearly equal between the two jellyrolls such that, unlike the hexon and P3, the polypeptide chain does not create a 'tower' for the Vp54 trimer. The helices following the F and F' β-strands (αF and αF', respectively) of both jellyrolls are orthogonal to and wedged between the jellyrolls.[3]

With the exception of Ser57, the glycosylation sites of Vp54 occur in the second jellyroll (Figure 1). The carbohydrate chains nearly cover the entire surface of the trimer and come close to interlocking with carbohydrates from the adjacent Vp54 chain. Glycosylation of Vp54 may play a role in stabilizing the PBCV-1 virion.[3] Interestingly, the carbohydrates decorating the polypeptide chain create the 'tower' for Vp54. The crystal structure fits into the PBCV-1 cryo-EM image reconstruction such that the Vp54 β-strands and symmetry axis are orthogonal to the icosahedral surface.

The *Sulfolobus* Turreted Icosahedral Virus (STIV)

The *Sulfolobus* turreted icosahedral virus (STIV) was isolated from a hot spring in Rabbit Creek Yellowstone National Park.[42] STIV infects the acidophilic archaea *Sulfolobus solfataricus* (phylum *Crenarchaeota*, domain archaea) that grows optimally at temperatures higher than 80 °C and in the pH range 2–4. Little is known about members of the archaea domain and there is great interest in the many unique biochemical properties of these organisms. The genome of *S. solfataricus* has been sequenced and it has become a model system for studying hyperthermophilic archaea.[43,44]

STIV has a mass of ~ 60 MDa, measures ~ 1000 Å from opposite vertices and has large pentameric turret-like appendages that protrude more than 130 Å from the shell surface. The 17 600 bp circular dsDNA genome encodes for ~ 36 predicted open reading frames (ORFS). The capsid shell is composed of the heterogeneously glycosylated major capsid protein B345 (38–45 kDa based on the extent of glycosylation) and proteins A223, C381 and C557 arranged on a pseudo T = 31d surface lattice. Five copies of each protein, A223 (24 kDa), C381 (42 kDa) and C557 (58 kDa), are believed to form each penton complex. Sandwiched between the capsid shell and the genome is a host-derived single layer of cyclic tetraether lipids. Possibly embedded in this lipid layer, *via* predicted transmembrane helices, are minor capsid proteins A55 (6 kDa) and B130 (14 kDa). Protein B164 (19 kDa) may be an ATPase responsible for packaging the STIV dsDNA genome into the capsid shell. Also associated with the purified virion is the virally encoded B109 (12 kDa) protein, a 7 kDa host-derived DNA-binding protein (SSO7D) and a 25 kDa host-derived protein (SSO0881) that has a VPS24 vacuolar sorting function.[42,45]

The Major Capsid Protein (B345)

The major capsid protein of STIV (B345) forms homo-trimeric capsomers that can be identified from cryo-EM image reconstruction of the intact virion.[42]

However, biochemical studies of recombinantly expressed B345 clearly indicate that it is monomeric, even at concentrations higher than 50 mg mL^{-1}.

The B345 crystal structure, solved to 2.0 Å resolution, identified all but the last 27 C-teminal residues. B345 crystallizes as a monomer and has a topology and fold that are nearly indistinguishable from those of the PRD1 P3 (Figure 1, Table 1). Once again, there are two homologous jellyrolls connected by an α-helix (αI). With the exception of loop l_{fg}, both the lengths and types of insertions in the loops connecting the jellyroll β-strands are nearly identical between B345 and the P3 (Figure 1). The α-helices following the F and F′ β-strands of both jellyrolls are orthogonal to the strands of each jellyroll and wedged between the two jellyrolls, in a similar manner to hexon, P3 and Vp54.

The most prominent differences between the P3 and B345 structures are the absence of the 17 residue N-terminal helix and the much longer l_{fg} loop of P3. In P3, the N-terminal helix is responsible for protein–membrane interaction and capsid assembly.[37] Whereas the crystal structure of B345 begins with Gly2 forming the B β-strand of the first jellyroll, the C-terminus ends with a stretch of 27 residues that could not be modeled owing to missing density. Using a combination of structure prediction, docking and difference mapping techniques, the C-terminus of B345 could be modeled to form an α-helix that could interact with the negatively charged viral membrane through a number of basic residues at its very C-terminus. Thus the C-terminus of B345 carries out the same capsid–membrane anchoring function as observed for the N-terminus of PRD1. The docking experiments also showed that B345 was oriented in the capsid such that its jellyroll β-strands are normal to the capsid surface.

In P3, the longer l_{fg} loop closely interacts with neighboring P3 molecules and is predominantly responsible for forming and stabilizing the P3 trimer (see above). Surface area calculations using the B345 trimer, modeled after the P3 trimer, indicated than ∼1140 Å^2 of surface area is buried at each of the B345 trimer interfaces. This is less than the 1600 ± 200 Å^2 reported for stable protein–protein interactions. The absence of such a loop, or an equivalent interaction, could explain why B345 is monomeric.[4] Interestingly, the five B345 capsomers in the iASU do not follow ideal threefold symmetry and are geometrically distinct from one another.[4] It may be possible that even in the context of the capsid shell the B345 capsomers are elastic.

Table 1 Structure and sequence comparison of the adenovirus lineage capsid proteins.[a]

	Hexon (type 2)	*P3*	*Vp54*	*B345*
Hexon (type 2)		21 (5)	21 (7)	25 (6)
P3	5.75 (318)		27 (9)	31 (11)
Vp54	5.13 (359)	4.03 (315)		27 (8)
B345	6.01 (304)	2.81 (296)	3.71 (293)	

[a]The diagonal on the right is the percentage sequence similarity calculated using the Blosum30 scoring matrix. Numbers in parentheses are percentage sequence identity. Structure-based sequence alignments were used for the scoring. The diagonal on the left is the r.m.s.d. for Cα positions using a 5 Å distance cut-off. Numbers in parentheses are the number of aligned residues.

The proposed evolutionary relationship between adenovirus, PRD1, PBCV-1 and STIV stems primarily from, but is not limited to, their homologous major capsid protein structure. Considering that an eight-stranded jellyroll can be formed *via* four different folds,[46] yet only one of these folds is observed for icosahedral virus capsids, strongly argues that the viruses, or at least their major capsid proteins, diverged from a common ancestor. What is unique to the adenovirus lineage is the perpendicular orientation of the jellyroll β-strands with respect to the icosahedral surface and the pattern of loop lengths connecting the β-strands. The jellyroll β-strands of other icosahedral viruses are oriented parallel to the icosahedral capsid surface and the pattern of loop lengths connecting their β-strands vary.

These viruses also share particularly similar capsid architectures. The pseudo-hexameric and pentameric capsomers are composed from different gene products. Long protrusions decorate the vertices of adenovirus, PRD1 and STIV, whereas PBCV-1 has a single similar protrusion on the surface of each iASU.[69] Beneath the capsid shell of adenovirus, PRD1 and STIV are minor capsid proteins surrounding the base of the vertex pentamer.[4,25,37] A tape measure-like protein identified from the crystal structure of PRD1 also appears to be present in adenovirus (Figure 1).[25,37]

3 The HK97 Lineage

The HK97 lineage likely involves the largest biomass on the planet, with an estimated 10^{31} particles infecting organisms in the archaea, prokarya and eukarya domains of life.[47] Members of this group (*e.g.* bacteriophages lambda, P22, T7, phi29 and Mu-1) are indispensable tools to the field of molecular biology – having paved the way for studying genetics, protein folding, protein–protein interaction and dynamics of macromolecular machines. The capsids of these virions assemble as spherical immature proheads that subsequently expand to a mature icosahedral head. The expansion is believed to accompany and accommodate the packing of the genome *in vivo*.

The Bacteriophage HK97

HK97 is a tailed, linear dsDNA temperate lambdoid coliphage that was isolated from pig dung in Hong Kong.[48] Although similar to lambda, HK97 has a slightly larger capsid (diameter of 660 Å from opposite vertices) and a longer tail (~ 1700 Å).[5] The $\sim 39\,800$ bp genome encodes for more than 60 proteins, of which ~ 10 are incorporated into the infectious virion. HK97 infects *Escherichia coli*, in the domain prokarya.

The icosahedral head includes the coat protein (gp5) and the portal protein (gp3) arranged on a $T = 7l$ surface lattice.[5] The tail includes gp7, gp12, gp16, gp24, gp28 and possibly gp72. Recombinant expression of gp5 leads to the spherical Prohead I form of the capsid. Expression of gp5 and the HK97 encoded protease (gp4) leads to a capsid that can be expanded *in vitro* to the

Head II form that is morphologically identical with the mature capsid from the infectious virion, with the exception of the missing portal.[49] Expansion of the Prohead I to Head II capsid involves a number of intermediates that have been captured and studied using cryo-EM, X-ray crystallography and solution X-ray scattering. This expansion involves rotational and translational movements of the major capsid protein and large changes in the subunit tertiary structure.

The Major Capsid Protein (gp5)

The crystal structure of the Head II capsid, solved to 3.5 Å resolution, identified a new protein fold.[5] The Head II structure has a surprisingly thin shell (18 Å) and is composed of planar hexamers and concave pentamers cross-linked to their neighbors *via* an isopeptide bond. The bonds between neighboring subunits occur along each hexamer–hexamer and hexamer–pentamer interface. These interactions form a chain-mail lattice such that the cross-linked subunits encircle the pentamers or hexamers, respectively (Figure 2). The cross-linking is believed to stabilize and promote capsid expansion.[50,51]

The gp5 structure is composed of two domains formed by non-contiguous portions of the polypeptide chain. The axial domain (domain A) is a six-stranded β-sheet, decorated by three α-helices, that sits adjacent to the hexamer and pentamer axes and protrudes slightly from the capsid. The peripheral domain (domain P) is a long α-helix packed against a long three-stranded anti-parallel β-sheet that extends from a two-stranded β-sheet decorated by an extended loop (E-loop) and forms the edges of the hexamers and pentamers. Both of the cross-linking sites are in domain P, with one near the three-stranded β-sheet (Asn356) and the other in the E-loop (Lys169).[5]

At the N-terminus of gp5 is a 103-residue scaffolding domain, referred to as the delta-domain, which is necessary for capsid assembly. The delta-domain also restrains the capsid to the Prohead I state and must be removed, by the HK97-encoded protease (gp4), for maturation to begin.[49] In the Head II structure, the N-terminus makes extensive interactions with neighboring subunits and terminates in the capsid shell plane, such that there is no room for the delta-domain. Structure prediction of the delta-domain and the cryo-EM image reconstruction of Prohead I together suggest that the delta-domain forms a coiled coil fold that resides inside the particles.[52]

The Herpesvirus

Members of the *Herpesviridae* family infect many species throughout the animal kingdom (domain eukarya). Infections are life long, latent, recurring and cannot be cleared from the system. Herpesviruses are enveloped viruses with large icosahedral capsids. Embedded into the envelope are a number of viral glycoproteins responsible for cellular attachment and viral entry. Sandwiched between the viral envelope and the capsid shell are tegument proteins responsible for hijacking the cellular machinery and propagating the viral lifecycle.

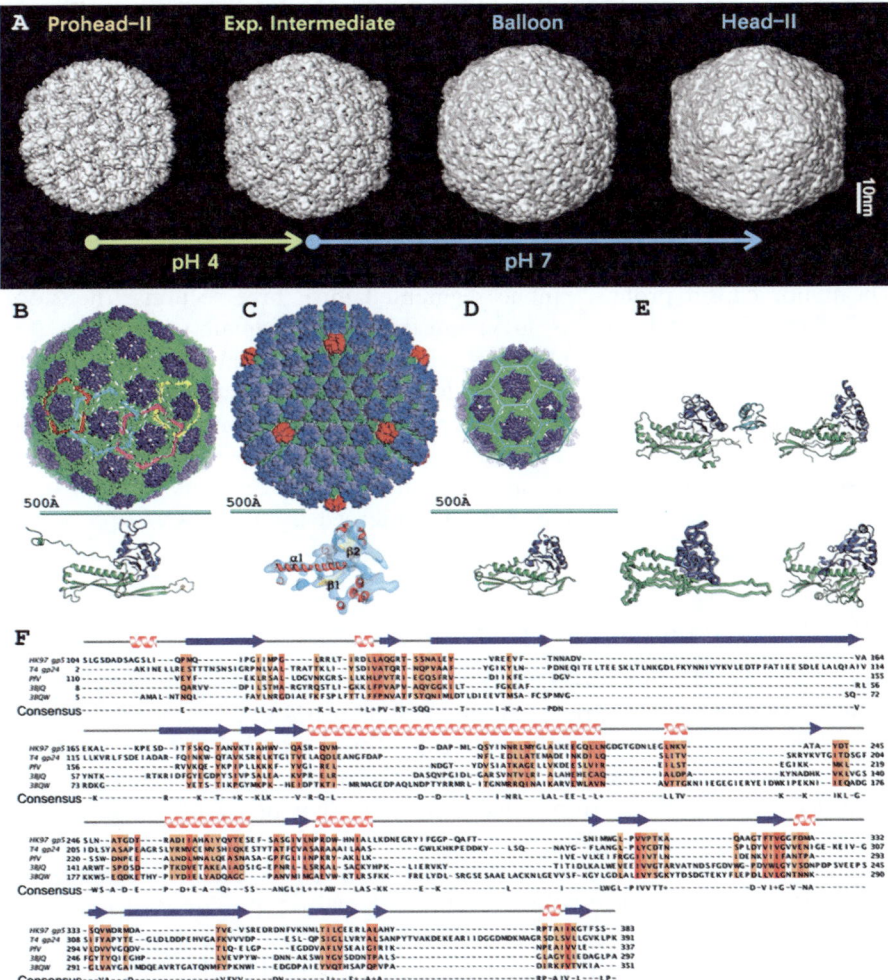

Figure 2 The HK97 lineage. (A) Cryo-EM image reconstructions of the different states of the HK97 expansion process.[66] (B) Tube representation of the HK97 Head II crystal structure. Cross-linking residues are shown as black circles. Beneath is the ribbon representation of the HK97 subunit. Domains A and P are shown in blue and green, –respectively (PDB entry 1OHG). (C) Surface representation of the 8.5 Å HSV-1 cryo-EM image reconstruction.[67] Beneath is the segmented floor domain of HSV-1 VP5, cyan mesh, showing putative helices and β-sheets.[6] (D) Tube representation of the PfV crystal structure. Beneath is the ribbon representation of the PfV subunit (PDB entry 2E0Z). (E) Ribbon representation of the T4 gp24 (PDB entry 1YUE), PDB entries 3BJQ and 3BQW and Cα trace of the epsilon 15 gp7 (PDB entry 3C5B). (F) Structure-based sequence alignment of the HK97 lineage. The secondary structure elements of HK97 are shown at the top of each row. Colored columns indicate greater than 50% physicochemical properties conserved.[68] Consensus is indicated by greater than 40% sequence identity per column.

The icosahedral capsids ($\sim$180 MDa) are composed of hexamers and pentamers arranged on a T = 16 surface lattice enclosing a linear dsDNA genome. The genome ranges in size, depending on the family member, from 80 000 to 250 000 bp. The hexamers and pentamers are composed of multiple copies of a single gene product. A number of minor capsid proteins or cementing proteins and a portal associate with the major capsid protein to make the capsid shell. The number and positioning of the minor capsid proteins differ slightly between the members of the family. Although the capsid shells also vary in thickness and size, they are close to 160 Å thick and $\sim$1250 Å from opposing vertices. The major capsid protein can be segmented into three sections: the 'floor' ($\sim$50 Å thick), the 'middle' ($\sim$30 Å) and the 'upper' domain ($\sim$85 Å).[53]

Herpesvirus capsids assemble as spherical procapsids that irreversibly mature into icosahedral-shaped capsids. Time-lapse cryo-EM studies of HSV-1 capsids undergoing this maturation *in vitro* have revealed the conformational change to involve rotational and translational movement of the 'floor' domain that progresses into the 'triplex' and 'upper' domains.[54] The motions observed in the 'floor' domain are akin to the motions observed for the HK97 gp5 protein during capsid expansion.

The HSV-1 major capsid protein (VP5)

The major capsid protein of herpes simplex virus 1 (HSV-1) is 149 kDa. The crystal structure of the 'upper' domain, solved to 2.9 Å resolution, shows a novel fold that is predominantly α-helical.[55] Structural data for the 'floor' and 'middle' domains are available from an 8 Å cryo-EM map of the HSV-1 capsid. Analysis of the VP5 electron density suggests that many of the HK97 gp5 secondary structural elements, along with its size and capsomer organization, are preserved in the 'floor' domain (Figure 2).[6] Moreover, the dynamics of gp5, in the context of the capsid assembly and maturation, are recurrent in the HSV-1 capsid.

Herpesvirus capsid proteins require a virally encoded scaffolding protein for assembly.[56] The scaffolding protein recruits a virally encoded protease into the capsid to proteolyze the scaffolding protein and initiate viral expansion. This mechanism is similar to that for the delta-domain and protease of HK97.

The *Pyrococcus furiosus* Virus (PfV)

Pyrococcus furiosus (phylum *Euryarchateota*) is a hyperthermophilic archaea that was discovered in geothermally heated marine sediments at the beach of Porto di Levante, Vulcano, Italy.[57] *P. furiosus* is a curious organism as it grows optimally at 100 °C and is highly resistant to radiation damage.

Analysis of *P. furiosus* lysate by EM identified 30 nm spherical particles resembling virus-like particles (VLPs). These particles, referred to as *P. furiosus* virus-like particles (PfVs), are composed of multiple copies of a 39 kDa protein that is encoded by the *P. furiosus* genome and has strong sequence similarity to a number of other archaeal encoded proteins.[58] Expression of the gene in *E. coli* resulted in particles with a similar morphology. The 3.6 Å crystal structure of the recombinantly expressed PfV revealed a 7 MDa spherical particle composed of a 39 kDa protein arranged on a T = 3 icosahedral surface lattice.[7] The dimension of the particles from opposite vertices is ∼ 360 Å and the shell is 30 Å thick (Figure 2).

The Major Capsid Protein

The PfV subunit structure is amazingly similar to the HK97 gp5 and is also composed of two domains (Figure 2, Table 2). Domain A sits at the center and domain P forms an edge of each capsomer complex. The most prominent differences between the two structures include the N-terminal region and the E-loop.

The N-terminus of the PfV subunit contains a 109-residue region that could not be modeled due to indiscernible electron density, but it appears to be located within the particle.[7] Secondary structure predictions suggest that this region forms a helix–coil–helix structure. The location and predicted fold for this region strongly imply that it may be the delta-domain of PfV. This agrees with the spherical shapes of PfV and Prohead I of HK97, where the delta-domain is fused to and resides within the capsid.

The E-loop of PfV is 17 Å shorter than that of gp5 –where one of the gp5 cross-linking residues resides. In general, hyperthermophilic proteins have shorter loops and extensions to increase protein stability.[59] The HK97 cross-linking residues are not conserved in PfV and no cross-links can be seen in the PfV structure.

The HK97 Fold

Two additional bacteriophage major capsid proteins (gp24 of T4 and gp7 of epsilon 15) and three bacterial genomic proteins have homologous structures to the HK97 gp5.[60–64] These structures are slightly larger than HK97 and contain

Table 2 Structure and sequence comparison of the HK97 lineage capsid proteins.[a]

	HK97 gp5	*T4 gp24*	*3BJQ*	*3BQW*	*PfV*
HK97 gp5		30 (11)	22 (8)	21 (5)	34 (8)
T4 gp24	4.12 (220)		26 (8)	21 (6)	30 (7)
3BJQ	3.68 (217)	4.64 (243)		31 (7)	35 (10)
3BQW	4.00 (216)	4.57 (246)	3.91 (253)		34 (6)
PfV	2.81 (202)	3.54 (224)	3.42 (220)	3.58 (210)	

[a]The diagonal on the right is the percentage sequence similarity calculated using the Blosum30 scoring matrix. Numbers in parentheses are percentage sequence identity. Structure-based sequence alignments were used for the scoring. The diagonal on the left is the r.m.s.d. for Cα positions using a 5 Å distance cut-off. Numbers in parentheses are the number of aligned residues.

a number of additional decorations to the HK97 fold. The most prominent differences between the structures include the N-termini and the E-loops. The N-termini adopt alternative conformations and the E-loops vary in size. There is a domain insertion in the E-loop of the T4 gp24.

The discovery of the HK97-fold in archaeal and bacterial genomes may have been explained as a remnant of viral infection and not relevant to the biology of the host (*e.g.* PfV). However, the identification of a bacterial microcompartment using the HK97-like fold suggests otherwise. *Thermotoga maritima* employs the HK97-like fold to build $T = 1$ icosahedral particles, termed encapsulin, that serve as chambers for ferritin-like proteins (Flps). Encapsulin is prevalent in a number of other prokaryotes and genomic analysis suggests that it may package different enzymes. The recruitment and encapsulation of Flp occur *via* the N-terminal domain of encapsulin.[64] This is akin to the delta-domain of HK97 gp5, which recruits the HK97 protease (gp4). Interestingly, the capsid protein that once served to protect the pathogen genome is now incorporated into the host genome, where it serves to fulfill important metabolic functions for the host.

4   Viral Capsid Evolution

Comparative genomics has been, and remains, an important tool for studying phylogenetic relationships among viruses. However, such relationships cannot be identified between related but highly diverged genomes. Structural information is crucial for developing new methods to detect distant evolutionary relationships. Members in the above-described lineages display genomes that share little to no sequence homology among structural proteins. The homologous capsid protein structures imply that, at least for these gene products, there is a shared common ancestor. Closer inspection of the viral architectures (*e.g.* shapes of capsomers and capsomer–capsomer interactions) provides additional similarities that are unlikely to have risen independently. Superimposed on these 'fixed' points of particle structure are a dazzling array of genes hijacked from the hosts and genes that effect interactions between hosts. Among the tailed phages, the function of these non-structural genes include control of bacterial latency, cellular and immune responses to bacterial infections and even proteins associated with autoimmune diseases such as lupus in humans.[65] The fact that one coat protein dominates the adenovirus capsid class of dsDNA viruses and another dominates the dsDNA tailed bacteriophage and herpes viruses suggests that the coat protein and gene products such as cementing proteins, portals and DNA packaging proteins function as a module that is structurally fixed, while Nature experiments with a huge variety of additional genes that are incorporated through illegitimate DNA recombination events. Indeed, the two classes of capsid proteins associated with the dsDNA viruses described are the 'test-tubes' within which Nature has performed extraordinary experiments in evolution.

References

1. F. K. Athappilly, R. Murali, J. J. Rux, Z. Cai and R. M. Burnett, *J. Mol. Biol.*, 1994, **242**, 430.
2. S. D. Benson, J. K. Bamford, D. H. Bamford and R. M. Burnett, *Cell*, 1999, **98**, 825.
3. N. Nandhagopal, A. A. Simpson, J. R. Gurnon, X. Yan, T. S. Baker, M. V. Graves, J. L. Van Etten and M. G. Rossmann, *Proc. Natl. Acad. Sci. USA*, 2002, **99**, 14758.
4. R. Khayat, L. Tang, E. T. Larson, C. M. Lawrence, M. Young and J. E. Johnson, *Proc. Natl. Acad. Sci. USA*, 2005, **102**, 18944.
5. W. R. Wikoff, *Science*, 2000, **289**, 2129.
6. M. L. Baker, W. Jiang, F. J. Rixon and W. Chiu, *J. Virol.*, 2005, **79**, 14967.
7. F. Akita, K. T. Chong, H. Tanaka, E. Yamashita, N. Miyazaki, Y. Nakaishi, M. Suzuki, K. Namba, Y. Ono, T. Tsukihara and A. Nakagawa, *J. Mol. Biol.*, 2007, **368**, 1469.
8. S. C. Harrison, A. J. Olson, C. E. Schutt, F. K. Winkler and G. Bricogne, *Nature*, 1978, **276**, 368.
9. C. Abad-Zapatero, S. Abdel-Meguid, J. E. Johnson, A. Leslie, I. Rayment, M. G. Rossmann, D. Suck and T. Tsukihara, *Nature*, 1980, **286**, 33.
10. J. M. Hogle, M. Chow and D. J. Filman, *Science*, 1985, **229**, 1358.
11. M. G. Rossmann, E. Arnold, J. W. Erickson, E. A. Frankenberger, J. P. Griffith, H. J. Hecht, J. E. Johnson, G. Kamer, M. Luo and A. G. Mosser, *Nature*, 1985, **317**, 145.
12. W. P. Rowe, R. J. Huebner, L. K. Gillmore, R. H. Parrott and T. G. Wart, *Proc. Soc. Exp. Biol. Med.*, 1953, **84**, 570.
13. M. R. Hilleman and J. H. Werner, *Proc. Soc. Exp. Biol. Med.*, 1954, **85**, 183.
14. R. J. Huebner, W. P. Row, T. G. Ward, R. H. Parrott and J. A. Bell, *N. Engl. J. Med.*, 1954, **251**, 1077.
15. E. Jawetz, *Br. Med. J.*, 1958, **i**, 873.
16. T. H. Flewett, A. S. Bryden, H. Davies and C. A. Morris, *Lancet*, 1978, **i**, 4.
17. R. K. Yolken, *J. Petiatr.*, 1982, **101**, 21.
18. R. W. Horne, S. Brenner, A. P. Waterson and P. Wildy, *J. Mol. Biol.*, 1959, **1**, 84.
19. H. S. Ginsberg, H. G. Pereira, R. C. Valentine and W. C. Wilcox, *Virology*, 1966, **28**, 782.
20. J. van Oostrum and R. M. Burnett, *J. Virol.*, 1985, **56**, 439.
21. R. M. Burnett, M. G. Grutter and J. L. White, *J. Mol. Biol*, 1985, **185**, 105.
22. P. L. Stewart, R. M. Burnett, M. Cyrklaff and S. D. Fuller, *Cell*, 1991, **67**, 145.
23. M. G. Grutter and R. M. Franklin, *J. Mol. Biol*, 1974, **89**, 163.
24. V. Mautner and H. G. Pereira, *Nature*, 1971, **230**, 456.
25. C. M. Fabry, M. Rosa-Calatrava, J. F. Conway, C. Zubieta, S. Cusack, R. W. Ruigrok and G. Schoehn, *EMBO J.*, 2005, **24**, 1645.
26. G. Akusjarvi, P. Alestrom, M. Pettersson, M. Lager, H. Jornvall and U. Petterssonn, *J. Biol. Chem.*, 1984, **259**, 13976.

27. M. V. Nermut, *Virology*, 1975, **65**, 480.

28. I. Nasz, and E. Adam, *Acta Micobiol. Hung.*, 1983, **30**, 169.

29. J. J. Rux, D. Pascolini and R. M. Burnett, *Large-scale Purification and Crystallization of Adenovirus Hexon*, Humana Press, Totowa, NJ, 1999.

30. M. M. Roberts, J. L. White, M. G. Grutter and R. M. Burnett, *Science*, 1986, **232**, 1148.

31. R. H. Olsen, J.-S. Siak and R. H. Gray, *J. Virol.*, 1974, **14**, 69.

32. S. J. Butcher, D. H. Bamford and S. D. Fuller, *EMBO J.*, 1995, **14**, 6078.

33. J. K. Bamford, J. J. Cockburn, J. Diprose, J. M. Grimes, G. Sutton, D. I. Stuart and D. H. Bamford, *J. Struct. Biol.*, 2002, **139**, 103.

34. L. Mindich, D. H. Bamford, T. McGraw and G. Mackenzie, *J. Virol.*, 1982, **44**, 1021.

35. L. Lo Conte, C. Chothia and J. Janin, *J. Mol. Biol.*, 1999, **285**, 2177.

36. C. San Martin, J. T. Huiskonen, J. K. Bamford, S. J. Butcher, S. D. Fuller, D. H. Bamford and R. M. Burnett, *Nat. Struct. Biol.*, 2002, **9**, 756.

37. N. G. Abrescia, J. J. Cockburn, J. M. Grimes, G. C. Sutton, J. M. Diprose, S. J. Butcher, S. D. Fuller, C. San Martin, R. M. Burnett, D. I. Stuart, D. H. Bamford and J. K. Bamford, *Nature*, 2004, **432**, 68.

38. L. J. Van Etten, R. H. Meints, D. Kuczmarski, D. E. Burbank and K. Lee, *Proc. Natl. Acad. Sci. USA*, 1982, **79**, 3867.

39. X. Yan, N. H. Olson, J. L. Van Etten, M. Bergoin, M. G. Rossmann and T. S. Baker, *Nat. Struct. Biol.*, 2000, **7**, 101.

40. J. L. Van Etten, *Annu. Rev. Genet.*, 2003, **37**, 153.

41. M. P. Skrdla, D. E. Burbank, Y. Xia, R. H. Meints and J. L. Van Etten, *Virology*, 1984, **135**, 308.

42. G. Rice, L. Tang, K. Stedman, F. Roberto, J. Spuhler, E. Gillitzer, J. E. Johnson, T. Douglas and M. Young, *Proc. Natl. Acad. Sci. USA*, 2004, **101**, 7716.

43. F. Pfeifer, P. Palm and K. H. Schleifer, (eds), *Molecular Biology of Archeae*, Gustav Fischer, Stuttgart, 1994.

44. Q. She, R. K. Singh, F. Confalonieri, Y. Zivanovic, G. Allard, M. J. Awayez, C. C. Chan-Weiher, I. G. Clausen, B. A. Curtis, A. De Moors, G. Erauso, C. Fletcher, P. M. Gordon, I. Heikamp-de Jong, A. C. Jeffries, C. J. Kozera, N. Medina, X. Peng, H. P. Thi-Ngoc, P. Redder, M. E. Schenk, C. Theriault, N. Tolstrup, R. L. Charlebois, W. F. Doolittle, M. Duguet, T. Gaasterland, R. A. Garrett, M. A. Ragan, C. W. Sensen and J. Van der Oost, *Proc. Natl. Acad. Sci. USA*, 2001, **98**, 7835.

45. W. S. Maaty, A. C. Ortmann, M. Dlakic, K. Schulstad, J. K. Hilmer, L. Liepold, B. Weidenheft, R. Khayat, T. Douglas, M. J. Young and B. Bothner, *J. Virol.*, 2006, **80**, 7625.

46. T. J. Gibson and P. Argos, *J. Mol. Biol.*, 1990, **212**, 7.

47. K. E. Wommack and R. R. Colwell, *Microbiol. Mol. Biol. Rev.*, 2000, **64**, 69.

48. E. K. Dhillon, T. S. Dhillon, A. N. Lai and S. Linn, *J. Gen. Virol.*, 1980, **50**, 217.

49. R. W. Hendrix and R. L. Duda, *Adv. Virus Res.*, 1998, **50**, 235.

50. P. D. Ross, N. Cheng, J. F. Conway, B. A. Firek, R. W. Hendrix, R. L. Duda and A. C. Steven, *EMBO J.*, 2005, **24**, 1352.
51. K. K. Lee, L, Gan, H. Tsuruta, C. Moyer, J. F. Conway, R. L. Duda, R. W. Hendrix, A. C. Steven and J. E. Johnson, *Structure*, 2008, **16**, 1492.
52. J. F. Conway, R. L. Duda, D. H. Chen, R. W. Hendrix and A. C. Steven, *J. Mol. Biol.*, 1995, **253**, 86.
53. L. Wu, P. Lo, X. Yu, J. K. Stoops, B. Forghani and Z. H. Zhou, *J. Virol.*, 2000, **74**, 9646.
54. J. B. Heymann, N. Cheng, W. W. Newcomb, B. L. Trus, J. C. Brown and A. C. Steven, *Nat. Struct. Biol.*, 2003, **10**, 334.
55. B. R. Bowman, M. L. Baker, F. J. Rixon, W. Chiu and F. A. Quiocho, *EMBO J.*, 2003, **22**, 757.
56. V. G. Preston, J. A. Coates and F. J. Rixon, *J. Virol.*, 1983, **45**, 1056.
57. G. Fiala and K. O. Stetter, *Arch. Microbiol.*, 1986, **145**, 56.
58. K. Namba, K. Hagiwara, H. Tanaka, Y. Nakaishi, K. T. Chong, E. Yamashita, G. E. Armah, Y. Ono, Y. Ishino, T. Omura, T. Tsukihara and A. Nakagawa, *J. Biochem. (Tokyo)*, 2005, **138**, 193.
59. R. J. Russell, J. M. Ferguson, D. W. Hough, M. J. Danson and G. L. Taylor, *Biochemistry*, 1997, **36**, 9983.
60. A. Fokine, P. G. Leiman, M. M. Shneider, B. Ahvazi, K. M. Boeshans, A. C. Steven, L. W. Black, V. V. Mesyanzhinov and M. G. Rossmann, *Proc. Natl. Acad. Sci. USA*, 2005, **102**, 7163.
61. W. Jiang, M. L. Baker, J. Jakana, P. R. Weigele, J. King and W. Chiu, *Nature*, 2008, **451**, 1130.
62. R. Zhang, C. Hatzos, J. Abdulla and A. Joachimiak, to be published..
63. Joint Center for Structural Genomics, To be published.
64. M. Sutter, D. Boehringer, S. Gutmann, S. Gunther, D. Prangishvili, M. J. Loessner, K. O. Stetter, E. Weber-Ban and N. Ban, *Nat. Struct. Mol. Biol.*, 2008, **15**, 939.
65. M. L. Pedulla, M. E. Ford, J. M. Houtz, T. Karthikeyan, C. Wadsworth, J. A. Lewis, D. Jacobs-Sera, J. Falbo, J. Gross, N. R. Pannunzio, W. Brucker, V. Kumar, J. Kandasamy, L. Keenan, S. Bardarov, J. Kriakov, J. G. Lawrence, W. R. Jacobs Jr., R. W. Hendrix and G. F. Hatfull, *Cell*, 2003, **113**, 171.
66. A. C. Steven, J. B. Heymann, N. Cheng, B. L. Trus and J. F. Conway, *Curr. Opin. Struct. Biol.*, 2005, **15**, 227.
67. Z. H. Zhou, M. Dougherty, J. Jokana, J. He, F. J. Rixon and W. Chiu, *Science*, 2000, **288**, 877.
68. C. D. Livingstone and G. J. Barton, *Comput. Appl. Biosci.*, 1993, **9**, 745.
69. X. Yan, V. Bowman, N. H. Olsen, J. R. Gurnon, J. L. Van Etten, M. G. Rossmann and T. S. Baker, *Microsc. Microanal*, 2005, **11**, 1056.

Mechanisms of Icosahedral Virus Assembly

ADAM ZLOTNICK[a] AND BENTLEY A. FANE[b]

[a] Department of Biology, Indiana University, Bloomington, IN 47405, USA
[b] Division of Plant Pathology and Microbiology, Department of Plant Sciences and The BIO5 Institute, University of Arizona, Tucson, AZ 85721, USA

1 Introduction

In the simplest icosahedral virus assembly systems, morphogenesis is a function of a single viral coat protein. The simplicity of these systems may be a direct consequence of the evolutionary niche that these viruses occupy. Many of these systems are small plant and enveloped animal viruses, systems in which coat proteins have not been constrained by the evolution of a receptor-binding domain or the recruitment of minor capsid proteins. One view is that as coat proteins evolved to perform additional functions or package a larger genome, a one-coat protein assembly mechanism was not feasible and additional adaptations were required. For example, even the ostensibly $T = 1$ parvovirus capsids are composed of two or three variants of a single protein. Although capsids can be formed from the 60 copies of the major coat protein variant,[1] it is possible that maximum infectivity and/or fitness will require the presence of the minor variants. $T = 3$ viruses, such as the nodaviruses and caliciviruses, provide exceptions to test this generalization. However, nodaviruses of vertebrates[2] and caliciviruses[3] have large exterior domains that provide additional complexity that may play important roles in receptor binding and immune evasion. The $P = 3$ picornavirus capsids may represent a more complex adaptation. Morphogenesis requires three different coat proteins of similar structure

RSC Biomolecular Sciences No. 21
Structural Virology
Edited by Mavis Agbandje-McKenna and Robert McKenna
© Royal Society of Chemistry 2011
Published by the Royal Society of Chemistry, www.rsc.org

and most likely a common evolutionary origin.[4,5] Other adaptations have required coat proteins to interact with minor capsid components, such as portal complexes and spikes that decorate icosahedral vertices. As genomes enlarged to encode these additional proteins, capsid T numbers increased. However, the coat proteins of many large viruses have retained the ability to form capsids with smaller T numbers, ones that cannot fully accommodate the volume of the genome. Recruiting other structural proteins during assembly and also fidelity and proper size formation may have driven the evolution of scaffolding proteins, which have been demonstrated to affect all of these phenomena.

Many approaches have been used to study icosahedral capsid assembly. X-ray and cryo-electron microscopy (cryo-EM) structures have provided a wealth of information regarding macromolecular interactions found in both mature particles and assembly intermediates. However, assembly is a dynamic process. Thus many morphogenetic interactions may be transitory and therefore not reflected in the structure of the final product. Genetic approaches have elucidated some of these transient interactions and defined the functions of proteins and protein domains *vis-à-vis* assembly. Biochemical and biophysical methodologies have illustrated the kinetics of virus assembly, defining rate-limiting nucleation events and subsequent elongation phases. The collective data for these varied approaches have recently led to the development of mathematical models. All of these approaches are reviewed below in the context of assembly systems of varying complexity, from single coat protein systems to those dependent on two scaffolding proteins.

2 Assembly in Viruses Without Scaffolding Proteins

The Basic Problem

Even the simplest of viruses have numerous pieces that must be assembled accurately and on a biologically limited time-scale. Furthermore, for a productive infection, the host must produce numerous progeny. These reactions can be recapitulated with purified proteins, demonstrating that self-assembly is 'programmed' into the capsid proteins in the same way that folding is 'programmed' into a protein's amino acid sequence.[6] However, a protein sequence is a covalent chain. Mechanistically, how can self-assembly of many components be achieved/explained? Even though a high-order reaction, where many components come together at the same instant, may seem an attractive answer, high-order reactions are confounded by the difficulty of getting many freely diffusing reactants to collide simultaneously while in exactly the right orientation. Alternatively, theoretical and experimental results support the hypothesis that virus capsids are assembled by a cascade of second-order reactions.

Most complex reactions can only be interpreted with the aid of a model (kinetics always requires a model). For this reason, it is instructive to describe theoretical studies as a starting point in a discussion of virus assembly.

Consider assembly of a population of 'capsids' from geometric 'assembly units' (AUs). Assembly proceeds by adding one AU at a time to a growing complex. The AUs are polyvalent and each contact adds one unit of association energy. Given this very basic model, assembly simulations can be calculated from a list of equations describing the concentration of intermediates,[7,8] which can readily be defined from a single assembly path.[9–11] Similar results are reached by a statistical mechanical approach that considers individual events[12] or coarse-grained dynamic simulations of Brownian collisions of AUs.[13–16] Simple models (*e.g.* geometric solids for coarse-grained simulations) can be amplified by adding protein-like details into the AUs[9,17] and with distinct forcefields quantifying association.[15,16] Remarkably, these relatively simple models lead to testable predictions.

In silico assembly simulations show a consistent picture of successful assembly reactions, reactions that lead to capsid in minimal time with minimal kinetically trapped intermediate. These reactions approach equilibrium rapidly and show sigmoidal kinetics for the appearance of capsid. The equilibrium for a capsid of n AUs is described by the law of mass action:

$$440n(\text{AU}) \quad \text{capsid} \tag{1}$$

$$K_{\text{capsid}} = [\text{capsid}]/[\text{AU}]^n \tag{2}$$

Assembly gives the appearance of a critical concentration. Because n can be a large number, there is a steep cutoff between concentrations of AU that are insufficient to support detectable assembly and those where almost all excess AU assembles into capsid. This behavior has been observed in every virus where there has been an effort to quantify assembly (BMV,[18] CCMV,[19,20] HBV,[21,22] MS2,[23,24] HPV[24]). However, examination of Equation (2) shows that, at equilibrium, even at very low concentrations of AU, there will be some capsid and that at high concentration there is no absolute limit to the concentration of free AU, though it may require physically impossible concentrations of protein. Thus, the cutoff concentration is a pseudo-critical concentration that is not truly constant. The value of K_{capsid} is in unwieldy units which can be readily be dissected into components.[7,25] A dissociation constant per AU ($K_{\text{Dapparent}}$) approximates the pseudo-critical concentration. Also convenient is the equilibrium constant per pairwise interaction between two AUs (K_{contact}), which is useful for relating assembly to structure.

The sigmoidal kinetics of assembly superficially resemble the formation of a crystal starting with a single nucleus. However, a typical *in vitro* capsid assembly reaction that results in the formation of 10^{12} capsids (0.1 μM in 100 μL) requires at least 10^{12} nuclei. In simulations, nucleation reactions occur continuously during the reaction. The lag phase reflects the time required to build up an assembly line of intermediate structures. These intermediates are expected to be at very low concentrations. However, this steady state of intermediates is required for efficient assembly in any stepwise reaction.

Successful reactions have a certain similarity; unsuccessful *in silico* assembly reactions also have a limited repertoire. Unsuccessful reactions follow two general descriptions: thermodynamic and kinetic. Thermodynamically, when association between AUs is very strong, reactions behave irreversibly, resulting in association of all AUs to form partial capsids. The kinetic basis of unsuccessful assembly also involves too many starting points and yields a similar result of trapped incomplete fragments, although with residual free AUs. If the association energy is not too strong, AUs can dissociate from some fragments and reassociate with others eventually to yield a capsid. If the association energy is too strong, these fragments are trapped. The thermodynamic–kinetic basis of unsuccessful assembly has a tendency to trap and to be exacerbated by errors in intersubunit geometry, which can result in 'monsters'. Thus, unsuccessful assembly resembles diffusion-limited aggregation that leads molecules to form fractals.[26]

In vitro assembly reactions are fairly robust whereas simple calculated reactions have a very narrow window of AU concentration, association rate and association energy in which they are successful.[12,13,16,21] However, a single regulatory step at the beginning of the assembly cascade makes assembly simulation robust over a very broad range of conditions. *In silico*, regulation can be imposed by initiating assembly at some seed, analogous to an exogenous protein or nucleoprotein complex or with a nucleation step. Nucleation may be differentiated from 'elongation' by weaker association energy and/or a slower association rate for the first two or more association reactions. There is no need to assume that nucleation occurs by a high-order reaction; as with actin, nucleation can also be achieved by a series of low-order reactions.[27]

A Simplest Case: HBV and Its Implications

Hepatitis B virus (HBV) has been an extremely successful test bed for analyzing *in vitro* assembly. HBV has a complicated lifecycle as a DNA virus with an RNA intermediate.[52] *In vivo*, the virus core protein (Cp) assembles in the cytoplasm around a complex of HBV reverse transcriptase, viral RNA and several host proteins. After Cp assembles to a T = 4 icosahedral shell, the DNA genome is synthesized within. The DNA-filled core then interacts with host partners to be directed from the cell, gaining a protein-studded envelope on the way or back to the nucleus, presumably to maintain infection. The requirements for Cp are strict: it must assemble, display signals to the host (many of which are buried) and eventually uncoat to release its genome.

In vitro capsid assembly studies have focused on the 149-residue assembly domain of the Cp, Cp149. Dimers can be purified from an *Escherichia coli* expression system and reassembled in response to increased ionic strength. These reactions are readily observed by light scattering, fluorescence and size-exclusion chromatography. The concentration dependence of assembly shows the predicted pseudo-critical concentration. The calculated association energy per contact is -3 to $-4 \, \text{kcal mol}^{-1}$.[28] The kinetics of HBV capsid assembly shows the predicted sigmoidal shape and concentration dependence;[21] a rigorous analysis derived from model studies[8] suggests that assembly is nucleated by a trimer of dimers.[21]

The physical chemical results have very distinct implications for biology. The driving force for assembly of the beautifully organized capsid, as predicted, was entropy[28] which is consistent with the largely hydrophobic surface that is buried at the contact between dimeric AUs.[29] The association energy was strong enough to drive assembly but low enough to minimize issues with kinetic trapping. The weak interactions led to the testable and confirmed prediction of a highly dynamic structure[30] that can, for example, readily respond to changes in its packaged nucleic acid. Weak association interactions seem to be a common feature of virus assembly,[25] although this may be limited to systems that conveniently self-assemble *in vitro*. Simulations also led to the testable and confirmed prediction that there is a substantial hysteresis to dissociation[31] that is believed to derive from both (i) the multivalent AUs in the context of closed capsid, which makes removal of the first few subunits energetically challenging, and (ii) the depletion of intermediates at late times during assembly, which collapses the equilibrium between association and dissociation. Finally, the observation that mutations, divalent metals and some small molecules all enhance assembly have led to the hypothesis that HBV dimers undergo a transition from an assembly-active to an assembly-inactive state.[22,32–35] Regulation of assembly by an activity transition of the AU has also been proposed for retroviruses based on biochemical evidence[36] and as a general mechanism to further minimize kinetic trapping.[37]

A Not So Simple Case: CCMV and BMV

Bromoviruses are simple plant viruses. They consist of a 28 nm, T = 3, 90-dimer capsid that assembles around viral RNA. The bromovirus cowpea chlorotic mottle virus (CCMV) was the first spherical virus reassembled *in vitro*.[38] Companion studies with the type member of the family, brome mosaic virus (BMV), whose coat shares about 80% sequence identity, confirmed the basic description of assembly.[39] Purified dimeric AUs reassembled into empty virus-like particles at acidic pH. RNA-filled particles, indistinguishable from native virions, can be reassembled at neutral pH by mixing AUs and RNA[40] and structure.[41] Bromoviruses have been one of the important systems for understanding virus assembly (see below) and also for using viruses in nanotechnology.[42,43]

In spite of the great similarity between BMV and CCMV, the assembly mechanisms show remarkable differences. The association energies are similar on comparing results from Zlotnick's group and other published accounts[19] (Table 1) with the (pseudo-) critical concentration observed by Adolph and Butler.[39] However, the results of kinetic studies indicate a very different assembly path. CCMV assembly is characterized by accumulations of pentamers of AUs [*i.e.* pentamers of dimers (PODs)].[19,20] An excess of PODs leads to

Table 1 Association behavior for different viruses.[a]

Parameter	HBV[b]	CCMV[c]	HPV16[d]	Phage P22[e]	Phage HK97[f]
AU geometry	Dimer	Dimer	Pentamer	Monomer	Pentamer and hexamer
$\Delta G_{contact}$	−3.1 to −4.1	−3.1 to −3.7	–	–	−2.0
ΔG per AU	−6.5 to −8.6	−6.5 to −7.8	−8.1 to <−9.8	−7.2	−7.0
$K_{Dcontact}$ (mM)	4.4 to 1.3	5.0 to 1.8	–	–	30
K_D per AU (μM)	9.9 to 0.9	10 to 1.6	0.5 to <0.03	~10	6.7

[a]The K_D per AU is approximately the pseudo-critical concentration of assembly. Calculations are as per Zlotnick (2003).[25] Italicized values were calculated for this chapter from published data. Except for HK97, association energy was determined from assembly experiments.
[b]Data from Ceres *et al.* (2002).[28]
[c]Data from Johnson *et al.* (2005).[19]
[d]Data from Mukherjee *et al.* (2008).[24] Distorted quasi-equivalence precluded per contact analysis.
[e]Data from Parent *et al.* (2006).[154] Scaffolding protein also contributes to K_{capsid} in bacteriophage P22.
[f]Data from Ross *et al.* (2006).[155] $\Delta G_{contact}$ for coat protein was determined by calorimetry. The AU for HK97 is a mixture of pentamers and hexamers.

their assembly to generate 60-dimer icosahedra.[44] These 60-dimer structures are not native but have been observed in over-expression of BMV in yeast[45]and in assembly of a CCMV N-terminal-domain deletion mutant[44] (structurally, these resemble the capsids of dsRNA viruses such as L-A[46]). POD formation creates a characteristic early phase of increased light scattering in CCMV kinetics. In contrast, BMV has sigmoidal kinetics with a pronounced lag phase.[18] Critical analysis of BMV kinetics indicates that the lag cannot be explained by the time to build up a steady state of intermediates, but can readily be modeled as a slow transition of the free AU from an assembly-inactive to an assembly-active state.[18]

Assembly-active and -inactive Species: Allostery and Autostery

Regulation of virus assembly prevents the formation of kinetic traps and also prevents the virus from assembling at the wrong time and place. Assembly can be controlled physicochemically (nucleation and activation of subunits) and/or by external factors. Nucleation has been observed *in vitro* for the formation of empty capsids.[19,21,47] Likewise, AU transitions from an assembly-inactive to an assembly-active state have been observed *in vitro* for HBV and BMV (as described above). Kinetic studies on pappillomavirus also demonstrated an activation-related lag.[129] Retroviral Gag proteins undergo a large assembly-associated conformational change[48] which may be activated by dimerization[36,49] and/or by binding to exogenous molecules such as phosphoinositol.[50,51] Presciently, Caspar predicted that autostery, a conformational change in free AUs induced by binding to a growing capsid already in the active state, is another mechanism for preventing inappropriate initiation of assembly.[37]

In addition to these physicochemical mechanisms, biological systems routinely use other proteins (or nucleic acids) to regulate assembly. This has the advantage of localizing important players in the virus lifecycle. This mechanism is largely outside the aims of this chapter. However, it is worth listing a few examples, each with radically different architecture. *In vivo*, HBV assembly is initiated by a nucleoprotein complex of the viral reverse transcriptase and the RNA form of the viral genome.[52] Closely related bacteriophages R17 and MS2 initiate assembly by specific interaction between viral RNA and coat protein.[53–56] Turnip crinkle virus coat protein binds to a specific RNA structure and size.[57–60] Of course, a well-developed example of this mechanism is the role of scaffolding proteins as described elsewhere in this review.

3   Post-nucleation Effects on Assembly: Mutations and Drugs

Capsid morphology can be critical to several stages of the viral lifecycle, such as genome packaging, stability, receptor recognition and uncoating. For small capsids, up to $T = 7$, the interactions between subunits may be sufficient to control geometry. The geometry of interaction can be expressed mathematically as local rules.[61] For large T numbers ($T \approx 25$), the rules become too degenerate and a distribution of sizes is expected[62] unless, for example, there is tape measure protein.[63] However, even with smaller capsids, it is possible to have heterogeneous assembly products. For example, assembly conditions can result in a surprising diversity in CCMV with $T = 1$, $T = 3$, 17 nm and 30 nm diameter rods and multilamellar forms.[38,64] HBV forms 28 nm ($T = 3$) and 35 nm ($T = 4$) diameter particles *in vivo*[65] and *in vitro*.[66] As there is no concentration dependence of the $T = 4$ to $T = 3$ ratio, it was suggested that both forms start with a similar nucleus and the choice between forms is made early during elongation.[67] Interestingly, C-terminal truncation of the HBV capsid protein leads to a shift in the assembly ratio to favor more $T = 3$.

Heterogeneity in size often results from such truncations, supporting the hypothesis that there are intrinsic protein switches that regulate intersubunit geometry.[68] This behavior is particularly well described in plant viruses. A proteolytic truncation of southern bean mosaic virus coat protein led to $T = 1$ particles instead of the usual $T = 3$ capsid.[69] Pleiomorphic assembly in the closely related *Sesbania* mosaic virus (SeMV) has been described in detail. SeMV is a $T = 3$ RNA virus assembled from a dimeric AU; an interesting feature of the capsid is that N-termini of three 'C' subunits twist around each other to form a 'β-annulus'. SeMV capsid protein formed $T = 3$ capsids even when the N-terminal 22 residues, including the RNA-binding domain, were removed; $T = 1$ and pseudo-$T = 2$ capsids resulted when 36 residues were deleted; when 65 residues were removed, including the segment required for the β-annulus, only $T = 1$ particles were observed.[70] Here there is a clear interplay between structural studies and deducing an assembly pathway. The pseudo-$T = 2$ structures, like aberrant 60-dimer BMV and CCMV capsids[44,45] and all known dsRNA icosahedral viruses,[46,71,72] are built from 12 PODs. (More exactly, these are T = 1 structures with two proteins per icosahedral asymmetric unit.) Because it is difficult to imagine how a POD could arise as the capsid is growing, it is highly likely that the pentamers assembled first and then assembled into a capsid. A pentamer-based assembly pathway is found in many viruses. For example, poliovirus assembles from a pentamer of hetero-trimers,[4,73] polyoma- and papillomaviruses are complexes of pentamers in a T = 7 lattice[74] and, as described later in this chapter, φX174 assembles from scaffolding-stabilized pentamers. Given the structural differences between the proteins, this convergence of mechanisms likely derives from the thermo-dynamic advantage of a polyvalent AU capable of many weak interactions.[25]

Although SeMV and CCMV are structurally dissimilar *vis-à-vis* subunit orientation,[75,76] their assembly pathways and size regulatory mechanisms may be very similar. Cleavage of the N-terminus of the CCMV coat protein led to a mixture of T = 1, pseudo-T = 2 and T = 3 capsids.[19] Accumulation of PODs was a hallmark of CCMV assembly and, even with intact capsid protein, correlated with pseudo-T = 2 formation under conditions where PODs were prevalent.[20]

Thus, what is the molecular determinant, or switch, for SEMV and CCMV capsid size? It was hypothesized that the SEMV β-annulus would be critical to T = 3 capsid formation because 30-dimer T = 1 particles form in the absence of the annulus and protein–protein interactions are the major contributor to virus stability.[77] However, when a peptide segment containing the annulus was deleted from the coat protein, leaving the rest of the N-terminus intact, a perfectly normal capsid was formed.[78] In contrast, CCMV stability is domi-nated by protein–RNA interaction.[38,79] However, like SeMV, normal T = 3 capsids were observed when the β-hexamer,[76] analogous to the β-annulus, was deleted.[80] The search for a switch thus leads to the poorly ordered regions of the N-terminus that appear to have little structure, probably little contribution to capsid stability and probably their greatest influence on the choice of kinetic pathways of assembly. As stated previously, many critical interactions directing assembly may be transitory and thus may not be reflected in the structure of the final product.

Mutations provide one method for defining assembly pathways. Another approach is to use small molecules that interfere with assembly, sometimes described as chemical genetics.[81] The first example of small molecules that affected virus stability, to the authors' knowledge, were the 'WIN' compounds that stabilize picornaviruses and prevent timely uncoating.[82,83] Focusing on assembly, the fluorophore bis-ANS prevented assembly of bacteriophage P22 by interacting with the scaffolding[84] and/or coat protein.[85] Bis-ANS also affects HBV assembly by inhibiting association and by misdirecting assembly to generate large 'non-capsid polymers'.[86] Conceptually, misdirected assembly may consume many AUs non-productively, which gives it an advantage over simple association inhibitors that can only affect one AU at a time.[86,87]

Another example of assembly misdirection is provided by the HAP molecules (heteroaryldihydropyrimidines) that affect HBV assembly. HAPs were first discovered as suppressors of viral HBV replication that targeted the capsid protein.[88] When added to *in vitro* reactions, HAPs were shown to increase both the rate and extent of association,[33,89] leading to very large sheets of HBV capsid protein.[33,90] A crystal structure of an HBV–HAP complex showed that HAPs fit into a gap at the AU–AU interface. This flattens the quasi-sixfold axes, explaining both the stronger association energy and the formation of hexagonal sheets.[91] As high concentrations of HAP are required for mis-direction,[33,89] it is unlikely that the misdirected assembly is the primary mechanism of HAP antiviral activity. A quantitative analysis of a series of HAPs demonstrated (i) activity at concentrations far below the level required for misdirection and (ii) a remarkable correlation between replication of HBV in culture and the enhancement of assembly kinetics. This implies that the primary anti-viral activity operates on the level of assembly initiation.

4 Scaffolding Protein-mediated Morphogenesis, P22 a Model System

Nucleation and Elongation

Two critical observations from the early studies of P22 phage assembly provided the initial insights into the kinetics of morphogenesis and the functions of scaffolding proteins. (1) Partially formed capsids were rarely observed in wild-type infections. Once formation began, it ran to completion. (2) In the complete absence of the scaffolding protein, coat proteins do not self-associate in an otherwise wild-type infection. However, if cell lysis was inhibited, coat protein-containing particles formed, indicating that the lag phase before particle production was substantially lengthened.[92] While the aberrant nature of the particles suggested other scaffolding proteins functions, such as size determination and morphogenetic fidelity, the delayed lag phase indicated that scaffolding proteins stimulated assembly by lowering the overall intracellular coat protein concentration required for association. Thus, the scaffolding protein was a key reactant in a rate-limiting nucleation reaction, which would be followed by rapid lower order reactions ultimately ensuring a homogeneous population of completed particles. The results of a seminal set of experiments verified this model *in vitro*.[93]

The subsequent use of mutant scaffolding proteins and altered *in vitro* assembly conditions have provided many of the molecular details of capsid nucleation and elongation (summarized in Figure 1). The C-terminal residues of the P22 scaffolding protein constitute the coat protein-binding domain,[94,95] a general feature that has been observed in many viral assembly systems as diverse as HSV and φX174.[96] Coat–scaffolding interactions are weak and governed by electrostatic interactions. Thus, *in vitro* assembly is sensitive to ionic solutions.[97] Low salt concentrations, corresponding to strong interactions between scaffold and coat proteins, led to kinetically trapped partially formed capsids, which can complete assembly by increasing salt concentrations.[97] In addition to demonstrating that the edges of the partially formed capsids

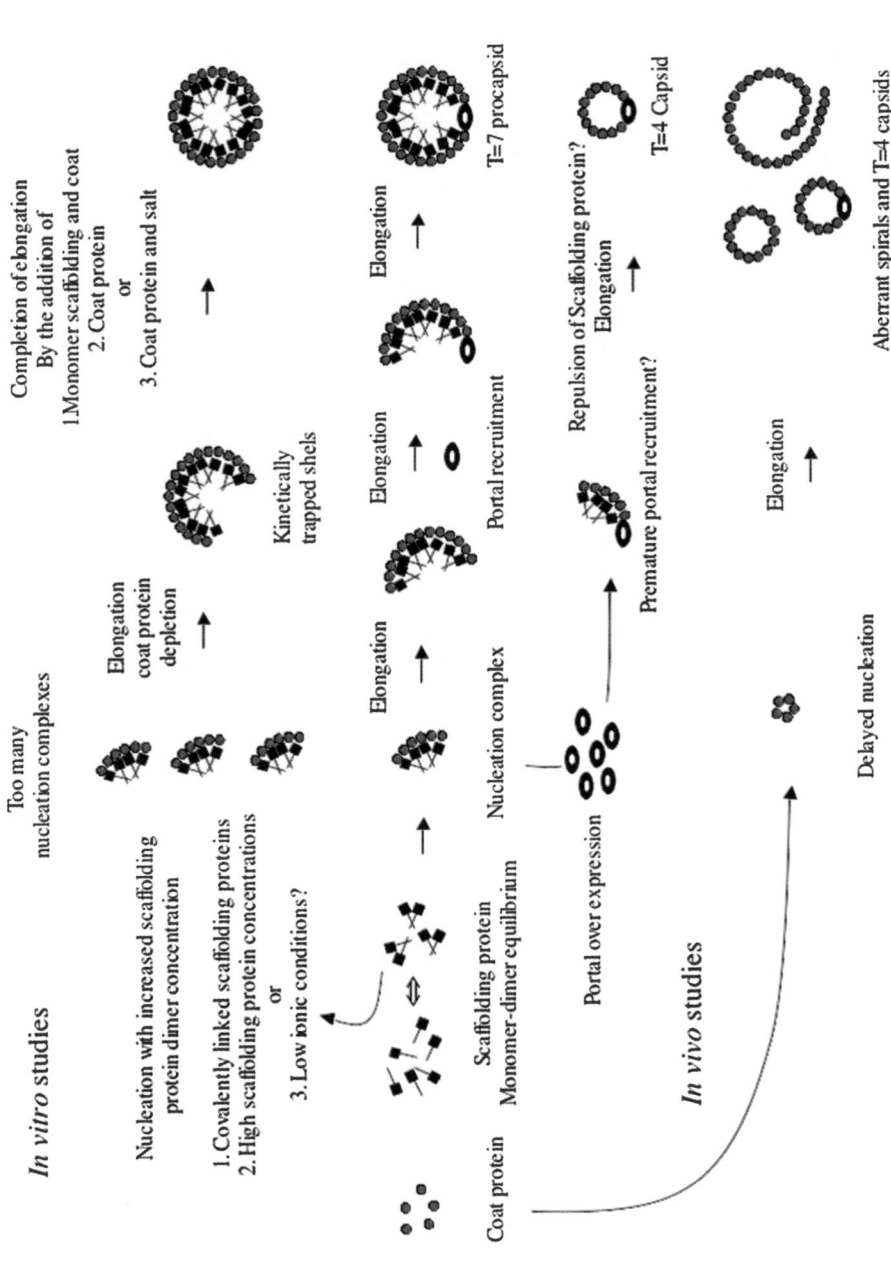

Figure 1 Summary of P22 assembly studies.

remain assembly competent, it indicates that low ionic strengths lower thermodynamic nucleation barriers. Consequently, the abundant nucleation complexes deplete components while the strong association energy prevents scaffolding from dissociation and reuse in the subsequent addition of coat protein. Excess scaffolding protein also leads to a preponderance of kinetically trapped shells, indicating that the nucleation–elongation balance is sensitive to scaffolding protein concentrations. *In vivo*, the concentration of the P22 scaffolding protein is tightly regulated on the post-transcriptional level,[98–100] strongly suggesting that the effects of excess scaffolding protein is not an artifact of *in vitro* assembly systems.

In solution, the P22 scaffolding protein exists in a monomer–dimer–tetramer equilibrium,[101] which likely governs the nucleation–elongation balance. Covalently linked scaffolding protein dimers accelerate and favor nucleation reactions.[102] These dimers or tetramers may act as an entropy sink, creating foci containing critical concentrations of coat protein. The kinetically trapped partially built shells formed in these reactions can be completed by the addition of scaffolding protein monomers and coat protein. The behavior of the covalently linked scaffolding protein may also explain the predominance of partially formed capsids in reactions with excess scaffolding protein or low salt concentrations.[97] High concentrations of scaffolding protein would increase the pool of scaffolding dimers, leading to an overabundance of nucleation complexes. Moreover, it is possible that low ionic conditions favor scaffolding dimerization. *In vitro*, a steady-state phase follows the nucleation phase, during which the rate of new nucleation events equals that of shell completion.[8,102] This stage continues until the coat protein falls below the concentration required to drive further nucleation reactions. However, elongation continues: those capsids in the process of forming finish assembly.

Portal and Minor Protein Recruitment

During elongation, other scaffolding protein functions, namely size determination, fidelity and the recruitment of portals and minor capsid proteins, become apparent. The first 140 amino acids of the P22 scaffolding protein affect the fidelity of morphogenesis. *In vitro* assembly reactions with proteins lacking this domain predominantly yield aberrant particles. These aberrant abominations form at a faster rate than wild-type particles,[103] suggesting that overly rapid assembly is error prone, perhaps creating incorrect binding orientations that lead to kinetically trapped shells. The timing of portal protein incorporation may also affect fidelity. Procapsid-like particle formation *in vivo* is not dependent on the portal protein.[104,105] Moreover, portal-less particles cannot incorporate portals after the completion of shell assembly. Thus, the portal is recruited during elongation. The timing of portal recruitment may be critical to size determination and fidelity as four types of particles were observed in cells over-expressing a cloned portal gene (see below).

Portal recruitment and minor protein incorporation are a function of the scaffolding protein.[106,107] This may be a general feature of scaffolding proteins as similar observations have been made in φ29, T4 and HSV.[108–111] Two genetic methodologies have defined the region of the scaffolding protein required for portal recruitment. Scaffolding protein missense mutants that fail to recruit portals were first isolated and mapped to a specific region of the scaffolding protein.[106,112] Subsequent deletion analyses further defined this domain,[107] which could directly interact with the portal protein and/or influence the ability of the coat protein to recruit the portal. Deletion analyses have also defined a portal recruitment domain in the HSV-1 scaffolding protein.[113] As stated above, the over-expression of the portal protein leads to a variety of aberrant structures *in vivo* and is therefore detrimental to the production of infectious progeny. One scaffolding protein mutation, located in the coat protein-binding domain, confers a portal protein over-expression resistance phenotype that restores assembly fidelity, suppressing the production of aberrant particles.[104,105]

5   Size Determination and Fidelity

Bacteriophage P22

Aberrant assembly, or a lack of fidelity, is often used to describe any defect that leads to the production of an abnormal particle, be it an unclosed spiral structure or a capsid of altered dimensions or T number. However, the molecular events that lead to a closed capsid of altered T number or lengthwise dimension may be fundamentally different from those producing aberrant structures. There are many examples of mutant coat proteins of prolate viruses forming isometric shells,[108,114–116] but few, if any, have been shown to affect the T number. In contrast, mutant T4 scaffolding proteins can lead to smaller T numbers.[117] Isometric virion coat proteins will form capsids with smaller T numbers in wild-type infections and bacteriophage λ coat protein mutations[118] been shown to affect the balance of T = 4 and T = 7 products *in vivo*.

The absence of scaffolding also affects the balance of T = 4 and T = 7 P22 particles.[119] Moreover, the T = 4 particles generated in cells over-expressing the portal protein are devoid of scaffolding protein.[104,105] Excess portals may alter the timing of portal recruitment. If recruitment occurs too early during elongation, scaffolding proteins may be excluded from the reaction. Comparisons between T = 4 and T = 7 P22 cryo-EM structures indicate unique T = 4 trimeric clusters of coat protein, which may be prevented by the scaffolding protein.[120] There may be a general correlation between very low scaffolding protein concentrations and the formation of smaller capsids. Limiting concentrations of the herpes simplex virus scaffolding protein leads to the formation of small T = 9 capsids, as opposed to the wild-type T = 16 core.[121]

Phages P2 and P4, Reprogramming Size Determination

The P2/P4 bacteriophage system represents a biologically programmed size re-determination. P2 is a T = 7 bacteriophage and, like P22, the P2 coat protein is capable of forming T = 4 capsids. The satellite bacteriophage P4 parasitizes this flexibility. Encoding an external scaffolding protein sid (*size determination*), it reprograms P2 coat and internal scaffolding assembly for efficient T = 4 capsid morphogenesis, a capsid size that can only accommodate the smaller P4 genome. The sid protein interacts with the coat protein, forming a cage around it. The lattice contains 12 large pentagons surrounding the 12 vertices and bifurcating the five surrounding hexameric capsomers.[122–124] In the five places where three hexamers meet, the C-termini of sid form trimeric structures. A distinct dimer conformation interconnects the trimers.[124] The dimers and trimers appear to interact *via* N-termini sid–sid interactions.

Genetic analyses have identified a critical coat protein hinge region required to build the P4 T = 4 lattice. Mutations in this hinge region, *sir* mutations (*sid responsiveness*), are resistant to the size-altering effects of the sid protein.[125] Extragenic second-site suppressors of *sir* mutations, *nms* mutations (wild-type N *mutation sensitive*; protein N is the viral coat protein), cluster to the C-terminus of sid[126] and most likely result in stronger sid trimers and dimers. This could produce a more rigid external lattice that overrides the less flexible hinge regions created by *sir* mutations. Alternatively, *sir* and *nms* mutations may favor different sets of binding angles or kinetically redirect assembly pathways.

Although the exact details of nucleating P2 and P4 capsids are unknown, a speculative model based on the cryo-EM reconstructions can be proposed. The nucleation complex of both capsids would involve the P2 coat and internal scaffolding proteins. Afterwards, assembly units, composed of coat and scaffolding proteins, would add to this structure. In the absence of the P4 sid protein, a T = 7 capsid would form. In a cell co-infected with P4, there would be a second nucleation event, one to form the external sid protein lattice. This could involve sid trimer–P2 coat protein interactions where three P2 coat hexamers interact, an arrangement found in both T = 4 and T = 7 structures. However, concurrent elongation of the less flexible and size-constraining sid lattice would direct assembly into the smaller T = 4 capsid.

6   ϕX174, a Two-scaffolding Protein System

Unlike the systems discussed above, ϕX174 assembly proceeds through pentameric intermediates. Assembly through capsomers has been seen in other viral systems. Polyomaviruses such as SV40 and structurally similar papillomaviruses assemble from 72 coat protein pentamers.[127–129] Bacteriophage HK97 assembly involves both hexameric and pentameric capsomers.[130,131] Twelve pentamers containing five copies each of the three major structural proteins associate during the morphogenesis poliovirus, a T = 3 capsid.[4,73]

Bacteriophage ϕX174 assembly is dependent on two scaffolding proteins, an internal and an external species. To understand the need for two scaffolding proteins in a small, but not necessarily simple, $T = 1$ assembly system, the evolutionary context in which it arose must be considered. The *Microviridae* consist of two distinct subfamilies, the microviruses, which infect free-living bacterial hosts and the Gokushoviruses, which primarily infect obligate intra-cellular parasitic bacteria such as *Chlamydia* and *Bdellovibrio*.[132] Microviruses make up only a small fraction of the phages infecting free-living bacteria, competing with a vast pool of large dsDNA viruses. In contrast, all known chlamydiaphages are Gokushoviruses, which do not encode an external scaffolding protein, a more recently acquired gene.[133] Two scaffolding proteins have allowed the evolution of a very fast viral lifecycle. New progeny can be detected as early as 5 min post-infection,[134] a time when many dsDNA phages are still in middle gene expression. In contrast, Gokushovirus progeny are detected approximately 48 h post-infection,[135] an affordable luxury in a niche without competition.

Early Assembly, the Internal Scaffolding Protein

In the complete absence of the internal scaffolding B protein, only 9S coat protein and 6S major spike protein pentamers form (Figure 2). Thus, the binding of the B protein to the underside of 9S particles induces a conformational switch, allowing the upper surface to interact with the major spike protein complex. The suppressors of a partially functional B protein are located on the coat protein's outer surface in three distinct sequences of considerable homology to each other.[136,137] These sequences are all found in loop regions, as opposed to the β-barrel core and thus may identify key hinge points. Further genetic analyses have been hindered by the lack of missense mutations with defective phenotypes, which may reflect a structure that tolerates alterations. Indeed, internal scaffolding proteins from related viruses cross-complement despite only 30% homology,[138] explaining why point mutations rarely confer defects. The behavior of chimeric internal scaffolding proteins demonstrates that optimal fitness is a function of the C-terminus being of the same origin as the viral coat protein.[139] The C-terminal 24 amino acids are highly ordered in the procapsid crystal structure and constitute a coat protein binding domain.[140,141] The protein electron density map becomes more diffuse as it extends toward the N-terminus, suggesting that N-terminal interactions are variable or that the N-terminus displays mobility. These characteristics of the N-terminus may be a general phenomenon: similar observations have been made in both P22 and HSV.[142–145]

The Relationship Between the Two Scaffolding Proteins

In contrast to the internal scaffolding protein, the external scaffolding protein is highly ordered and extremely sensitive to mutation,[140,141,146–148] suggesting that the external scaffolding protein is more critical for morphogenesis. After the acquisition of the external scaffolding protein, the B protein may have evolved

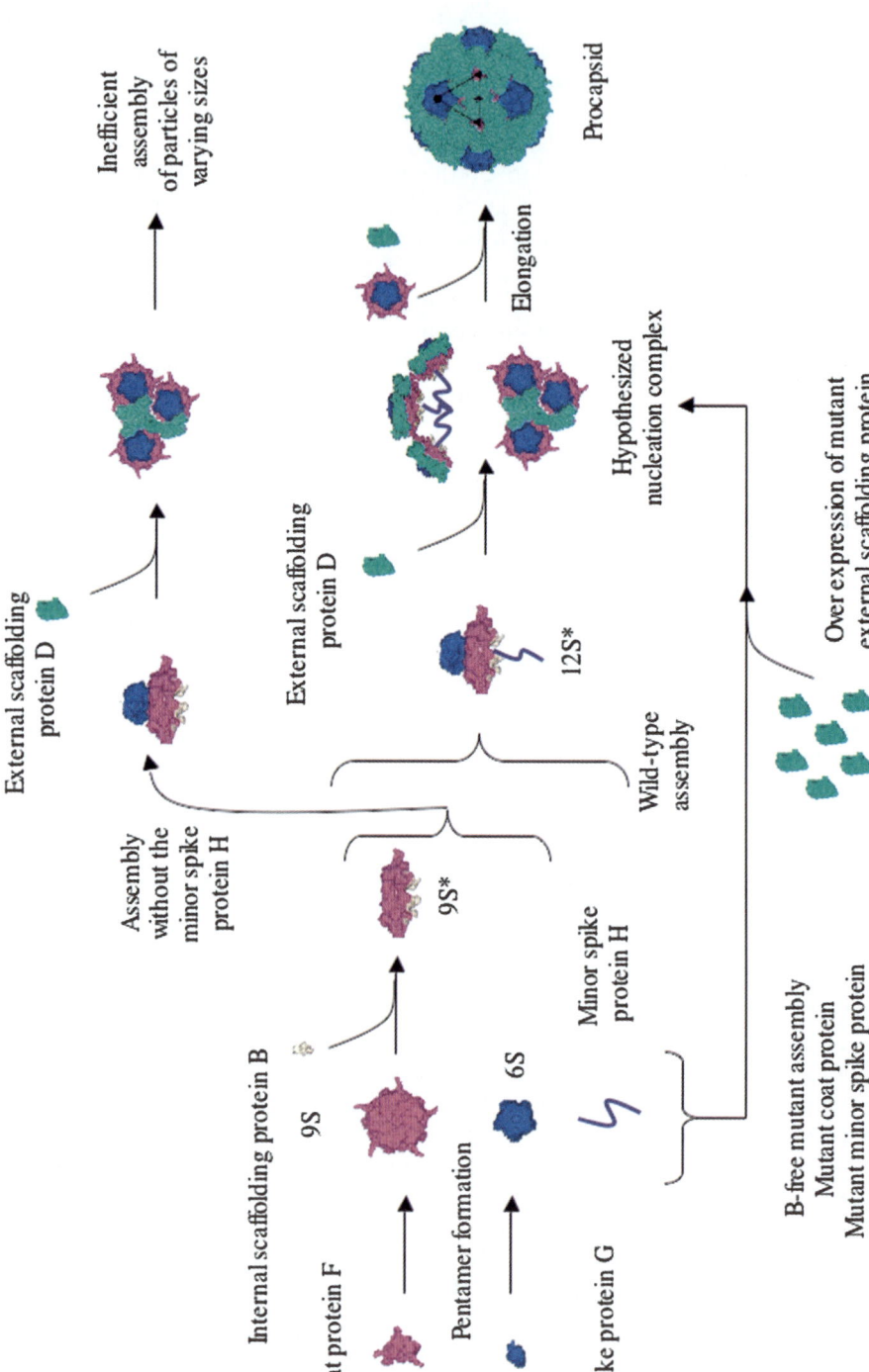

Figure 2 Summary of φX174 assembly studies.

into a general 'efficiency protein', aiding several morphogenetic processes, but not strictly required for any one reaction. If correct, an inherent plasticity would allow other proteins to compensate for reduced or absent B protein function. A series of targeted selections were designed to lessen sequentially the requirement for B protein until it was no longer needed. A sextuple mutant was eventually isolated.[134] The primary adaptive mechanism involved mutations that strengthened external scaffolding–coat protein interactions and grossly over-expressed the external scaffolding protein. The results of kinetic analyses indicated that only the over-expression of the mutant external scaffolding protein was required to form virions. However, the lag phase before progeny production *in vivo* was 50 min, long after programmed cell lysis, and therefore could only be detected in φX174 lysis-resistant cells. With the addition of the subsequent mutations, which included more promoter mutations, lag phases became progressively shorter. Hence one of the primary functions of the internal scaffolding protein is to lower the critical concentration of the external scaffolding protein required to nucleate procapsid morphogenesis.

Functions of the External Scaffolding Protein

In the complete absence of functional external scaffolding protein, 12S* particles accumulate in infected cells.[134] The results of genetic analyses conducted with chimeric external scaffolding proteins, in which the first α-helices between related viruses were interchanged, have defined the first helix as a coat protein substrate specificity domain, which may nucleate procapsid morphogenesis.[147,149,150] While the chimeric scaffolding proteins cannot support plaque formation, progeny are produced in lysis-resistant cells after an extended lag phase. Once progeny production initiates, virions appear to be produced at a wild-type rate. Plaque-forming suppressor mutations shorten the lag phase and map to a distinct α-helix in the viral coat protein. Strains with small deletions within the first helix also delay the timing of progeny production, which can be shortened by regulatory mutations that alter the level of D protein expression.[151]

The nature and existence of the next intermediate in the pathway remain obscure. Structural consideration suggest that it may contain the same components of the 12S* particle with the addition of 20 D protein subunits. An intermediate with these characteristics and stoichiometry, the 18S particle, has been detected. However, conditions to trap this intermediate kinetically or genetically have not been successful. It is either very short-lived or represents an off-pathway particle. If the 18S particle is a *bona fide* intermediate, the next step in the reaction would be rate limiting, involving at least three 18S particles, followed by the successive addition of pentamers to the growing shell. An alternative mechanism would involve the concurrent association of pentamers as the external scaffolding protein decorates 12S* particles. If this model is correct, the 18S particle must be demoted to the status of an off-pathway or a kinetically trapped abomination.

Regardless of the exact nature of the next intermediate, elements of the conformational switching of the external scaffolding protein have been elucidated by genetic analyses and structural studies. In the atomic structure of the viral procapsid, there are four structurally distinct external scaffolding proteins arranged as dimers of asymmetric dimers (D_1D_2, D_3D_4), per viral coat protein. The two subunits in the assembly naïve D_AD_B dimer structure are poised to occupy these four positions.[140,141,150]. The D_A subunit has a structure most similar to D_1 and D_3, whereas the D_B subunit shares a fold more closely related to D_2 and D_4. To achieve this unique arrangement, one monomer in each asymmetric dimer $(D_1, D_3$ or $D_A)$ must be bent 30°. This occurs at glycine residue 61 in α-helix 3. Substitutions for G61 result in dominant lethal phenotypes.[146] The inhibitory proteins alone appear to have no function, allowing the formation of the 12S* particle, the intermediate directly preceding the first D protein mediated step in the pathway, but not supporting further progress. However, if both wild-type and inhibitory G61 mutants are present during assembly, 12S* particles no longer accumulate and appear to be removed from the assembly pathway. Hence wild-type subunits must be present for the mutant subunits to interfere with morphogenesis, suggesting that heterodimers are the inhibitory species. Theoretically, heterodimers should not be able to form 18S particles. Therefore, the removal of 12S* particles suggests that elongation may not involve fully formed 18S particles and hence assembly proceeds *via* the concurrent association of pentamers as the external scaffolding protein decorates 12S* particles.

The lethal dominant D proteins can be viewed as antiviral agents that specifically target virus assembly. Viruses resistant to the effects of the inhibitory proteins were selected. The resulting resistance mutations were not allele specific, conferring resistance to numerous proteins with different substitutions for glycine 61. They cluster to the C-terminus of the internal scaffolding protein and to a distinct region on the upper surface of the viral coat protein. Both this region and the C-terminus of the internal scaffolding protein reside directly beneath the D_3 subunit, suggesting that these changes either exclude heterodimers from the D_3D_4 position or accommodate them in that location. A multiple mutant resistant strain was generated by propagating cultures for over 180 viral lifecycles. It displays optimal fitness in cells expressing the inhibitory protein, suggesting that the evolution of resistance also selected for a level of dependence (J. E. Cherwa and B. A. Fane, unpublished results).

Size Determination and the Minor Vertex Protein H

The minor vertex protein H may play an active role in morphogenesis on the level of size determination. 12S* particles efficiently form in the absence of H protein, but capsid yield is reduced.[152] Unlike wild-type ϕX174 capsids, H-less particles vary considerably in diameter, ranging from wild-type to significantly larger dimensions (M. G. Rossmann, personal communication). In the 12S* particle there appears to be one copy of protein H; however, in solution the protein forms oligomers (G. Cingolani, personal communication). It is conceivable that H protein oligomerization may occur during the association of pentamers and set the curvature of the capsid. Whereas it is known that SV40, a capsid composed of 72 pentamers, can form T = 1 structures,[127] size variation in a T = 1 particle has never been observed.

7 Summary

Virus assembly must be controlled to ensure correct geometry and assembly at the right time and place. Assembly can be regulated by the physical chemistry of the coat protein but frequently involves other components such as scaffolding proteins and/or nucleic acids. Additional regulation can be supplied by host factors/partners, macromolecules or small molecules. Disturbing assembly regulation is an important approach to understanding virus lifecycles and may be a strategy for antiviral therapeutics.

References

1. E. Hernando, A. L. Llamas-Saiz, C. Foces-Foces, R. McKenna, I. Portman, M. Agbandje-McKenna and J. M. Almendral, *Virology*, 2000, **267**, 299.
2. L. Tang, C. S. Li n, N. K. Krishna, M. Yeager, A. Schneemann and J. E. Johnson, *J. Virol.*, 2002, **76**, 6370.
3. R. Chen, J. D. Neill, M. K. Estes and B. V. Prasad, *Proc. Natl. Acad. Sci. USA*, 2006, **103**, 8048.
4. Y. Verlinden, A. Cuconati, E. Wimmer and B. Rombaut, *J. Gen. Virol.*, 2000, **81**, 2751.
5. J. M. Hogle, M. Chow and D. J. Filman, *Science*, 1985, **229**, 1358.
6. C. B. Anfinsen, *Science*, 1973, **181**, 223.
7. A. Zlotnick, *J. Mol. Biol.*, 1994, **241**, 59.
8. D. Endres and A. Zlotnick, *Biophys. J.*, 2002, **83**, 1217.
9. V. S. Reddy *et al.*, *Biophys. J.*, 1998, **74**, 546.
10. D. Endres, M. Miyahara, P. Moisant and A. Zlotnick, *Protein Sci.*, 2005, **14**, 1518.
11. T. Keef, C. Micheletti and R. Twarock, *J. Theor. Biol.*, 2006, **242**, 713.
12. T. Zhang and R. Schwartz, *Biophys. J.*, 2006, **90**, 57.
13. D. C. Rapaport, *Phys. Rev. E*, 2004, **70**, 051905.
14. D. C. Rapaport, J. E. Johnson and J. Skolnick, *Comp. Phys. Commun.*, 1999, **121**, 231.
15. M. F. Hagan and D. Chandler, *Biophys. J.*, 2006, **91**, 42.
16. H. D. Nguyen, V. S. Reddy and C. L. Brooks III, *Nano Lett.*, 2007, **7**, 338.
17. M. Hemberg, S. N. Yaliraki and M. Barahona, *Biophys. J.*, 2006, **90**, 3029.
18. C. Chen, C. Kao and B. Dragnea, *J. Phys. Chem. A*, 2008, **112**, 9405.
19. J. M. Johnson *et al.*, *Nano Lett.*, 2005, **5**, 765.

20. A. Zlotnick, R. Aldrich, J. M. Johnson, P. Ceres and M. J. Young, *Virology*, 2000, **277**, 450.
21. A. Zlotnick, J. M. Johnson, P. W. Wingfield, S. J. Stahl and D. Endres, *Biochemistry*, 1991, **38**, 14644.
22. S. J. Stray, P. Ceres and A. Zlotnick, *Biochemistry*, 2004, **43**, 9989.
23. A. E. Ashcroft *et al.*, *J. Nanosci. Nanotechnol.*, 2005, **5**, 2034.
24. S. Mukherjee, M. V. Thorsteinsson, L. B. Johnston, P. DePhillips and A. Zlotnick, *J. Mol. Biol.*, 2008, **381**, 229.
25. A. Zlotnick, *Virology*, 2003, **315**, 269.
26. T. A. Witten Jr and L. M. Sander, *Phys. Rev. Lett.*, 1981, **47**, 1400.
27. P. Matsudaira, J. Bordas and M. H. Koch, *Proc. Natl. Acad. Sci. USA*, 1987, **84**, 3151.
28. P. Ceres and A. Zlotnick, *Biochemistry*, 2002, **41**, 11525.
29. S. A. Wynne, R. A. Crowther and A. G. Leslie, *Mol. Cell*, 1999, **3**, 771.
30. J. K. Hilmer, A. Zlotnick and B. Bothner, *J. Mol. Biol.*, 2008, **375**, 581.
31. S. Singh and A. Zlotnick, *J. Biol. Chem.*, 2003, **278**, 18249.
32. P. Ceres, S. J. Stray and A. Zlotnick, *J. Virol.*, 2004, **78**, 9538.
33. S. J. Stray, C. R. Bourne, S. Punna, W. G. Lewis, M. G. Finn and A. Zlotnik, *Proc. Natl. Acad. Sci. USA*, 2005, **102**, 8138.
34. C. Bourne, S. Lee, B. Venkataiah, A. Lee, B. Korba, M. G. Finn and A. Zlotnick, *J. Virol.*, 2008, **82**, 10262.
35. F. M. Suk, M. H. Lin, M. Newman, S. Pan, S. H. Chen, J. D. Lui and C. Shih, *J. Virol.*, 2002, **76**, 12069.
36. M. C. Johnson, H. M. Scobie, Y. M. Ma and V. M. Vogt, *J. Virol.*, 2002, **76**, 11177.
37. D. L. Caspar, *Biophys. J.*, 1980, **32**, 103.
38. J. B. Bancroft, *Adv. Virus Res.*, 1970, **16**, 99.
39. K. W. Adolph and P. J. Butler, *Philos. Trans. R. Soc. London, Ser. B*, 1976, **276**, 113.
40. J. B. Bancroft and E. Hiebert, *Virology*, 1967, **32**, 354.
41. X. Zhao, J. M. Fox, N. H. Olson, T. S. Baker and M. J. Young, *Virology*, 1995, **207**, 486.
42. T. Douglas and M. Young, *Science*, 2006, **312**, 873.
43. J. Sun, C. DuFort, M. C. Daniel, A. Murali, C. Chen, K. Gopinath, B. Stein, M. De, V. M. Rotello, A. Holzenburg, C. C. Kao and B. Dragnea, *Proc. Natl. Acad. Sci. USA*, 2007, **104**, 1354.
44. J. Tang, J. M. Johnson, K. A. Dryden, M. J. Young, A. Zlotnick and J. E. Johnson, *J. Struct. Biol.*, 2006, **154**, 59.
45. M. A. Krol, N. H. Olson, J. Tate, J. E. Johnson, T. S. Baker and P. Ahlquist, *Proc. Natl. Acad. Sci. USA*, 1999, **96**, 13650.
46. R. H. Cheng, J. R. Caston, G. J. Wang, F. Gu, T. J. Smith, T. S. Baker, R. F. Bozarth, B. L. Trus, N. Cheng and R. B. Wickner, *J. Mol. Biol.*, 1994, **244**, 255.
47. P. E. Prevelige, D. Thomas and J. King, *Biophys. J.*, 1993, **64**, 824.
48. S. A. Datta, J. E. Curtis, W. Ratcliff, P. K. Clark, R. M. Crist, J. Lebowitz, S. Krueger and A. Rein, *J. Mol. Biol.*, 2007, **365**, 812.

49. Y. M. Ma and V. M. Vogt, *J. Virol.*, 2002, **76**, 5452.
50. S. Campbell, R. J. Fisher, E. M. Towler, S. Fox, H. J. Issaq, T. Wolfe, L. R. Phillips and A. Rein, *Proc. Natl. Acad. Sci. USA*, 2001, **98**, 10875.
51. S. A. Datta, Z. Zhao, P. K. Clark, S. Tarasov, J. N. Alexandratos, S. J. Campbell, M. Kvaratskhelia, J. Lebowitz and A. Rein, *J. Mol. Biol.*, 2007, **365**, 799.
52. R. Bartenschlager and H. Schaller, *EMBO J.*, 1992, **11**, 3413.
53. J. Carey, V. Cameron, P. L. de Haseth and O. C. Uhlenbeck, *Biochemistry*, 1983, **22**, 2601.
54. D. Beckett, H. N. Wu and O. C. Uhlenbeck, *J. Mol. Biol.*, 1988, **204**, 939.
55. K. Valegard, J. B. Murray, P. G. Stockley, N. J. Stonehouse and L. Liljas, *Nature*, 1994, **371**, 623.
56. W. T. Horn, M. A. Convery, N. J. Stonehouse, C. J. Adams, L. Liljas, S. E. Phillips and P. G. Stockley, *RNA*, 2004, **10**, 1776.
57. P. K. Sorger, P. G. Stockley and S. C. Harrison, *J. Mol. Biol.*, 1986, **191**, 639.
58. F. Qu and T. J. Morris, *J. Virol.*, 1997, **71**, 1428.
59. N. Wei, L. A. Heaton, T. J. Morris and S. C. Harrison, *J. Mol. Biol.*, 1990, **214**, 85.
60. J. C. Carrington, T. J. Morris, P. G. Stockley and S. C. Harrison, *J. Mol. Biol.*, 1987, **194**, 265.
61. B. Berger, P. W. Shor, L. Tucker-Kellogg and J. King, *Proc. Natl. Acad. Sci. USA*, 1994, **91**, 7732.
62. J. Lidmar, L. Mirny and D. R. Nelson, *Phys. Rev. E*, 2003, **68**, 051910.
63. N. G. Abrescia, J. J. Cockburn, J. M. Grimes, G. C. Sutton, J. M. Diprose, S. J. Butcher, S. D. Fuller, C. San Martin, R. M. Burnett, D. I. Stuart, D. H. Bamford and J. K. Bamford, *Nature*, 2004, **432**, 68.
64. P. Grayson, A. Evilevitch, M. M. Inamdar, P. K. Purohit, W. M. Gelbart, C. M. Knobler and R. Phillips, *Virology*, 2006, **348**, 430.
65. L. M. Stannard and M. Hodgkiss, *J. Gen. Virol.*, 1979, **45**, 509.
66. R. A. Crowther, N. A. Kiselev, B. Bottcher, J. A. Berriman, G. P. Borisova, V. Ose and P. Pumpens, *Cell*, 1994, **77**, 943.
67. A. Zlotnick, N. Cheng, J. F. Conway, F. P. Booy, A. C. Steven, S. J. Stahl and P. T. Wingfield, *Biochemistry*, 1996, **35**, 7412.
68. J. E. Johnson, *Proc. Natl. Acad. Sci. USA*, 1996, **93**, 27.
69. J. W. Erickson, A. M. Silva, M. R. Murthy, I. Fita and M. G. Rossmann, *Science*, 1985, **229**, 625.
70. G. L. Lokesh, T. D. Gowri, P. S. Satheshkumar, M. R. Murthy and H. S. Savithri, *Virology*, 2002, **292**, 211.
71. C. L. Hill *et al.*, *Nat. Struct. Biol.*, 1999, **6**, 565.
72. M. Yeager, K. A. Dryden, N. H. Olson, H. B. Greenberg and T. S. Baker, *J. Cell Biol.*, 1990, **110**, 2133.
73. Y. Ghendon, E. Yakobson and A. Mikhejeva, *J. Virol.*, 1972, **10**, 261.
74. D. M. Belnap, N. H. Olson, N. M. Cladel, W. W. Newcomb, J. C. Brown, J. W. Kreider, N. D. Christensen and T. S. Baker, *J. Mol. Biol.*, 1996, **259**, 249.

75. V. Sangita, G. L. Lokesh, P. S. Satheshkumar, C. S. Vijay, V. Saravanan, H. S. Savithri and M. R. Murthy, *J. Mol. Biol.*, 2004, **342**, 987.
76. J. A. Speir, S. Munshi, G. Wang, T. S. Baker and J. E. Johnson, *Structure*, 1995, **3**, 63.
77. P. S. Satheshkumar, G. L. Lokesh, M. R. Murthy and H. S. Savithri, *J. Mol. Biol.*, 2005, **353**, 447.
78. A. Pappachan, C. Subashchandrabose, P. S. Satheshkumar, H. S. Savithri and M. R. Murthy, *Virology*, 2008, **375**, 190.
79. K. W. Adolph and P. J. G. Butler, *Nature*, 1975, **255**, 737.
80. D. Willits, X. Zhao, N. Olson, T. S. Baker, A. Zlotnick, J. E. Johnson, T. Douglas and M. J. Young, *Virology*, 2003, **306**, 280.
81. R. L. Strausberg and S. L. Schreiber, *Science*, 2003, **300**, 294.
82. T. J. Smith, M. J. Kremer, M. Luo, G. Vriend, E. Arnold, G. Kamer, M. G. Rossmann, M. A. McKinlay, G. D. Diana and M. J. Otto, *Science*, 1986, **233**, 1286.
83. Y. Zhang, A. A. Simpson, R. M. Ledford, C. M. Bator, S. Chakravarty, G. A. Skochko, T. M. Demenczuk, A. Watanyar, D. C. Pevear and M. G. Rossmann, *J. Virol.*, 2004, **78**, 11061.
84. C. M. Teschke, J. King and P. E. Prevelige Jr, *Biochemistry*, 1993, **32**, 10658.
85. P. E. Prevelige Jr, J. King and J. L. Silva, *Biophys. J.*, 1994, **66**, 1631.
86. A. Zlotnick, P. Ceres, S. Singh and J. M. Johnson, *J. Virol.*, 2002, **76**, 4848.
87. P. E. Prevelige Jr, *Trends Biotechnol.*, 1998, **16**, 61.
88. K. Deres, C. H. Schroder, A. Paessens, S. Goldmann, H. J. Hacker, O. Weber, T. Kramer, U. Niewohner, U. Pleiss, J. Stoltefuss, E. Graef, D. Koletzki, R. N. Masantschek, A. Reimann, R. Jaeger, R. Gross, B. Beckermann, K. H. Schlemmer, D. Haebich and H. Rubsamen-Waigmann, *Science*, 2003, **299**, 893.
89. S. J. Stray and A. Zlotnick, *J. Mol. Recognit.*, 2006, **19**, 542.
90. H. J. Hacker, K. Deres, M. Mildenberger and C. H. Schroder, *Biochem. Pharmacol.*, 2003, **66**, 2273.
91. C. Bourne, M. G. Finn and A. Zlotnick, *J. Virol.*, 2006, **80**, 11055.
92. S. Casjens and J. King, *J. Supramol. Struct.*, 1974, **2**, 202.
93. P. E. Prevelige Jr, D. Thomas and J. King, *Biophys J.*, 1993, **64**, 824.
94. R. Tuma, M. H. Parker, P. Weigele, L. Sampson, Y. Sun, N. R. Krishna, S. Casjens, G. J. Thomas Jr and P. E. Prevelige Jr, *J. Mol. Biol.*, 1998, **281**, 81.
95. Y. Sun, M. H. Parker, P. Weigele, S. Casjens, P. E. Prevelige Jr and N. R. Krishna, *J. Mol. Biol.*, 2000, **297**, 1195.
96. B. A. Fane and P. E. Prevelige Jr, *Adv. Protein Chem.*, 2003, **64**, 259.
97. K. N. Parent, S. M. Doyle, E. Anderson and C. M. Teschke, *Virology*, 2005, **340**, 33.
98. E. Wyckoff and S. Casjens, *J. Virol.*, 1985, **53**, 192.
99. S. Casjens and M. B. Adams, *J. Virol.*, 1985, **53**, 185.
100. S. Casjens, M. B. Adams, C. Hall and J. King, *J. Virol.*, 1985, **53**, 174.

101. M. H. Parker, W. F. Stafford III and P. E. Prevelige Jr, *J. Mol. Biol.*, 1997, **268**, 655.

102. R. Tuma, H. Tsuruta, K. H. French and P. E. Prevelige Jr, *J. Mol. Biol.*, 2008, **381**, 1395.

103. M. H. Parker, S. Casjens and P. E. Prevelige Jr, *J. Mol. Biol.*, 1998, **281**, 69.

104. S. D. Moore and P. E. Prevelige Jr, *J. Mol. Biol.*, 2002, **315**, 975.

105. S. D. Moore and P. E. Prevelige Jr, *J. Virol.*, 2002, **76**, 10245.

106. B. Greene and J. King, *Virology*, 1996, **225**, 82.

107. P. R. Weigele, L. Sampson, D. Winn-Stapley and S. R. Casjens, *J. Mol. Biol.*, 2005, **348**, 831.

108. P. X. Guo, S. Erickson, W. Xu, N. Olson, T. S. Baker and D. Anderson, *Virology*, 1991, **183**, 366.

109. V. V. Mesyanzhinov, B. N. Sobolev, E. I. Marusich, A. G. Prilipov and V. P. Efimov, *J. Struct. Biol.*, 1990, **104**, 24.

110. W. W. Newcomb, D. R. Thomsen, F. L. Homa and J. C. Brown, *J. Virol.*, 2003, **77**, 9862.

111. W. W. Newcomb, F. L. Homa and J. C. Brown, *J. Virol.*, 2005, **79**, 10540.

112. C. Bazinet and J. King, *J. Mol. Biol.*, 1988, **202**, 77.

113. J. B. Huffman, W. W. Newcomb, J. C. Brown and F. L. Homa, *J. Virol.*, 2008, **82**, 6778.

114. C. Y. Fu, M. C. Morais, A. J. Battisti, M. G. Rossmann and P. E. Prevelige Jr, *J. Mol. Biol.*, 2007, **366**, 1161.

115. K. Iwasaki, B. L. Trus, P. T. Wingfield, N. Cheng, G. Campusano, V. B. Rao and A. C. Steven, *Virology*, 2000, **271**, 321.

116. D. T. Mooney, J. Stockard, M. L. Parker and A. H. Doermann, *J. Virol.*, 1987, **61**, 2828.

117. E. Kellenberger, *Eur. J. Biochem.*, 1990, **190**, 233.

118. I. Katsura and H. Kobayashi, *J. Mol. Biol.*, 1990, **213**, 503.

119. W. Earnshaw and J. King, *J. Mol. Biol.*, 1978, **126**, 721.

120. P. A. Thuman-Commike, B. Greene, J. A. Malinski, J. King and W. Chiu, *Biophys. J.*, 1998, **74**, 559.

121. W. W. Newcomb, F. L. Homa, D. R. Thomsen and J. C. Brown, *J. Struct. Biol.*, 2001, **133**, 23.

122. O. J. Marvik, T. Dokland, R. H. Nokling, E. Jacobsen, T. Larsen and B. H. Lindqvist, *J. Mol. Biol.*, 1995, **251**, 59.

123. S. Wang, P. Palasingam, R. H. Nokling, B. H. Lindqvist and T. Dokland, *Virology*, 2000, **275**, 133.

124. T. Dokland, S. Wang and B. H. Lindqvist, *Virology*, 2002, **298**, 224.

125. E. W. Six, M. G. Sunshine, J. Williams, E. Haggard-Ljungquist and B. H. Lindqvist, *Virology*, 1991, **182**, 34.

126. K. J. Kim, M. G. Sunshine, B. H. Lindqvist and E. W. Six, *Virology*, 2001, **283**, 49.

127. D. M. Salunke, D. L. Caspar and R. L. Garcea, *Cell*, 1986, **46**, 895.

128. S. N. Kanesashi, K. Ishizu, M. A. Kawano, S. I. Han, S. Tomita, H. Watanabe, K. Kataoka and H. Handa, *J. Gen. Virol.*, 2003, **84**, 1899.

129. G. L. Casini, D. Graham, D. Heine, R. L. Garcea and D. T. Wu, *Virology*, 2004, **325**, 320.
130. Z. Xie and R. W. Hendrix, *J. Mol. Biol.*, 1995, **253**, 74.
131. R. L. Duda, K. Martincic, Z. Xie and R. W. Hendrix, *FEMS Microbiol. Rev.*, 1995, **17**, 41.
132. B. A. Fane, in *Virus Taxonomy. Eighth Report of the International Committee on Taxonomy of Viruses*, ed. C. M. Fauquet, M. A. Mayo, J. Maniloff, U. Desselberger and L. A. Ball, Elsevier Academic Press, Amsterdam, 2005, pp. 289–299.
133. D. R. Rokyta, C. L. Burch, S. B. Caudle and H. A. Wichman, *J. Bacteriol.*, 2006, **188**, 1134.
134. M. Chen, A. Uchiyama and B. A. Fane, *J. Mol. Biol.*, 2007, **373**, 308.
135. O. Salim, R. J. Skilton, P. R. Lambden, B. A. Fane and I. N. Clarke, *Virology*, 2008, **377**, 440.
136. B. A. Fane and M. Hayashi, *Genetics*, 1991, **128**, 663.
137. M. C. Ekechukwu and B. A. Fane, *J. Bacteriol.*, 1995, **177**, 829.
138. A. D. Burch, J. Ta and B. A. Fane, *J. Mol. Biol.*, 1999, **286**, ch. 95.
139. A. D. Burch and B. A. Fane, *Virology*, 2000, **270**, 286.
140. T. Dokland, R. A. Bernal, A. Burch, S. Pletnev, B. A. Fane and M. G. Rossmann, *J. Mol. Biol.*, 1999, **288**, 595.
141. T. Dokland, R. McKenna, L. L. Ilag, B. R. Bowman, N. L. Incardona, B. A. Fane and M. G. Rossmann, *Nature*, 1997, **389**, 308.
142. P. A. Thuman-Commike, B. Greene, J. A. Malinski, M. Burbea, A. McGough, W. Chiu and P. E. Prevelige Jr, *Biophys. J.*, 1999, **76**, 3267.
143. B. L. Trus, F. P. Booy, W. W. Newcomb, J. C. Brown, F. L. Homa, D. R. Thomsen and A. C. Steven, *J. Mol. Biol.*, 1996, **263**, 447.
144. Z. H. Zhou, S. J. Macnab, J. Jakana, L. R. Scott, W. Chiu and F. J. Rixon, *Proc. Natl. Acad. Sci. USA*, 1998, **95**, 2778.
145. W. W. Newcomb, B. L. Trus, N. Cheng, A. C. Steven, A. K. Sheaffer, D. J. Tenney, S. K. Weller and J. C. Brown, *J. Virol.*, 2000, **74**, 1663.
146. J. E. Cherwa Jr, A. Uchiyama and B. A. Fane, *J. Virol.*, 2008, **82**, 5774.
147. A. D. Burch and B. A. Fane, *J. Virol.*, 2000, **74**, 9347.
148. A. D. Burch and B. A. Fane, *Virology*, 2003, **310**, 64.
149. A. Uchiyama, M. Chen and B. A. Fane, *J. Virol.*, 2007, **81**, 8587.
150. A. Uchiyama and B. A. Fane, *J. Virol.*, 2005, **79**, 6751.
151. A. Uchiyama, P. Heiman and B. A. Fane, *Virology*, 2009, **386**, 303.
152. M. C. Morais, C. M. Fisher, S. Kanamaru, L. Przybyla, J. Burgner, B. A. Fane and M. G. Rossmann, *Mol. Cell*, 2004, **15**, 991.
153. K. R. Spindler and M. Hayashi, *J. Virol.*, 1979, **29**, 973.
154. K. N. Parent, A. Zlotnick and C. M. Teschke, *J. Mol. Biol.*, 2006, **359**, 1097.
155. P. D. Ross, J. F. Conway, N. Cheng, L. Dierkes, B. A. Firek, R. W. Hendrix, A. C. Steven and R. L. Duda, *J. Mol. Biol.*, 2006, **364**, 512.

CHAPTER 11
Mechanisms of Genome Packaging

MARK ORAM AND LINDSAY W. BLACK

Department of Biochemistry and Molecular Biology, University of Maryland School of Medicine, Baltimore, MD 21201, USA

1 Introduction

A strategy for viral maturation that is common to all viral classes is the assembly of progeny genome units within fresh envelope or capsid structures, once the nucleic acid genome is replicated during an infective cycle. Two basic modes of virion nucleic acid packaging are known: (1) co-condensation of the nucleic acid with viral capsid proteins to form a virus particle (employed by viruses such as HIV, TMV and M13) and (2) translocation of the nucleic acid into a preformed procapsid shell (found in eukaryotic viral families such as herpes and tailed dsDNA bacteriophages such as T4, lambda and φ29, among others). Packaging is not an energetically favorable process, in part as the negatively charged DNA can be packed to near crystalline density (~ 500 $mg\,mL^{-1}$) within the capsid, so that a powerful energy-transducing, molecular translocation system (or motor) is required to effect the condensation. This second strategy tends to be more frequently employed among viruses with larger, more complex and especially dsDNA-based genomes. Indeed, one advantage to a virus of active DNA translocation into the procapsid is the potential for repair of the nascent genome during packaging before it is sequestered for maturation. A detailed molecular picture of the organization and action of the DNA packaging machinery from the class I order Caudovirales (dsDNA bacteriophage genome) has emerged from biochemical, genetic and biophysical studies of representative systems. Similar results from studies

RSC Biomolecular Sciences No. 21
Structural Virology
Edited by Mavis Agbandje-McKenna and Robert McKenna
© Royal Society of Chemistry 2011
Published by the Royal Society of Chemistry, www.rsc.org

on model class I dsDNA eukaryotic viruses such as the herpes, CMV and poxvirus families, and also with class II (ssDNA genome) and class III (dsRNA genome) phage systems, highlight the conservation of many shared structural and mechanistic features.

The phage packaging systems are currently an area of active research, for several key reasons. The dsDNA phages are the most numerous ($\sim 10^{31}$) and energetic biological entities in the biosphere (since $\sim 1\%$ of these are thought to be undergoing assembly at any given moment and since 300 ATPs are consumed per $\sim 600\,\mathrm{bp\,s}^{-1}$ DNA encapsidated by the average terminase, then $\sim 10^9$ J or equivalently more than one tank-full of gasoline per second, are expended in the biosphere by phage packaging at any one time). The condensation of phage DNA into the prohead forms an ideal model system for DNA or chromosome condensation and the packasome itself serves as a paradigm for both molecular motors and also the related hexameric helicase superfamily. Potential nanotechnology or other medically related applications exploiting the phage packaging machinery are being actively followed.[1,2] We focus primarily on the biology and mechanism of dsDNA packaging in this chapter, while noting significant similarities and differences between other viral classes where appropriate. A number of comprehensive reviews of this topic have appeared recently and are a valuable summary of the extensive recent[3–5] or older[6] literature.

2   The Biological Context of Phage DNA Packaging

With numerous bacteriophages and also herpesviruses, the initial products of viral DNA synthesis in an infected host are 'endless' concatemeric (*i.e.* head to tail coupled) genomes that arise from multiple replication and/or recombination events (Figure 1A).

Individual genome units are cut and packaged from the concatemer to yield molecules carrying a full complement of viral genes, and also a terminal sequence redundancy in many cases. This feature allows the concatemer to be reconstituted by recombination after phage genome injection in a subsequent infected host. As their name implies, the enzymes primarily responsible for cutting both strands of the phage DNA to generate linear genome units are the terminases, common to all dsDNA viral families. These proteins also provide the DNA translocation activity (driven by ATP hydrolysis) once packaging has been initiated and act to make a second DNA cut after one genome equivalent has been translocated into the procapsid, the latter to release the packaged head for further phage maturation events. This second cut can arise in a sequence-dependent manner once the terminase recognizes a defined DNA sequence element during the latter stages of genome translocation. Alternatively, the cut is made non-specifically in the DNA substrate once the prohead is filled to capacity and a resultant 'headful' signal has been transduced to the terminase to activate the nuclease activity.

It is important to note that *in vivo* nucleic acid packaging occurs in a developmentally regulated infected cytoplasmic milieu that must balance concurrent and sometimes competing viral developmental pathways. To ensure that DNA cutting is both controlled and coupled to packaging, the nuclease activity of the terminase complex must be tightly integrated with other viral processes not obviously linked to packaging and that differ mechanistically from one virus to another. For example, phages T7 and T3 show an obligatory participation of the phage RNA polymerase to transcribe a late terminase promoter that both initiates packaging on the phage genome in a sequence-specific manner[7,8] and couples essential DNA end repair synthesis to the packaging reaction[9,10] (Figure 1B). Analogously, phage T4 regulates initiation of DNA packaging *via* an interaction of the terminase with components of the late transcriptional machinery,[11] more specifically to the gp55 late sigma factor complexed with the gp45 DNA sliding clamp (Figure 1C). Thus for phages T3, T7 and T4, DNA replication, late transcription and packaging are apparently linked. Additional cellular or viral components are also involved in the initiation of DNA packaging in other well-characterized phage systems. The φ29 phage of *Bacillus subtilis* exploits a small RNA molecule, termed the pRNA, as an essential component of the DNA packaging machinery,[12] while in phage lambda the host IHF protein is essential for formation of a terminase–DNA complex that is competent for packaging initiation.[13,14]

Further differences in genomic packaging pathways are also apparent in the mechanisms for the generation of the ends between representative systems (Figure 1A). Sequence-specific recognition of the DNA substrate at specific *cos* or *pac* sites can occur by action of the phage terminase, which is then followed by terminase-catalyzed DNA cleavage, either at a precise sequence (as in lambda, T3 or T7) or less specifically near the *pac* site (P1 or P22, for example). In the case of lambda, the second *cos* site at the end of the newly translocated genome is recognized and then cut in a sequence-specific manner by the terminase (provided that a near headful of DNA has been packaged), so that exactly one genome unit – that can re-circularize upon a fresh round of infection – is provided to each head. Indeed, deletions in phage lambda DNA lead by this mechanism to underpackaged phage heads.[15] By contrast, with phage terminases that utilize *pac* sites to generate the initial DNA end for packaging, around 102–105% of the primary genome sequence is packaged (thus ensuring terminal sequence redundancy) and the subsequent cuts made by the enzyme to liberate the filled head then occur in a sequence-independent, complete headful-dependent fashion. The T-even phages, as exemplified by T4, were initially believed to have completely non-specific DNA recognition and cleavage events, but more recent evidence points to a functional *pac* site present in T4 DNA (actually in the vicinity of the gene for the small terminase subunit; see below) that functions in the initial T4 DNA recognition event.[16,17] A further class of packaged structures is exemplified by φ29, where the substrate for packaging is generated as a genome monomer with the phage gp3 protein bound to each end.[18]

3 Components of the Packaging Machinery

Clearly, the initiation of phage packaging between different systems appears to employ varying structural and functional connections to the specific developmental pathway. Studies of the actual translocation mechanism itself, by contrast, reveal a highly conserved process that employs few and relatively similar components, both *in vivo* and more recently in defined *in vitro* packaging systems. In fact, only three components seem to be essential for active translocation once a functioning packasome (*i.e.* the protein complex engaged with both a specific prohead and a genomic substrate that is competent to fill the nascent head) has been assembled within the scope of the particular viral developmental pathway (Figure 2A).

These components are the DNA substrate itself, the terminase enzyme (the large subunit alone often suffices) and the portal or connector protein that occupies a unique vertex in the icosahedral procapsid shell. Although the

---

Figure 1   Strategies for phage genome packaging. (A) Phage DNA synthesis frequently yields concatemers (genome units linked head-to-tail) that must be matured by terminase cutting during packaging. The concatemer may also contain nicks, gaps and recombination-induced branches [see also (C)] that must be removed during processive head filling by the packasome (blue octagon). (1) For lambda-like phages, a unique 3′ recessed end of the first genome unit to be packaged is generated by terminase cutting at the *cos* site. Packaging proceeds until the next *cos* site is recognized and cut identically. (2) The ϕ29 genome is generated as separate single genome units with covalently linked gp3 proteins (green dot) required for DNA synthesis and packaging. (3) For phages that package DNA by a pure headful mechanism, the initial terminase recognition of concatemeric DNA occurs at a specific *pac* site, followed by imprecise cutting around *pac*. Headful packaging of over 100% of the genome length ensures terminal redundancy with subsequent terminase-dependent cuts occurring in a sequence and *pac*-independent processive fashion. (B) For T3/T7 phages, each packaged genome contains a short direct terminal repeat (TR) that must be resynthesized during packaging. (1) The *pac* site adjacent to the TR (pacB is shown) also contains a promoter (P_r) for the phage RNA polymerase (green oval). (2) Transcription from this promoter proceeds through the TR, but is blocked at a pause site (functionally indicated by vertical red bars). This allows the displaced DNA strand to be nicked (red triangle); in addition, the large terminase subunit (orange oval) is recruited by direct interaction with the RNA polymerase. (3) End repair by DNA polymerase (not shown) recreates a duplex form of the TR and provides a substrate for the ds nuclease activity of the terminase (red arrow). (4) The packasome is assembled with the terminase–DNA complex engaged by the portal protein (red annulus) of an empty prohead. (C) The T4 concatemer, complete with nicks and branches arising from replication, recombination and late transcription, is shown by dark grey lines. The T4 sliding clamp (gp45), loaded at a discontinuity in dsDNA, participates in replication (right), late transcription *via* an attached late sigma factor (gp55) (center) and packaging *via* gp45–gp55–gp17 (large terminase subunit) interaction (left). Part (C) reproduced with permission from reference 11.

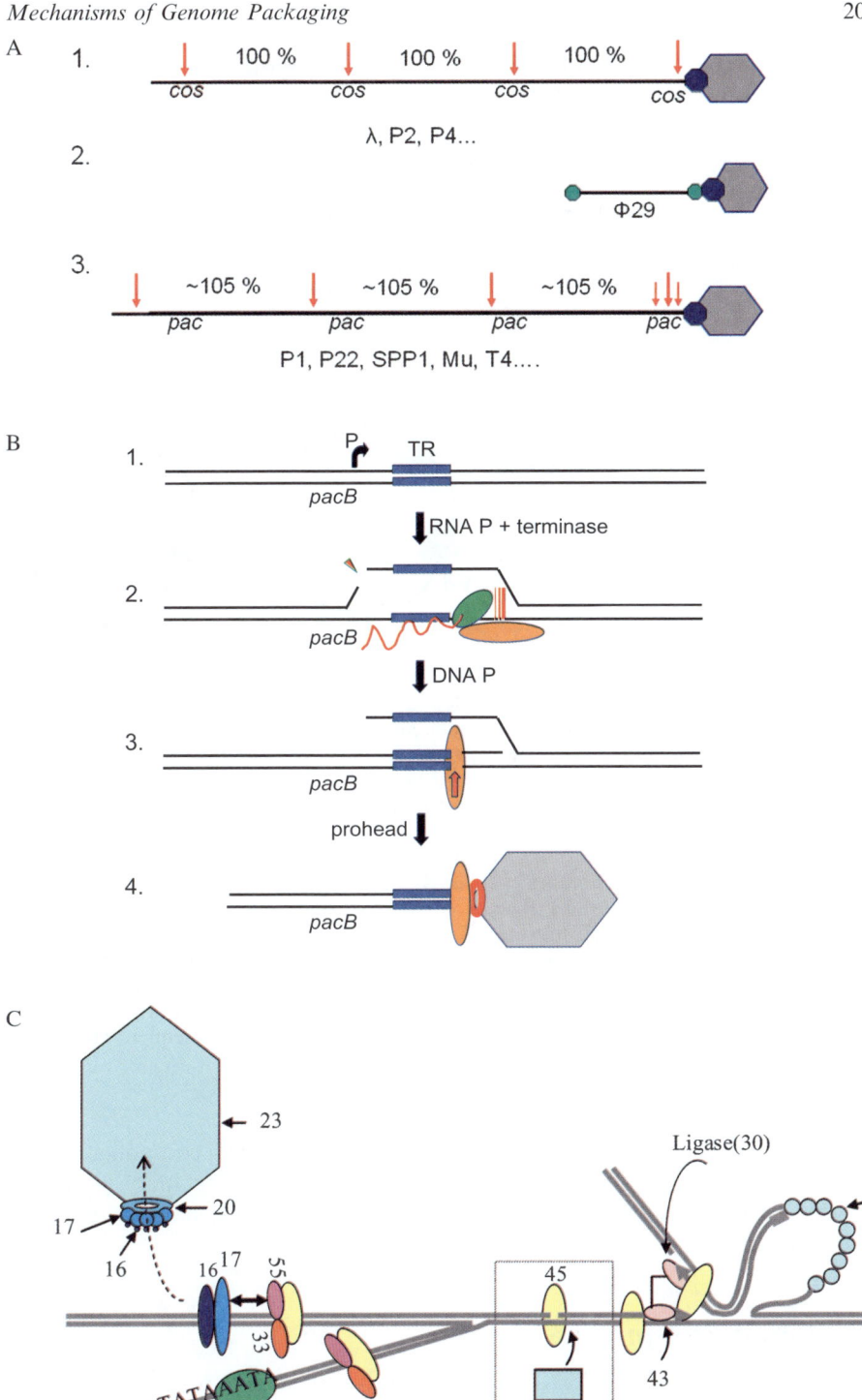

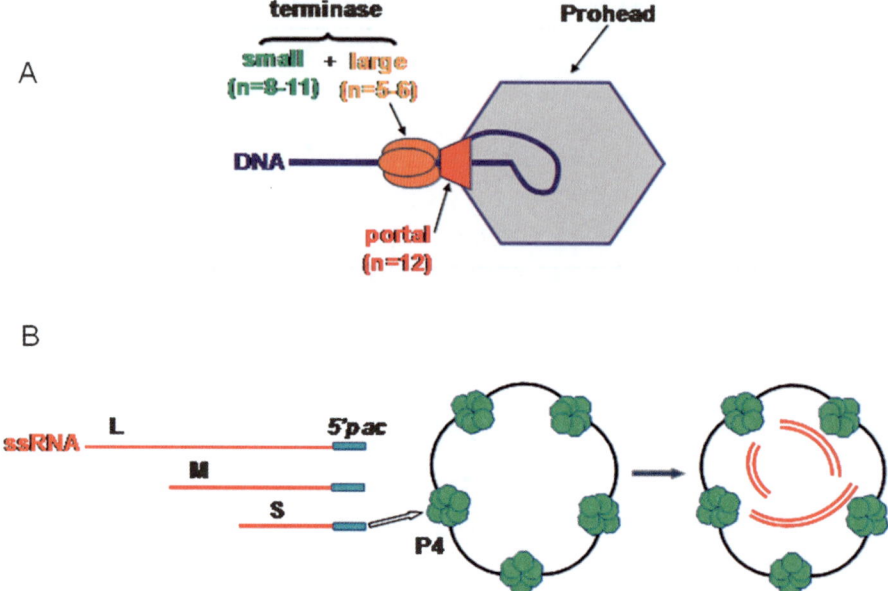

Figure 2   Overview of a packasome assembly in DNA or RNA phages. (A) Overview
of a 'generic' packasome formed by class 1 (dsDNA) phages. Linear genomic DNA undergoes an ATP-dependent translocation by a multimeric (5–6
copies) large terminase subunit protein engaged with the dodecameric portal
of the prohead. The small terminase subunit (8–11 copies) is required for
initiating packaging on the concatemer, but is frequently non-essential and
even inhibitory to active DNA translocation. (B) RNA packaging in class 3
(dsRNA) phages. The genome of *Cystoviridae* such as the *Pseudomonas*
phages φ6–φ13 comprises three RNA segments, named Small, Medium and
Large (S, M and L). The three RNA segments are generated in ss form
during phage maturation. A hexameric P4 phage protein permanently
occupies procapsid vertices, whereas the major procapsid structural protein
selects 5′ pac site ssRNAs for packaging. Packaging is catalyzed by the
NTP-driven P4 hexameric motor and synthesis of the second RNA strand of
each genome segment occurs within the head subsequent to RNA
packaging.

pRNA is also an essential if somewhat mysterious packaging component of the
φ29 system, the same tripartite DNA–terminase–portal machinery is utilized
by this phage also, underscoring the conservation of mechanism in phage
packaging.

The Portal Protein

The portal protein of dsDNA viruses is crucial to the processes of head
assembly and phage maturation and also the packaging reaction itself. Portal
structures in phages such as φ29, SPP1, P2, P22, ε15 and herpesvirus family

members have been determined using electron microscopy (EM), cryo-EM or X-ray crystallization methodologies.[19–26] These structural analyses concur in revealing a conserved dodecameric ring structure embedded at the pentameric prohead vertex. All the dodecameric portal structures that have been determined show a conserved wedge-like arrangement, with the larger (C-terminal) end of each monomer pointing inwards and the narrower N-terminal portion exiting the body of the capsid. The central channel of the portal ring is the path for entry of DNA during packaging and also for DNA exit upon a subsequent infection event.

Although the portal protein is essential for the packaging reaction to proceed, the way in which this is realized is not currently well understood. One long-standing model for DNA packaging proposed that a rotating portal, energetically facilitated by a symmetry mismatch between portal and pentameric capsid vertex, was necessary for DNA translocation.[19,27] However, the only significant experiments that specifically addressed this possibility have recently all but eliminated portal rotation as being a feature of the packaging process.[28,29] Nevertheless, it does seem that free relative movement of the portal subunits themselves is essential to the packaging reaction, as revealed by use of site-directed mutations in the SPP1 portal.[30] In this work, adjacent portal subunits were 'locked' together by engineered disulfide bridge formation, an action that concomitantly blocked active DNA translocation. Finally, genetic analyses show that the portal gauges a headful of DNA,[31,32] consistent with cryo-EM analyses which suggest that conformational changes in the portal, induced by the dense packing of DNA at the end of the packaging process once the prohead is filled to capacity, are essential in transducing the headful signal to the terminase protein.[33]

The Terminase Enzyme

The terminases of dsDNA phages, aside from their role in generating the ends of the packaged genome molecule, are also responsible for the ATP-dependent translocation reaction. Structurally, these enzymes typically possess a large and a small subunit, with a clear division of labor between each. Both terminase subunits generally form homo-multimers, in addition to interacting (often weakly) with each other to form an active terminase complex. The large terminase subunits possesses the ATPase sites that provide the biochemical energy for the DNA translocation process and also provide the terminase–nuclease activity. The small subunits play a key role in DNA recognition and, consistent with this, an NMR structural analysis of the lambda Nu1 small terminase subunit revealed a winged helix–turn–helix DNA binding motif.[34] Homology searches suggest that other small terminase subunits have a comparable structural feature. In addition, the small terminase subunit often enhances the ATPase activity of the large subunit. The genes for the terminase subunits are usually adjacent – even overlapping – in many phage genomes, and in addition the *cos* or *pac* sites used in varying phages map close to or even within the gene for the small subunit itself. Given the role of the small subunit in DNA recognition, the proximity of a *cos* or a *pac* element to the site of synthesis of the protein that it binds to may well have a key mechanistic role in efficient assembly of a packaging initiation complex.

Several experimental approaches have attempted to define the stoichiometry of the terminase complex of representative phages, in part as this may be a crucial structural feature of their mode of action. The small terminase subunits often form an annular ring structure and varying copy number for the protein from differing systems have been reported in the literature. In the lambda and SPP1 phages, a 2:1 ratio of small subunit to large subunit has been established,[35,36] and in addition the active form of the SPP1 small subunit was reported as a decamer. By contrast, sedimentation analyses of an active lambda terminase complex did not provide an unambiguous stoichiometry, although a pentamer of a 2:1 small to large heterotrimer (hence 10 copies of the small subunit in total) could account for the data observed.[37] About eight copies per oligomer – and twice that of a dimeric ring oligomer – of the T4 small terminase subunit (gp16) were determined by scanning transmission electron microscopy[17] and nine copies of the P22 small terminase subunit have been detected by electron microscopy.[38,39]

The DNA Substrate and Packasome Assembly

Irrespective of the actual number of subunits, the small terminase annulus seems to function to assemble the DNA within the large subunit oligomer to initiate assembly of a complex competent to begin genome translocation. It also appears that the large subunit itself is often relatively unable to oligomerize until the small terminase ring is present to facilitate this step. Nucleolytic processing of the genomic DNA at a *cos* or *pac* site by the small and large terminase subunits leads to the formation of a terminase initiation complex bound to the free end of a linear genome unit and engagement of this entity with the portal protein of an empty capsid completes assembly of the packasome or DNA packaging motor complex. The outer domains of the portal protein (*i.e.* those accessible from the capsid exterior) interact with the large terminase subunits and intimate portal–terminase interaction is also very likely mandatory to the packaging reaction (Figure 2A). Translocation of DNA is catalyzed by the ATP hydrolysis turnover at the large terminase subunits, acting to drive the phage genome into the phage head. After the second terminase cut has been made, once packaging is complete, the filled head is released for further phage maturation events. Concurrently, the newly liberated terminase–DNA structure, still attached to the next genome unit in the concatemer, can be recruited by another empty capsid for a fresh initiation and translocation event, thus enabling a processive repetition of the packaging process. At least three or more heads can be filled from the same initiating terminase–DNA complex in lambda and P22 phages in this fashion.[40]

4 Analysis of the Packasome as a Molecular Motor

The packaging reaction is now efficiently reproducible *in vitro*, in part as the purified components of the packasome can be readily obtained in significant quantities. Standard molecular biology techniques allow specific molecular modifications of any component to be generated essentially at will, enabling the mechanism to be probed structurally in a defined way. All of these approaches, in conjunction with mass spectrometry of megadalton protein complexes, and also with cryo-EM and crystallographic structure determination, have begun to dissect the packaging reaction in atomic detail.

Class 1 dsDNA Packaging

A series of elegant experiments by Smith's group, exploiting the use of optical tweezers, has started to explore the energetics and dynamics of the packing process itself.[41–45] The experimental system used in these studies typically consists of an immobilized prohead–terminase complex attached to one optical sphere, while the genomic substrate is attached at one end to a second, movable sphere. By carefully advancing the genome to the vicinity of the packasome, packaging initiation occurs at the free genome end. The movable sphere is then retracted and productive initiation and translocation events can then be analyzed individually by measuring the force on the second sphere as the reaction proceeds. Results obtained in this way with T4, lambda and φ29 also highlight significant differences in the DNA translocation rate – decreasing from T4 (171 kb genome: packaged at up to 2000 bp s^{-1}), to lambda (48 kb genome: 600 bp s^{-1}) to φ29 (19.3 kb genome: 100–165 bp s^{-1}) – with comparably high force generation being key features among these systems. The strengths generated by the packaging systems are noteworthy, in that the stalling forces can be greater than 100 pN, up to an order of magnitude greater than the forces generated by myosin or kinesin motors.[46,47] The interesting proposal that the significantly higher velocity achieved by the T4 packaging motor might be due to electrostatic enhancement to the motor from the T4 internal proteins[48] can also be tested by this experimental approach.

All three of the motors studied in this fashion showed some slips and pauses while packaging the genome, although the T4 motor in addition showed a significantly greater variability in translocation speed.[45] This mechanistic feature may be related to nicks and other discontinuities in the DNA itself. It has been known for some time that ligase activity is required for T4 DNA packaging *in vivo*,[49] and thus that discontinuities in the concatemer are strongly inhibitory to packaging. In a complementary approach, we recently explored the consequences of departures from B-form DNA in small defined substrates in an *in vitro* T4 packaging system, which also revealed the inhibitory effect of nicks in small ($\sim$ 100 bp) DNA substrates. To account for these observations, we also proposed that a linear lever arm motion of a terminase subunit imparts a spring-like compression force (due to transient DNA binding by the portal) in the substrate that is then released by the portal to power the translocation.[50]

Such a model would clearly demand a minimum length of DNA that can simultaneously contact binding sites in both the terminase and portal. The current minimum length of DNA capable of being translocated by any phage system remains unknown, although we also showed in this work that substrates as short as 20 bp were efficiently packaged by the T4 system (in that 100% of the duplexes in the reaction were sequestered into proheads, with multiple molecules per prohead accommodated).

Whether such a DNA compression model for the T4 system is unique or generally applicable, and also the structural basis for any lever-like motion of the motor subunit(s) in the packaging process, may well be answered by the generality of the only available terminase structure described at the present time. Rossmann and co-workers have determined crystal structures of the large T4 terminase (gp17) subunit[51] that together with SDM studies allow motor and nuclease (RNaseH-like) functions to be assigned to N- and C-terminal portions of the protein, respectively. They also proposed that a terminase pentamer docked on the portal translocates the DNA substrate by a helicase inchworm-like mechanism – one that results from large-scale motion of terminase subunit domains coupled to electrostatic interaction with the DNA substrate.[52] The identification of such structural elements in the enzyme provides invaluable information relating to how the action of each individual monomer relates to the overall packaging process. However, many questions still remain, such as how ATP binding and hydrolysis are coupled to the proposed domain movement, how this in turn engages with the DNA and how far a given substrate will be translocated per power stroke. In two cases (ϕ29 and T3), measurements in bulk suggested that an average of two DNA base pairs are packaged per ATP molecule hydrolyzed;[10,53] however, the individual DNA step size of any of these molecular motors remains to be determined.

Nucleic Acid Packaging Systems of Other Viral Classes

Although some class 2 ssDNA phages such as M13 apparently co-condense capsid proteins and DNA, others such as ϕX174 virus (discussed in Chapter 10) do package their ssDNA genome into a preformed capsid particle. Unlike the class 1 systems described above, however, there is no dedicated motor protein or complex that is required for packaging. Instead, it seems that replication is concurrent with entry of the ss genome into the capsid (*i.e.* packaging) and that the DNA polymerase, complexed with the capsid at a given vertex acts both to provide the DNA synthesis and translocation activities.[54]

In comparison with the dsDNA phage terminase–portal motors or ssDNA polymerase packaging systems, the ϕ6 ssRNA packaging motor may be a simpler single-component motor and its understanding is currently mostly complete (Figure 2B). Following binding of a 5' *pac*-site containing ssRNA by the ϕ6 major procapsid structural protein, the RNA packaging specificity determinant, the RNA is 'handed off' to an opened hexameric P4 motor protein; this subsequently powers translocation of the ssRNA into the procapsid through NTP hydrolysis[5] (Figure 2B). The precise and sequential packaging of each of three single-stranded RNA segments, followed by coordinated synthesis within the procapsid of the corresponding dsRNAs, occurs by a fascinating molecular mechanism that is far from simple.[55] However, with respect to the motor component, complementary oligonucleotide displacement helicase assays establish that some of the hexameric P4 motor proteins can function *in vitro* as RNA helicases in isolation from the prohead.[56,57] Crystal structures of several NTP hydrolysis intermediates,[58] H–D exchange studies of polypeptide mobility and SDM studies of P4 support a swiveling arm mechanism for the P4 power stroke with relatively little domain movement during translocation.[59–61] The structure of the translocating RNA segment engaged with the motor arm has not so far been revealed. It should be noted that the RNA translocation velocity is 20–60 times lower than that of the dsDNA phage packaging motors; hence the degree to which the φ6 mechanism will prove to be comparable to the dsDNA two-component portal–terminase motors remains to be established.

5 Fluorescence Approaches to Packaging Motor Dynamics

Fluorescence methods are now routinely employed to probe the dynamics and structure of both the protein complexes and substrate during many DNA transactions,[62] as in the analysis of the dynamic properties of the DNA sliding clamp in replication[63] or in the identification of RNA polymerase as a DNA scrunching machine.[64,65] Such approaches are equally applicable to the phage packaging reaction and may potentially discriminate among several of the proposed DNA packaging mechanisms discussed above. Small fluorophores can be covalently linked to the DNA substrate and GFP can be fused to virtually any part of the protein architecture of the packing complex, as non-invasive tools to probe the dynamics of the packaging process. For example, we recently employed fluorescence correlation spectroscopy[66] (FCS) to demonstrate T4 DNA packaging into procapsids in real time, by quantifying the change in diffusion of 100 bp dye-labeled DNAs as they entered the slower procapsids (Figure 3).

Additionally, their apparent decrease in number in this experimental system demonstrated that multiple DNAs entered a single procapsid.[67] FCS–FRET was also used to demonstrate that the dye-labeled DNAs entered the T4 prohead in proximity to packaged internal protein–GFP fusions (Figure 3D–F). Comparable packaging competent T4 procapsids have been constructed with portal fusion proteins containing the GFP fusion portion located inside (C-terminal) or outside (N-terminal) the procapsid, and the T4 large terminase protein can be labeled with a fluorescent dye and retain packaging activity (Figure 4A,B)

Large-scale motion of terminase domains to propel DNA into the prohead can potentially be detected by FRET changes between terminase and portal, whereas by comparable measurements compression or other alteration of the B-form DNA duplex by the motor force might be detected by FRET changes in double dye-labeled Y-DNA (Figure 4C). In fact, mixing of Texas Red dye-labeled Y-DNA and portal–GFP proheads demonstrates that the T4 terminase propels an arrested Y-DNA packaging complex into FRET proximity to the GFP located on the C-terminal portal protein within the procapsid (Figure 4D; unpublished observations). Thus several of the components of the packasome should be accessible to FRET determination of their molecular distances and, most significantly, to dynamic changes during translocation.

6   Summary

Viral nucleic acid is packaged into procapsid containers by conserved mechanisms based on one- or two-component multimeric motors of the helicase–translocase family.[68] Bacteriophages have evolved powerful nucleic acid translocation motors that function to fill a preformed viral capsid with genomic DNA or RNA during phage maturation. The phage motors are precisely regulated *in vivo* to prevent destruction of the replicated genome, yet are remarkably efficient and powerful, capable of generating forces up to an order of magnitude greater than the myosin/actin filament or kinesin motor, for example. How these motor proteins are adapted to serve specific viral needs

---

Figure 3   Fluorescence correlation spectroscopic (FCS) analysis of the phage T4 packaging reaction. (A) FCS detects fluorescence from a femtoliter reaction chamber excited by a laser. (B, C) Only a limited number of fluorophores are analyzed, so stochastic fluctuations in their number are apparent. Correlating the variations in fluorescent intensity over time (on the millisecond–second time-scale) generates an autocorrelation function, $G(\tau)$, that represents the decay of the fluorescence correlation. In this manner, the diffusion of fluorophores can be determined, since larger fluorophores will diffuse more slowly through the volume element and hence show an increased fluorescent correlation (*i.e.* a more slowly decaying autocorrelation function). In particular, the packaging of faster-diffusing fluorescent dye-tagged DNA substrates into much larger, slower diffusing proheads can be followed by FCS analysis. Parts (B) and (C) reproduced with permission from reference 67. (D) Packaging of Texas Red (TR)-tagged 100 bp DNAs (vertical black bars with red dots) into a T4 prohead that had also encapsidated GFP (green spheres), in the presence of ATP and T4 terminase (not shown) allows a non-invasive analysis of the packaging reaction. (E) Correlation curves from a combined FCS–FRET analysis of TR–DNA substrates packaged into a GFP-encapsidated prohead. Excitation at the GFP absorbance maximum of a completed packaging reaction yields correlation curves for the proheads (green dots) that have the diffusion constant of T4 proheads. In addition, FRET to the encapsidated TR substrate gives correlation curves for the TR–DNA (red crosses) that match the fluorescence decay of the prohead. This observation, together with the FRET from the proximity of both GFP and TR–DNA by co-encapsidation in the same prohead, confirms that packaging has occurred. (F) The dependence of these features on ATP excludes passive diffusion or spontaneous association of the TR–DNA with the GFP prohead as an explanation for the data obtained.

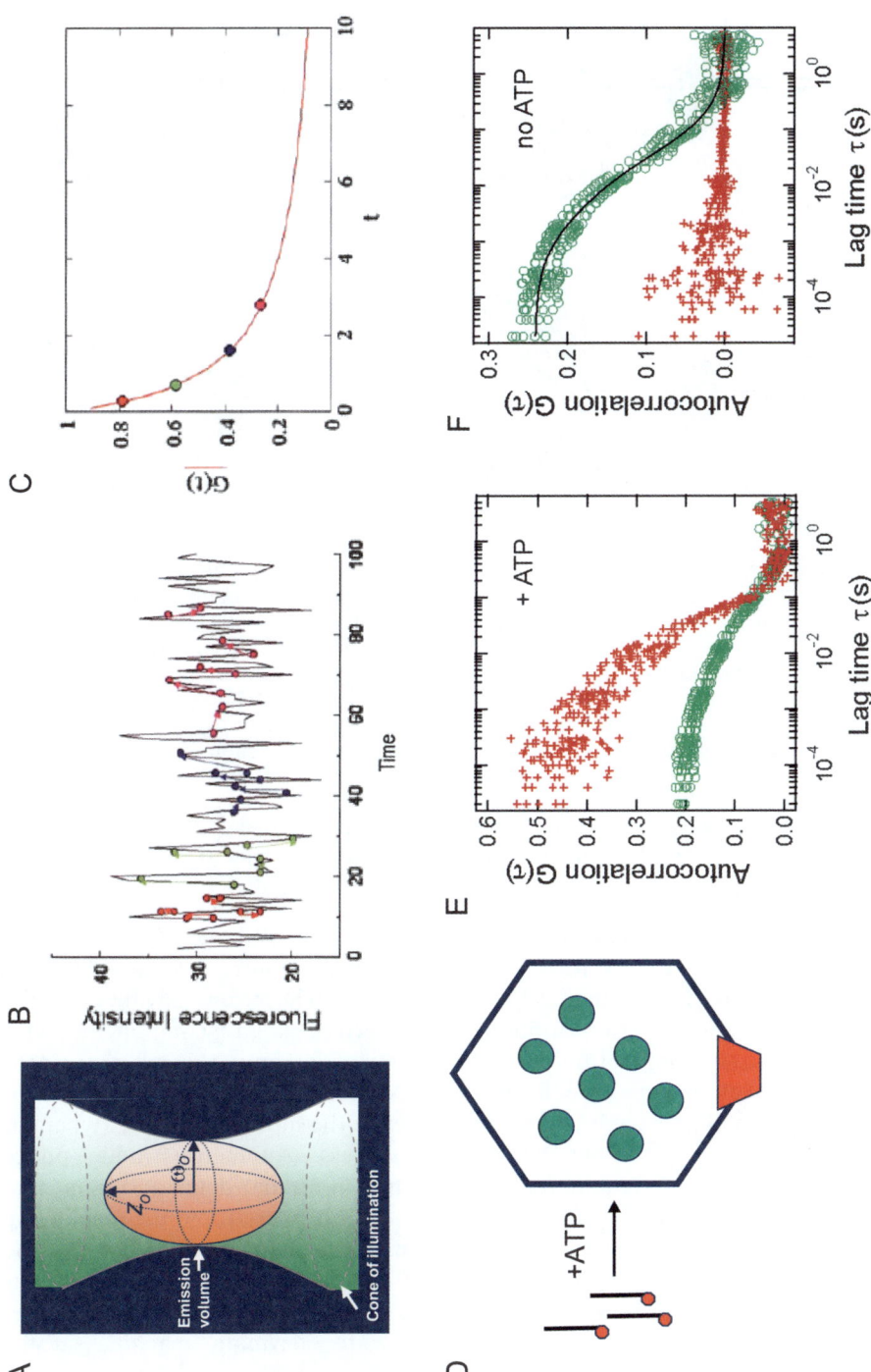

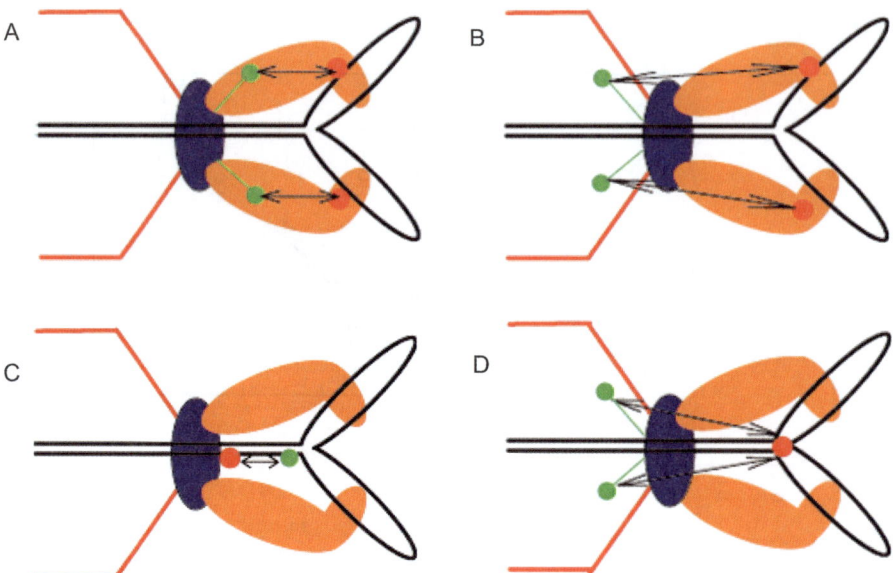

Figure 4   Dynamic analyses of the packasome employing fluorescence methods. Shown are four examples, based on T4 packasome components that could allow the relative movement and dimensions of packasome subunits to be probed by FCS, single-molecule platforms, FRET or other fluorescent approaches. The ability to attach fluorophores to either the DNA substrate, the outside face of the portal protein (by use of N-terminal GFP-portal fusions[28]) (A), the inside face of the portal protein (by use of C-terminal GFP-portal fusions[28]) (B) and the large terminase subunit (A and B) allows dynamic analysis of packaging proteins. FRET–dye pairs in packaging arrested Y-DNAs engaged with the portal (C) or single Y-DNA dye in proximity to portal-GFP (D) in packaging arrested Y-DNAs could reveal changes in the dimensions of the substrate during packaging.

(*e.g.* measurement of the amount and integrity, of the genome to be packaged) and how the high rate and force of the translocation motor are achieved remain to be understood at a detailed molecular level. The recent advances in the development and detection of fluorescent-tagged packaging components (viral packasomes), in conjunction with the structural determination of motor proteins and development of biochemical assays for DNA packaging, are forging a new biophysical approach to dissect and define the mechanisms and dynamics of these key biological systems. Recent progress in numerous model phage systems has begun to illuminate the reaction in molecular detail and point to a relatively well-conserved structure and function shared by the packaging enzymes of bacteriophages and related eukaryotic viruses. The pivotal role played by phage research in establishing the field of molecular genetics is well documented: it now seems that continued biophysical study of phage paradigms will yield fundamental insights into enzymatic machines at the molecular level.

Acknowledgments

We thank J.-X. Ma, University of Maryland Medical School, and V.B. Rao, Catholic University, for helpful comments on the manuscript.

References

1. P. Guo, *J. Nanosci. Nanotechnol.*, 2005, **5**, 1964.
2. P. Guo, *Methods Mol. Biol.*, 2005, **300**, 285.
3. V. B. Rao and M. Feiss, *Annu. Rev. Genet.*, 2008, **42**, 647.
4. P. Guo and T. J. Lee, *Mol. Microbiol.*, 2007, **64**, 886.
5. D. E. Kainov, R. Tuma and E. J. Mancini, *Cell. Mol. Life Sci.*, 2006, **63**, 1095.
6. L. W. Black, *Annu. Rev. Microbiol.*, 1989, **43**, 267.
7. C. Hashimoto and H. Fujisawa, *Virology*, 1992, **187**, 788.
8. C. Hashimoto and H. Fujisawa, *Virology*, 1992, **191**, 246.
9. X. Zhang and F. W. Studier, *J. Mol. Biol.*, 2004, **340**, 707.
10. H. Fujisawa and M. Morita, *Genes Cells*, 1997, **2**, 537.
11. L. W. Black and G. Peng, *J. Biol. Chem.*, 2006, **281**, 25635.
12. P. Guo, S. Erickson and D. Anderson, *Science*, 1987, **236**, 690.
13. I. Mendelson, M. Gottesman and A. B. Oppenheim, *J. Bacteriol.*, 1991, **173**, 1670.
14. M. E. Ortega and C. E. Catalano, *Biochemistry*, 2006, **45**, 5180.
15. M. Feiss, R. A. Fisher, M. A. Crayton and C. Egner, *Virology*, 1977, **77**, 281.
16. H. Lin and L. W. Black, *Virology*, 1998, **242**, 118.
17. H. Lin, M. N. Simon and L. W. Black, *J. Biol. Chem.*, 1997, **272**, 3495.
18. R. P. Mellado, M. A. Penalva, M. R. Inciarte and M. Salas, *Virology*, 1980, **104**, 84.
19. A. A. Simpson, Y. Tao, P. G. Leiman, M. O. Badasso, Y. He, P. J. Jardine, N. H. Olson, M. C. Morais, S. Grimes, D. L. Anderson, T. S. Baker and M. G. Rossmann, *Nature*, 2000, **408**, 745.
20. A. Guasch, J. Pous, A. Parraga, J. M. Valpuesta, J. L. Carrascosa and M. Coll, *J. Mol. Biol.*, 1998, **281**, 219.
21. J. M. Valpuesta, J. J. Fernandez, J. M. Carazo and J. L. Carrascosa, *Structure*, 1999, **7**, 289.
22. R. Lurz, E. V. Orlova, D. Gunther, P. Dube, A. Droge, F. Weise, M. van Heel and P. Tavares, *J. Mol. Biol.*, 2001, **310**, 1027.
23. E. V. Orlova, B. Gowen, A. Droge, A. Stiege, F. Weise, R. Lurz, M. van Heel and P. Tavares, *EMBO J.*, 2003, **22**, 1255.
24. D. N. Doan and T. Dokland, *J. Struct. Biol.*, 2007, **157**, 432.
25. J. Chang, P. Weigele, J. King, W. Chiu and W. Jiang, *Structure*, 2006, **14**, 1073.
26. B. L. Trus, N. Cheng, W. W. Newcomb, F. L. Homa, J. C. Brown and A. C. Steven, *J. Virol.*, 2004, **78**, 12668.
27. R. W. Hendrix, *Proc. Natl. Acad. Sci. USA*, 1978, **75**, 4779.

28. R. G. Baumann, J. Mullaney and L. W. Black, *Mol. Microbiol.*, 2006, **61**, 16.

29. T. Hugel, J. Michaelis, C. L. Hetherington, P. J. Jardine, S. Grimes, J. M. Walter, W. Falk, D. L. Anderson and C. Bustamante, *PLoS Biol.*, 2007, **5**, e59.

30. A. Cuervo, M. C. Vaney, A. A. Antson, P. Tavares and L. Oliveira, *J. Biol. Chem.*, 2007, **282**, 18907.

31. S. Casjens, E. Wyckoff, M. Hayden, L. Sampson, K. Eppler, S. Randall, E. T. Moreno and P. Serwer, *J. Mol. Biol.*, 1992, **224**, 1055.

32. P. Tavares, M. A. Santos, R. Lurz, G. Morelli, H. de Lencastre and T. A. Trautner, *J. Mol. Biol.*, 1992, **225**, 81.

33. G. C. Lander, L. Tang, S. R. Casjens, E. B. Gilcrease, P. Prevelige, A. Poliakov, C. S. Potter, B. Carragher and J. E. Johnson, *Science*, 2006, **312**, 1791.

34. T. de Beer, J. Fang, M. Ortega, Q. Yang, L. Maes, C. Duffy, N. Berton, J. Sippy, M. Overduin, M. Feiss and C. E. Catalano, *Mol. Cell*, 2002, **9**, 981.

35. M. A. Tomka and C. E. Catalano, *J. Biol. Chem.*, 1993, **268**, 3056.

36. A. G. Camacho, A. Gual, R. Lurz, P. Tavares and J. C. Alonso, *J. Biol. Chem.*, 2003, **278**, 23251.

37. N. K. Maluf, Q. Yang and C. E. Catalano, *J. Mol. Biol.*, 2005, **347**, 523.

38. D. Nemecek, E. B. Gilcrease, S. Kang, P. E. Prevelige Jr, S. Casjens and G. J. Thomas Jr, *J. Mol. Biol.*, 2007, **374**, 817.

39. D. Nemecek, G. C. Lander, J. E. Johnson, S. R. Casjens and G. J. Thomas Jr, *J. Mol. Biol.*, 2008, **383**, 494.

40. M. Feiss, J. Sippy and G. Miller, *J. Mol. Biol.*, 1985, **186**, 759.

41. D. E. Smith, S. J. Tans, S. B. Smith, S. Grimes, D. L. Anderson and C. Bustamante, *Nature*, 2001, **413**, 748.

42. D. N. Fuller, J. P. Rickgauer, P. J. Jardine, S. Grimes, D. L. Anderson and D. E. Smith, *Proc. Natl. Acad. Sci. USA*, 2007, **104**, 11245.

43. J. P. Rickgauer, D. N. Fuller, S. Grimes, P. J. Jardine, D. L. Anderson and D. E. Smith, *Biophys. J.*, 2008, **94**, 159.

44. D. N. Fuller, D. M. Raymer, J. P. Rickgauer, R. M. Robertson, C. E. Catalano, D. L. Anderson, S. Grimes and D. E. Smith, *J. Mol. Biol.*, 2007, **373**, 1113.

45. D. N. Fuller, D. M. Raymer, V. I. Kottadiel, V. B. Rao and D. E. Smith, *Proc. Natl. Acad. Sci. USA*, 2007, **104**, 16868.

46. J. T. Finer, R. M. Simmons and J. A. Spudich, *Nature*, 1994, **368**, 113.

47. C. M. Coppin, D. W. Pierce, L. Hsu and R. D. Vale, *Proc. Natl. Acad. Sci. USA*, 1997, **94**, 8539.

48. T. Y. Yu and J. Schaefer, *J. Mol. Biol.*, 2008, **382**, 1031.

49. A. Zachary and L. W. Black, *J. Mol. Biol.*, 1981, **149**, 641.

50. M. Oram, C. Sabanayagam and L. W. Black, *J. Mol. Biol.*, 2008, **381**, 61.

51. S. Sun, K. Kondabagil, P. M. Gentz, M. G. Rossmann and V. B. Rao, *Mol. Cell*, 2007, **25**, 943.

52. B. Draper and V. B. Rao, *J. Mol. Biol.*, 2007, **369**, 79.
53. P. Guo, C. Peterson and D. Anderson, *J. Mol. Biol.*, 1987, **197**, 229.
54. R. A. Bernal, S. Hafenstein, N. H. Olson, V. D. Bowman, P. R. Chipman, T. S. Baker, B. A. Fane and M. G. Rossmann, *J. Mol. Biol.*, 2003, **325**, 11.
55. L. Mindich, *Microbiol. Mol. Biol. Rev.*, 1999, **63**, 149.
56. D. E. Kainov, J. Lisal, D. H. Bamford and R. Tuma, *Nucleic Acids Res.*, 2004, **32**, 3515.
57. D. E. Kainov, M. Pirttimaa, R. Tuma, S. J. Butcher, G. J. Thomas Jr, D. H. Bamford and E. V. Makeyev, *J. Biol. Chem.*, 2003, **278**, 48084.
58. E. J. Mancini, D. E. Kainov, J. M. Grimes, R. Tuma, D. H. Bamford and D. I. Stuart, *Cell*, 2004, **118**, 743.
59. D. E. Kainov, E. J. Mancini, J. Telenius, J. Lisal, J. M. Grimes, D. H. Bamford, D. I. Stuart and R. Tuma, *J. Biol. Chem.*, 2008, **283**, 3607.
60. J. T. Huiskonen, F. de Haas, D. Bubeck, D. H. Bamford, S. D. Fuller and S. J. Butcher, *Structure*, 2006, **14**, 1039.
61. J. Lisal, T. T. Lam, D. E. Kainov, M. R. Emmett, A. G. Marshall and R. Tuma, *Nat. Struct. Mol. Biol.*, 2005, **12**, 460.
62. A. A. Deniz, M. Dahan, J. R. Grunwell, T. Ha, A. E. Faulhaber, D. S. Chemla, S. Weiss and P. G. Schultz, *Proc. Natl. Acad. Sci. USA*, 1999, **96**, 3670.
63. T. A. Laurence, Y. Kwon, A. Johnson, C. W. Hollars, M. O'Donnell, J. A. Camarero and D. Barsky, *J. Biol. Chem.*, 2008, **283**, 22895.
64. A. N. Kapanidis, E. Margeat, S. O. Ho, E. Kortkhonjia, S. Weiss and R. H. Ebright, *Science*, 2006, **314**, 1144.
65. A. Revyakin, C. Liu, R. H. Ebright and T. R. Strick, *Science*, 2006, **314**, 1139.
66. E. L. Elson, *Traffic*, 2001, **2**, 789.
67. C. R. Sabanayagam, M. Oram, J. R. Lakowicz and L. W. Black, *Biophys. J.*, 2007, **93**, L17.
68. J. M. Berger, *Cell*, 2008, **134**(5), 888.

CHAPTER 12

Attachment and Entry: Receptor Recognition in Viral Pathogenesis

DAMIAN C. EKIERT AND IAN A. WILSON

Department of Molecular Biology and Skaggs Institute for
Chemical Biology, The Scripps Research Institute, La Jolla, CA 92037, USA

1 Introduction – Viral Receptors Mediate Attachment and Entry

The entry of a virus into a target cell is a relatively slow process.[1,2] Endocytosis of a particle may take several minutes and membrane fusion or penetration requires the coordinated action of multiple copies of the viral surface proteins.[3,4] In order to remain in close proximity to a cell long enough to be endocytosed or penetrate the plasma membrane, a virus must anchor itself to the surface of the target cell. Initially, low-affinity, non-specific binding due to electrostatic or lectin–glycan interactions may keep a virus particle loosely associated with the cell surface.[5–8] Eventually, a more stable connection is accomplished through specific interaction of a protein on the surface of the virion and a receptor on the cell surface.

In many cases, receptor binding also serves as a signal that the virus has encountered a target cell so that initiation of the entry process is triggered. This mechanism helps to avoid premature firing of the membrane fusion or the penetration machinery. For example, water-borne viruses, such as hepatitis A and poliovirus, must remain infectious for days to months in contaminated water before being ingested by a new host.[9] In contrast, the use of receptors that are only expressed on particular tissues or cell types may confer distinct

RSC Biomolecular Sciences No. 21
Structural Virology
Edited by Mavis Agbandje-McKenna and Robert McKenna
© Royal Society of Chemistry 2011
Published by the Royal Society of Chemistry, www.rsc.org

advantages on certain viruses. HIV-1 escapes clearance by hijacking and suppressing the cellular immune system, which it accomplishes primarily due to its tropism for T-cells.[10] In the case of rabies virus, replication in muscle and neural tissues allows the establishment of a robust infection with little antiviral response.[11–13]

Finally, as many cell surface signaling molecules are activated by receptor clustering, the simultaneous interaction of a polyvalent virus with multiple receptors can mimic activation and trigger endocytosis of the virion. The virus acts as a molecular Trojan horse, mimicking food or extracellular debris and inducing the cell to internalize the virus particle. Once inside, the virus hijacks the cellular machinery in order to replicate its genome and produce a multitude of new, infectious progeny. This common theme is observed in many different virus families, and, therefore, it is not surprising that many of the viral receptors are involved in host cell signaling processes at the cell surface.[14]

2 Cell Surface Receptors

Any component of the cell surface could potentially serve as a receptor for viral attachment. The cell envelope consists of three main components: first, a mixture of phospholipids and cholesterols assemble to form the membrane itself; second, hundreds of membrane-associated proteins are distributed throughout the lipid bilayer and are anchored by membrane-spanning peptides and post-translational acylation; third, glycans (complex assemblies of sugar polymers) decorate many of the proteins and lipids in the membrane. The membrane lipids are largely hidden within the bilayer, leaving only their relatively small head groups exposed. Proteins and glycans, however, are well exposed and can project up to hundreds of Ångstroms away from their membrane anchors. Indeed, viruses have been found to use both resident membrane proteins and membrane glycolipids for attachment. Receptors can, therefore, be divided into two major classes: (1) glycans, which may be displayed on the target cell as protein glycosylations or as elaborations on membrane lipid head groups and (2) cell surface proteins. Some examples are depicted in Figure 1.

Due to the immense diversity of glycan structures and their often stochastic and irregular addition to target proteins and lipids, most viruses that engage glycan receptors recognize short oligosaccharide motifs rather than the entire complex glycan. The affinity of viral receptor binding proteins for glycan receptors can vary over six orders of magnitude. At one end of the spectrum is the interaction between the foot-and-mouth disease virus (FMDV) coat proteins VP1, VP2 and VP3 and heparan sulfate. The dissociation constant (K_d) for this interaction is on the order of 1 nM, comparable to the affinity of a fully matured antibody for its antigen.[15] In contrast, hemagglutinin, the receptor binding protein from the influenza A virus, binds sialylated glycans with a K_d of ~1–5 mM.[16–19] As a result, a single heparan sulfate-binding event is likely sufficient to anchor FMDV to a target cell, whereas the attachment of the

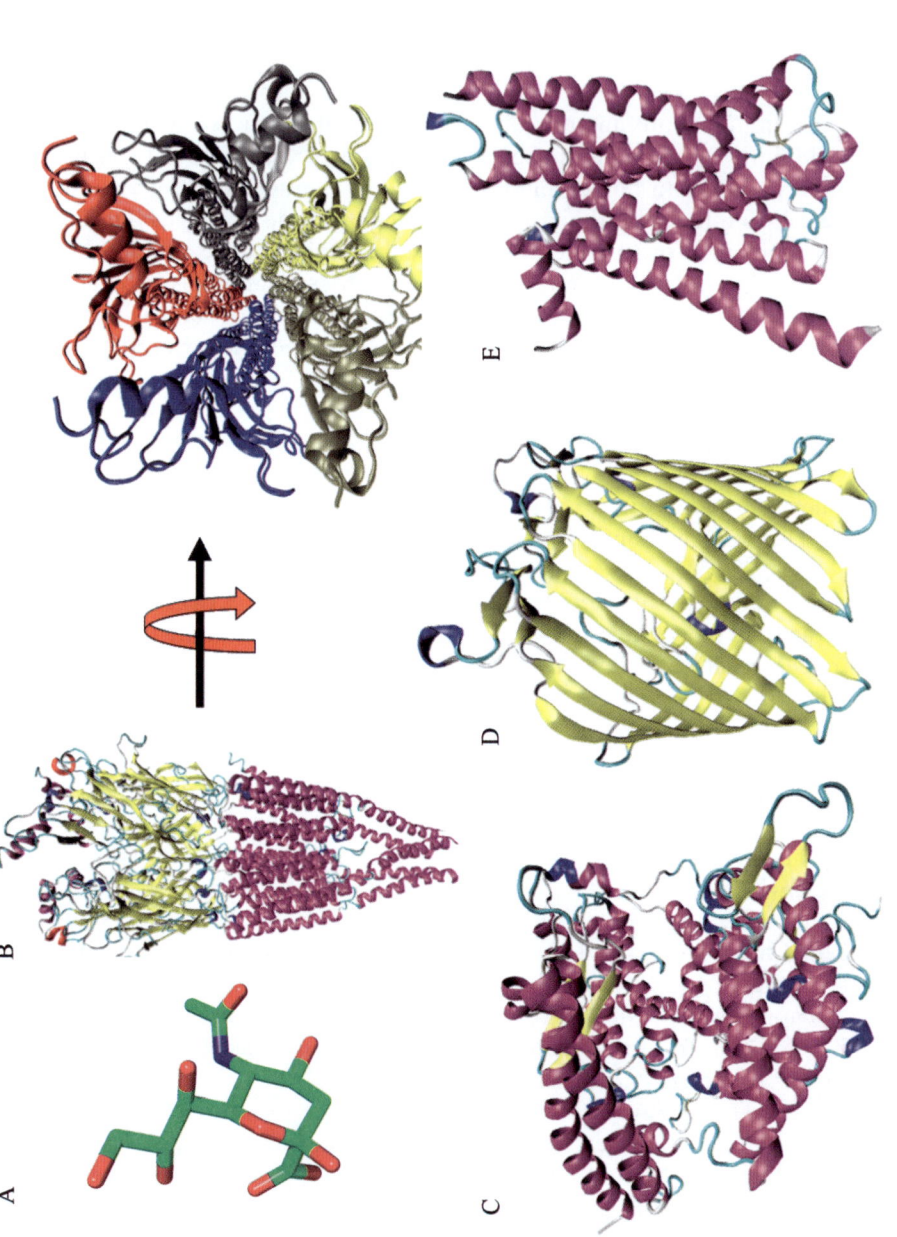

Figure 1 Diversity of viral receptors. (A) Sialic acid (influenza A virus and other respiratory viruses). (B) Left: the nicotinic acetylcholine receptor (rabies virus). The transmembrane helices are colored purple (bottom) and the extracellular domain is yellow (top). Right: alternative view down the length of the channel, from the extracellular face towards the cell interior; PDB code 2BG9. (C) ACE2; PDB code 2AJF (SARS-CoV). (D) Maltoporin; PBD code 1AF6 (bacteriophage lambda). (E) G-protein coupled receptors; PDB

influenza virus requires concerted interaction with multiple receptors in a multivalent interaction that substantially increases avidity.

Heparan sulfate (HS), one of the most commonly used glycan motifs, serves as a receptor or co-receptor for a number of viruses, including members of the *Herpesviridae* (*e.g.* HSV-1 and -2 and BHV), human papillomaviruses and some retroviruses (*e.g.* HTLV-1, FIV).[8,20–23] HS is a highly sulfated, linear glycopolymer built up from disaccharide units consisting of glucosamine and uronic acid. This polymer is anchored to the surface of animal cells through covalent linkages to resident plasma membrane proteins and is a major component of the extracellular matrix. In many cases, heparan sulfate appears to act as a *bona fide* receptor, as in the case of HSV-1 and FMDV.[15,20,24] In other situations, it may serve as a weaker, tethering interaction to hold the virus in close proximity to the cell surface until the primary receptor is engaged. Due to its high charge density, heparan sulfate has been suggested to contribute to non-specific virus binding through electrostatic interactions.[8]

Another dominant type of glycan receptor motif is *N*-acetylneuraminic acid, more commonly known as sialic acid. Sialic acid is a nine-carbon sugar containing an amino group at the C-5 position. In many higher animals, such as birds and mammals, N-linked glycosylations are frequently capped with sialic acid moieties. Influenza A is the prototypical virus that uses sialylated glycans for attachment and will be discussed later in more detail. Many other respiratory viruses, including many members of the *Paramyxoviridae* and *Coronaviridae*, also use sialic acids as their primary receptor.

It is also worth noting that the glycolipid lipopolysaccharide (LPS) that encapsulates Gram-negative bacteria is the receptor for many bacteriophages. Coliphages T3, T4, T7, P1 and P2 and *Salmonella* phages P22, Felix 0 and C21 all use different LPS substructures to anchor themselves for genome injection.[25] Similarly, a number of Gram-positive phages have been shown to use peptidoglycan motifs for attachment.[25]

Proteins on the surface of the target cell serve as the receptors for viruses in all three kingdoms of life. The diversity of cell surface molecules employed by viruses is equally broad and includes proteins from essentially all imaginable functional classes (see Figure 1). Bacteriophage lambda attaches to the surface of *Escherichia coli* cells *via* maltoporin, a β-barrel, porin-like protein involved in maltose transport across the outer membrane.[26,27] The SARS coronavirus binds to an enzyme ACE2, which is involved in the processing of peptidic signaling molecules.[28] The cell adhesion molecule, CAR, is the receptor for the Coxsackie and adenoviruses.[29] The rabies virus uses an ion channel, the nicotinic acetylcholine receptor.[30] Finally, HIV-1 requires two receptor-binding events: a primary interaction with CD4, which is normally involved in T-cell receptor signaling and activation, and a subsequent co-receptor interaction with a G-protein coupled chemokine receptor, CCR5 or CXCR4 that first requires a conformational change in gp120 from prior engagement with CD4.[31–36]

A common theme shared by both protein and glycan receptors is the need to abrogate receptor binding in the late stages of infection. For many viruses, down-regulation or inactivation of receptors after entry is critical to prevent superinfection and to achieve efficient viral egress and release. A classic example of this phenomenon is found for influenza A virus and other related viruses that use sialic acid receptors. In addition to its receptor binding protein, hemagglutinin (HA), influenza A virus also encodes a receptor-destroying enzyme, neuraminidase (NA). In the late stages of infection, cleavage of sialic acids from cellular glycans by NA prevents newly budded virions from being trapped on the cell surface through HA–receptor interactions and, hence, enables progeny escape. HIV-1 also actively suppresses its primary receptor, CD4, by several different mechanisms. HIV-1 encodes two accessory proteins that target CD4 for degradation. In the early stages of infection, Nef promotes the internalization of existing CD4 from the plasma membrane, in addition to the redirection of nascent CD4 from the *trans*-Golgi network.[37] Nef seems to act as an adapter molecule that links CD4 to components of the clathrin-dependent, vesicular trafficking pathways, such as AP-1, -2 and -3. Later, newly synthesized CD4 is sequestered in the ER by the viral glycoprotein, gp160.[37] Finally, the second accessory protein Vpu binds to the cytoplasmic tail of CD4 molecules retained in the ER by gp160, in addition to an E3 ubiquitin ligase complex.[37] These viral-encoded events result in the ubiquitination and proteosomal degradation of CD4.

Receptor down-regulation and destruction may extend all the way to the bacteriophages. For example, bacteriophage P22 encodes an *endo*-rhamnosidase activity that is localized near the tail spike and injection machinery.[38,39] Although such receptor-destroying enzymes may also aid in the entry process by allowing the phage particle to slide over the surface of a bacterium or by clearing away excess LPS, it seems likely that they may also play a role in release of progeny phage from bacterial debris after lysis.[39]

3 Viral Receptor Binding Proteins

Viral receptor binding domains (RBDs) are a highly variable class of molecules, reflecting the enormous diversity of virus lifestyles (see Figure 2). Some RBDs are an integral part of the protein capsid that surrounds the genome. For example, the poliovirus receptor CD155 binds deep in the 'canyon' formed around each fivefold axis by the major capsid proteins VP1, VP2 and VP3.[40,41] In membrane-bound viruses, on the other hand, the RBDs are found in the envelope proteins. In either case, RBDs are often part of a larger, multi-functional protein that plays a role in attachment, entry and membrane bypass, immune evasion, antiviral response suppression and budding of new virus particles.

Two of the best-studied receptor binding protein families are the hemagglutinin-esterase-fusion (HEF) proteins and the hemagglutinin–neuraminidases (HN/H/G). Due to the relatively high frequency of recombination and exchange of genomic segments among RNA viruses, members of the HEF and HN/H/G families are found in a wide range of respiratory viruses. A clear pattern of modular insertion and removal of various functional domains is

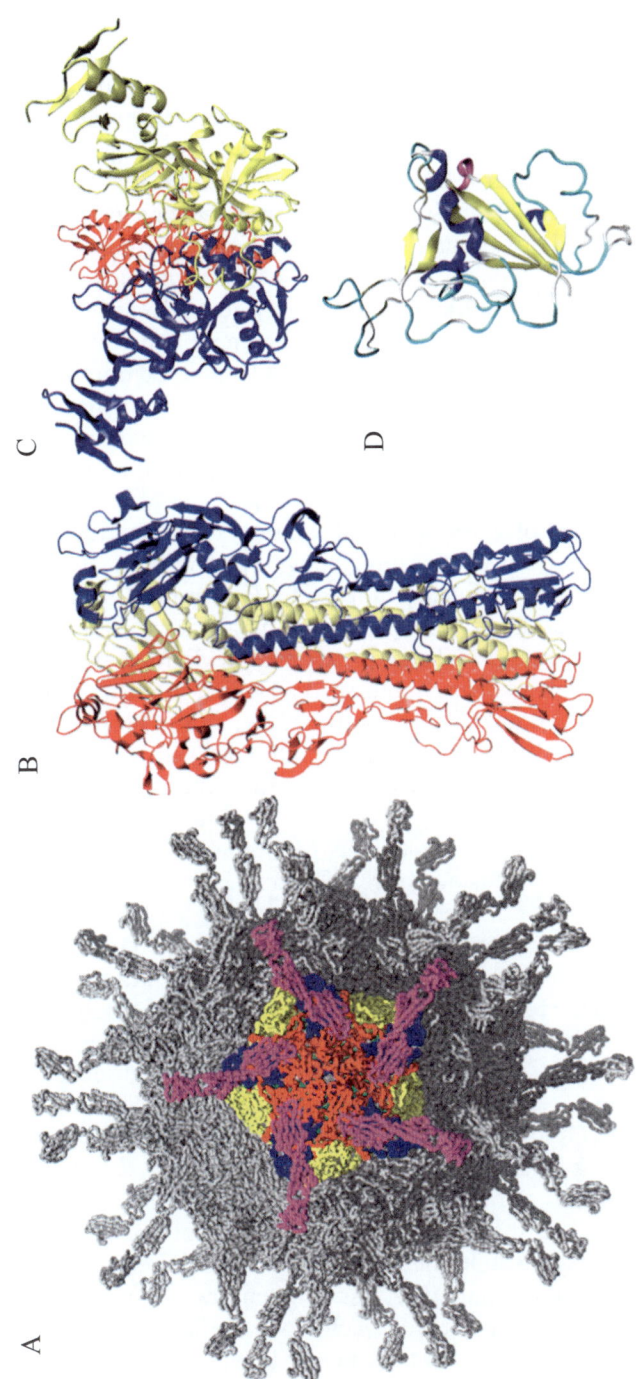

Figure 2 Viral receptor binding proteins. (A) Poliovirus particle decorated with soluble poliovirus receptor (magenta); PDB code 1NN8. The capsid proteins (VP1, VP2, VP3 and VP4) about one of the fivefold axes are colored red, yellow, blue and green, respectively (VP4 is largely buried). (B) The trimeric hemagglutinin protein from the influenza A virus, with the three protomers rendered in red, blue and yellow; PDB code 1RD8. (C) The core of the trimeric Ebola virus glycoprotein, GP, with the three protomers rendered in red, blue and yellow; PDB code 3CSY. The receptor for Ebola is unknown. (D) The receptor binding domain from the SARS-CoV spike protein S, PDB code 2AJF. SARS-CoV uses an enzyme, ACE2, for entry.

found on existing protein scaffolds, and also modification and tuning of these domains for new uses. Further, these two families can be found to overlap in some viruses, where HEF and HN/H/G family members are both present.

The prototypical members of the HEF family are the hemagglutinin proteins from the influenza A, B and C viruses. The influenza A HA is the predominant glycoprotein in the envelope of the influenza virus and was the first viral receptor binding protein to be characterized at atomic resolution.[42] A total of 16 HA subtypes have been identified based upon their antigenic properties.[43–47] Of these, only H1, H2 and H3 are found in *bona fide* human viruses. HA is a multifunctional protein that mediates attachment of virus particles to the surface of target cells through interaction with sialylated glycans on host proteins and glycolipids. In addition to its role in attachment and endocytosis, HA mediates the fusion of the viral envelope with endosomal membranes in a pH-dependent manner. HA is also the primary target of neutralizing antibodies of the host immune system. As a result, HA is the most critical determinant in the early stages of the virus life-cycle and in our ability to neutralize and clear influenza infections.

HA is a symmetric trimer of three identical chains. The immature HA0 chains are co-translationally translocated into the endoplasmic reticulum and each protomer is anchored in the membrane by a C-terminal, single-pass transmembrane segment. In addition, the cytoplasmic tail is acylated at one or more conserved cysteine residues.[48,49] These C-terminal acylations are critical for efficient membrane fusion and virus viability.[49–52] During insertion into the endoplasmic reticulum and trafficking through the Golgi to the plasma membrane, the HA trimer is multiply glycosylated and modified *via* glycosyltransferases and each HA0 protein is also proteolytically cleaved into two chains, HA1 and HA2. This cleavage leads to a metastable conformation of the protein that primes the HA for subsequent pH-induced conformational changes that lead to membrane fusion, once the virus has been taken into endosomes.

The mature HA trimer is an elongated, rod-shaped molecule (Figure 3).[42] From the viral envelope to the membrane distal tip, the HA trimer is roughly 135 Å long. The N-terminal HA1 chains form a globular head domain at the membrane distal end of the trimer. The head group consists of an eight-stranded, anti-parallel β-jelly roll surrounded by several short coils and α-helices. The HA2 chains fold to form a prominent three-helix bundle that aligns along the long, threefold axis. Three additional, shorter α-helices pack loosely around the triple-helical core near the viral envelope. Overall, the extracellular domain contains the tripod stem of HA2, with the three HA1 globular head domains packing together atop the trimeric stem and concealing much of the top of the tripod. While the bulk of HA1 lies at the membrane distal end of the trimer, a series of paired β-strands extending from the N- and C-termini snake their way nearly all the way back along the length of the stem to the base. The HA1 and HA2 chains are covalently linked together by a pair of disulfide bonds. The 'feet' of the tripod are tightly held together by the triple coiled coil and are immediately adjacent to the viral envelope. The whole

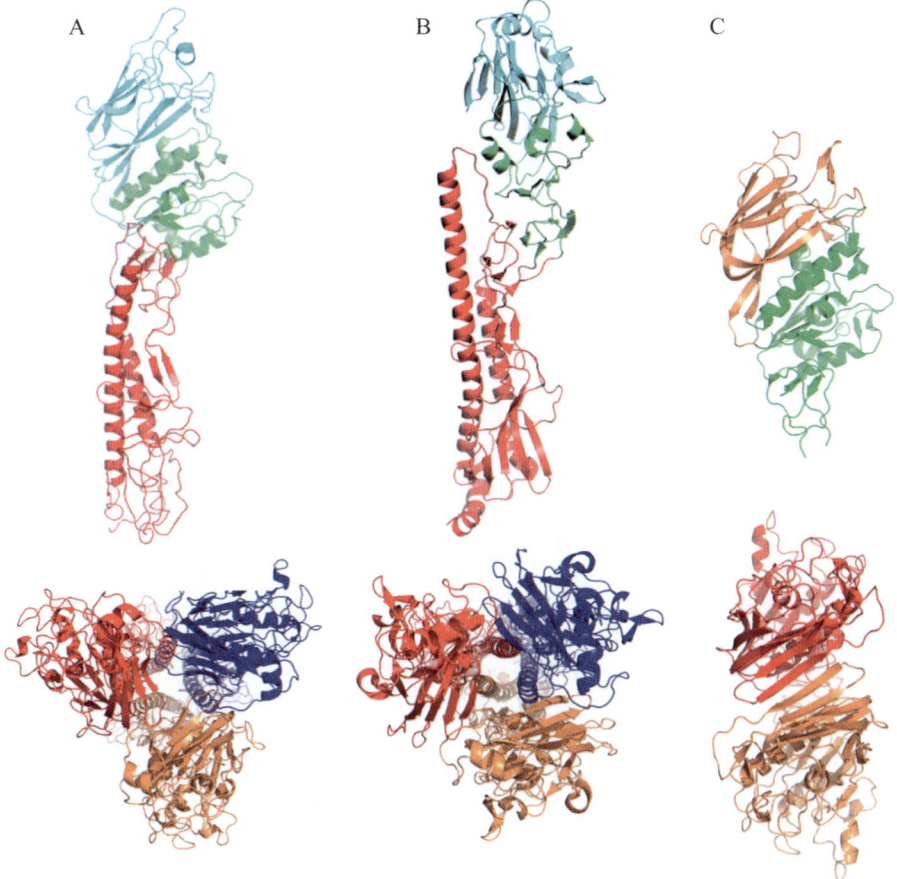

Figure 3 Domain shuffling played a major role in the evolution of respiratory virus glycoproteins. Recombination between ancestral receptor binding, esterase and fusion protein modules can be observed in the surface glycoproteins of many respiratory viruses. In the top row, the remnants of the receptor binding, esterase and fusion domains are colored in cyan, green and red, respectively, on the monomeric structures. In the bottom row, the same proteins are depicted in their native oligomeric form [trimeric for (A) and (C), dimeric for (B)], with coloring indicating each protomer of the oligomer. (A) The influenza C virus hemagglutinin–esterase-fusion (HEF) protein, which contains all three domains in functional forms; PDB code 1FLC. (B) The influenza A virus hemagglutinin protein, which was derived from an ancestral HEF protein like that in (A), has lost much of the esterase domain *via* a deletion [note the reduction in size and domain organization of the esterase module compared to (A) and (C)]; PDB code 1RD8. (C) The hemagglutinin–esterase (HE) protein from the bovine coronavirus, which closely resembles the HEF protein in (A), but lacks the helical fusion domain; PDB code 3CL5. Interestingly, a change in symmetry occurs between HE and HA/HEF. The former is twofold symmetric, whereas HAs and HEFs have a central threefold axis (bottom row).

assembly is anchored in the membrane *via* the transmembrane helices at the C-terminal ends of HA2.

A single receptor-binding site resides at the membrane distal end of each HA protomer, in HA1. The sialic acid receptor binds in a shallow, roughly triangular pocket lined by three secondary structural elements: the 130-loop, the 190-helix and the 220-loop.[53] The floor of the pocket is defined by one of the sheets that make up the HA1 β-sandwich. Sialic acids in N-linked glycans are found in two predominant forms that differ only in the linkage of the terminal sialic acid to the penultimate galactose: α(2,3) *versus* α(2,6). The crystal structures of several different HA variants bound to multiple receptor analogs have shed light on the mode and specificity of binding to glycan receptors.[19,53–58] First, despite the high degree of structural similarity between HAs of different subtypes, significant variation is observed in how each HA recognizes a given sialoglycan receptor. For example, structures of two human HAs that preferentially bind α(2,6) sialic acids over α(2,3) were determined in the presence of identical α(2,6) receptor analogs.[58] Whereas the terminal sialic acid moieties were very similarly positioned, the penultimate galactose residue was displaced by nearly 2 Å between the two structures. In the H1 HA structure, the galactose is bound lower in the receptor-binding pocket and makes additional hydrogen bonds with nearby amino acids. In contrast, the interaction with the same galactose in the H3 HA structure is dominated by hydrophobic contacts with a leucine residue not present in the H1 HA. Second, comparisons of human and avian viruses of the same subtype have led to the identification of key positions in the receptor binding site that determine specificity for α(2,3) *versus* α(2,6) sialoglycans.[54,59–64] These analyses support the notion that human pandemics arise when zoonotic influenza viruses gain the ability to use human-like, α(2,6) receptors.[65–69] For the human H1, H2 and H3 subtypes, as few as two mutations are sufficient to switch completely the receptor specificity of an HA from α(2,3) to α(2,6) or *vice versa*.[54,59,60,63,70] Perhaps not surprisingly, given the differing modes of receptor binding between subtypes, the specificity determining positions also differ from subtype to subtype. For example, positions 190 and 225 play the major role in the H1 subtype, whereas residues 226 and 228 are most important for H2 and H3. For other subtypes, such as H5, other combinations of residues seem to be involved, as introduction of mutations at the above positions only partially convert α(2,3)-specific H5 HAs to α(2,6) binders.[54,63,64,71]

HA2 carries the membrane fusion machinery that ultimately drives the translocation of the viral genome into the cytoplasm of a newly infected cell. After translation and translocation of the nascent polypeptide into the ER, the uncleaved HA0 chains fold to adopt the lowest energy conformation accessible at the time. Cleavage of HA0 to HA1 and HA2 allows the trimer to relax into an intermediate energy state and primes the complex for membrane fusion, as observed by comparing the similar, but subtly different, structures of the cleaved and uncleaved HA ectodomains.[55,72] However, the molecule is unable to relax into its most stable and lowest energy conformation – the post-fusion state.[73] This transition is suppressed in large part due to the steric constraints imposed by on HA2 by the receptor-binding domain, HA1. Upon exposure to low pH, these restraints are removed and HA2 undergoes major, irreversible rearrangements.[73–75] Consistent with this idea, HA2 expressed alone spontaneously folds into the post-fusion conformation, even at neutral pH.[76–78] These rearrangements expose the fusion peptide at the N-terminus of HA2, which was previously buried in the center of the coiled-coil helices, resulting in its insertion into a target membrane and its fusion with the viral envelope.

While the influenza B virus hemagglutinin is fairly similar to that of influenza A, the structure of the HEF protein from influenza C sheds light on the origins of this class of receptor binding protein.[79,80] The overall structure of the receptor binding and membrane fusion domains is largely conserved between the influenza A and C hemagglutinins. However, HEF differs by the insertion of an esterase domain in the membrane distal, HA1 (*cf.* Figure 3A *versus* 3B).[79] The esterase is structurally homologous to several non-viral esterases, suggesting that this pre-existing domain was incorporated into the viral glycoproteins.[79] Based upon the overall topology of the receptor-binding, esterase and fusion domains relative to one another, it seems that HEF was generated by at least two recombination events: the insertion of the receptor binding domain into the esterase domain and the insertion of the esterase domain into an ancestral fusion protein. The order of events is unclear. Many contemporary viruses are known to possess separate membrane fusion and receptor binding/ destroying proteins, yet no prominent examples of esterase-fusion proteins lacking receptor-binding activity are to be found in the literature. Hence it seems plausible that HEF-like proteins may have been generated by the insertion of a dual domain receptor-binding/esterase module into an existing fusion protein.[79] Comparison of the HA and HEF structures support the notion that domain insertion and deletion has driven the evolution of these proteins (Figure 3A–C).[42,79,80] The influenza A and B hemagglutinins lack any esterase activity and are missing much of the esterase domain observed in the HEF. However, clear remnants of a vestigial HEF esterase domain are present in the influenza A and B HA structures. The short helices and coils supporting the β-sandwich fold of the receptor-binding domain are derived from the core of the HEF esterase.[79] This structural equivalence then implies that the influenza A and B HAs are derived from a parental, HEF-like protein by the deletion of portions of the esterase domain.

A final twist on the recombination and reassortment of HEF-like proteins comes from recent work on a coronavirus receptor binding protein, hemagglutinin-esterase (HE).[81] In contrast to the HA and HEF proteins, the HE protein from bovine coronavirus contains no membrane fusion activity. The crystal structure reveals that this HE is wholly derived from an influenza C-like HEF precursor and comprises the complete membrane distal portion of HEF1/HA1, and also a substantial portion of the membrane proximal regions (Figure 3C). However, it completely lacks the membrane fusion domain normally formed from the C-terminal HEF2/HA2 portions of HEF and HA.[42,79,80] Remarkably, this rearrangement and deletion have converted the strictly trimeric HEF1/HA1 head groups into a twofold symmetric dimer.[81] This difference in oligomeric state results in the joining of the membrane distal faces of the β-sandwich receptor-binding domain to form an extended, eight-stranded sheet (Figure 3C). Subsequent evolution has led to a radical alteration of the mode of binding to *O*-acetylated sialic acid receptors. Most strikingly, the co-crystal structures with receptor analogs have revealed that the receptor binds the bovine coronavirus HE in almost completely the opposite orientation.[81] The entire receptor analog is rotated by nearly 150° compared with its position in HEF. In addition, the location of the sialic acid moiety is displaced by approximately 4 Å. From an evolutionary perspective, the existence of hemagglutinin-esterase proteins, such as coronavirus HE, confirms that the receptor-binding and esterase domains of this type are functional in the absence of the fusion domain and may have arisen before HEF. Thus, while the bovine coronavirus HE is believed to have been derived from an HEF on the basis of its membrane proximal regions, it is conceivable that an ancestral, HE-like protein was inserted into a membrane fusion protein to generate the first HEF.

Despite the substantial differences in the overall architectures between the HEF-type proteins and retroviral glycoproteins, some similarities can be found. Due to its medical relevance, the gp120/gp41 trimer from HIV-1 has been the main target of structural analysis from the *Retroviridae* family. While a high-resolution structure of the trimeric, prefusion gp120/gp41 remains elusive, crystal structures of a post-fusion gp41 core, and also the gp120 core in complex with the CD4 receptor and several neutralizing antibodies, have been determined (Figure 4A and B).[82–87] Through a combination of the crystal structures with several cryo-EM and cryo-tomographic reconstructions of gp120/gp41 trimers in solution and on the surface of virus particles, a picture of this envelope glycoprotein is emerging (Figure 4C and D).[88–91] Like influenza virus hemagglutinin, each Env protomer is synthesized as a single protein, in this case called gp160, which is subsequently glycosylated and cleaved by a furin protease. This proteolysis results in a larger, N-terminal receptor binding domain (gp120, analogous to HA1) and a smaller membrane-anchored, helical-stalk domain derived from the C-terminus that is implicated in membrane fusion (gp41, analogous to HA2). Based upon the apparent similarities between gp41, HA2 and other viral glycoproteins that use the zipping of trimeric coiled coils to drive membrane fusion, it has been noted that HIV *Env* may have been formed from the insertion of an independent receptor-binding domain into the gene of an ancient fusion protein.[79] This would imply that the first 62 and last 20 residues of gp120 are homologous to the extended N- and C-terminal segments from HA1, which were proposed to derive from the ancestral fusion protein on either side of the insertion of the receptor binding module. Overall, the structural and functional similarities suggest that the HEF and Env receptor binding proteins are probably distantly related members of a larger glycoprotein superfamily.

Although their name suggests otherwise, the hemagglutinin–neuraminidase proteins are structurally unrelated to HEF-type hemagglutinins. The HN/H/G family of receptor binding proteins is frequently divided into three groups based on their activity and receptor usage. HN proteins contain both hemagglutination and neuraminidase activities; H proteins act as hemagglutinins, but

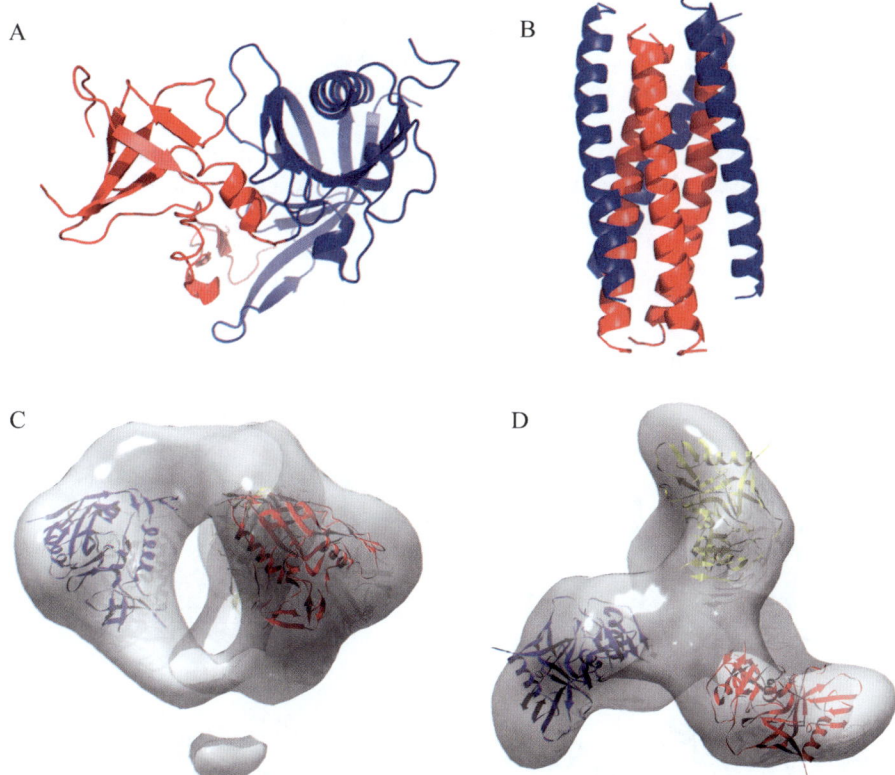

Figure 4   Structures of the HIV-1 envelope proteins gp120 and gp41. (A) Monomeric gp120 core. The inner and outer domains have been colored; PDB code 2NY7. (B) Trimeric helical core of gp41 in the post-fusion state; PDB code 1AIK. (C, D) Cryo-EM reconstructions of native HIV-1 env trimers. gp120 monomers from crystal structures (red, blue and yellow) have been fitted into the molecular envelope displayed in gray. (C) Side view of the trimer, with the location of the viral membrane and the gp41 stalk (not shown) at bottom. Note the loose association of gp120 monomers and the cavity at the center of the trimer, compared with the more compact structure of hemagglutinin (see Figure 3). (D) Top-down view of the top of the spike, along the threefold axis.

lack any receptor destroying activity, whereas the G-type proteins possess neither hemagglutinin nor neuraminidase functions. None of the HN/H/G proteins have any membrane fusion activity of their own. Instead, most viruses containing an HN/H/G receptor binding protein also contain a separate fusion protein called F. Unlike the pH-dependent fusogenic activity of the HEF-type proteins, fusion by F is pH independent. Through mechanisms that are still poorly understood, receptor binding by HN/H/G activates the F protein, presumably by a direct interaction between the two. All known members of the *Paramyxoviridae* possess an HN, H or G-type receptor binding protein.

In addition, several other virus families have been found to use proteins related to HN/H/G. Most notably, in addition to their hemagglutinin proteins, influenza A and B viruses possess neuraminidase proteins closely related to HN/H/G. In some cases, influenza A neuraminidases possess receptor-binding activity.[92,93]

The HN/H/G protein family members, along with neuraminidases from influenza viruses and many bacteria, share a distinctive β-propeller fold (Figure 5).[94–102] A monomer consists of six subdomains, each comprised of a

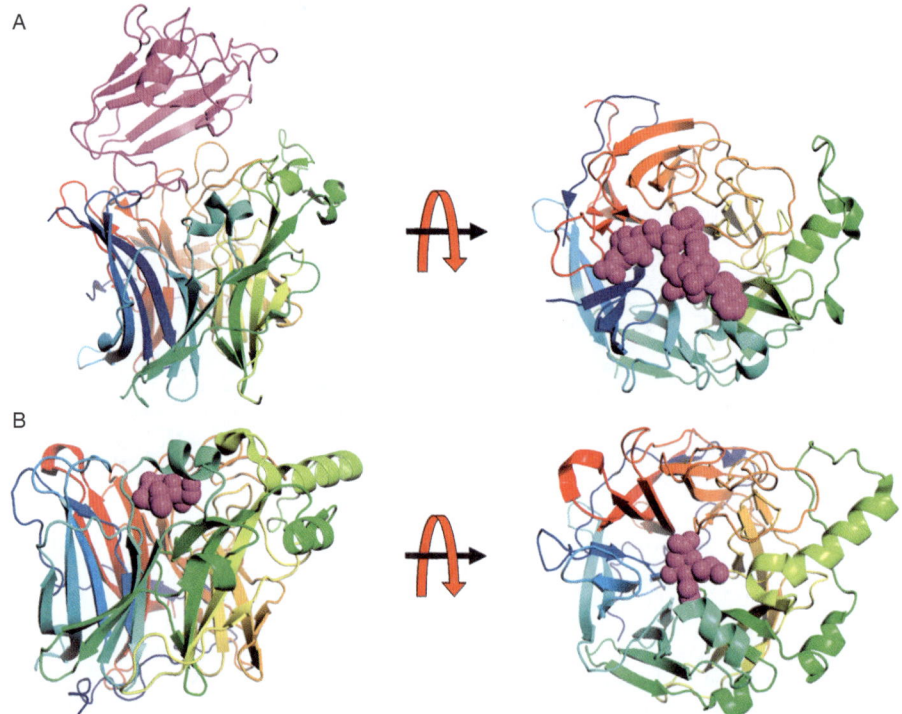

Figure 5   Receptor binding in H/HN/G proteins. (A) Interaction of the Hendra virus protein G with its ephrin-B2 receptor. Ephrin-B2 is colored in magenta, with the viral receptor binding domain rainbow colored by secondary structure progression. Left: side view of the HeV-G–Ephrin-B2 interaction; PDB code 2VSK. The viral envelope would be at bottom and the target cell at the top. Right: top-down view of the same structure. Most of Ephrin-B2 has been removed for clarity, except the loops involved in binding (magenta van der Waals spheres). (B) interaction of the Newcastle disease virus HN protein (NCDV-HN) with a sialo-glycan receptor; PDB code 1E8U. The glycan receptor is depicted in magenta with van der Waals spheres. The HN protein is colored and oriented approximately as in (A), above. Left: side view of the HN–sialo-glycan interaction, with the viral envelope at the bottom and the target cell at the top. Right: top-down view of the same structure. Note the similar modes of binding for protein and glycan receptors by HeV-G and NCDV-HN.

four-stranded anti-parallel β-sheet. The six sheets are arranged with pseudo sixfold rotational symmetry, with the edges of the N-terminal strand from each sheet running parallel and just adjacent to the symmetry axis. The sheets extend radially from the central axis with a 'right-handed' twist, giving the appearance of six propeller blades about a rotor. The monomers can oligomerize to form dimers and tetramers. There is some debate regarding the oligomeric state of native HN/H/G proteins on the virus surface. Many of the soluble ectodomains used for structural studies have been monomeric or dimeric, with little evidence of tetramers in solution or in the crystal.[99,102] In other cases, such as G from Hendra virus, similar constructs produce a mixture of tetramers, dimers and monomers.[96,97,103] Similarly, cross-linking studies on intact Sendai virus particles detected both tetramers and dimers of HN.[104] Thus, it is not yet clear whether different oligomeric states may be dominant in different viruses or if the relevant forms are labile and dissociate under some conditions. The orientation of the protomers in dimeric crystal structures can also vary substantially between different crystal forms of the same protein, suggesting a high degree of plasticity and flexibility at the oligomeric interface. It is worth noting that, although the tetrameric nature of the influenza virus neuraminidase has been firmly established by a variety of methods, including cryo-EM reconstruction of intact virus particles, the soluble ectodomains from several different isolates have been reported to be monomeric or dimeric in solution or in the crystal structure.[105–108] It has been suggested that the transmembrane segments and the linkers that tether the ectodomains to the viral envelope – both of which were removed from the soluble constructs – may also promote tetramerization *in vivo*.[109] Considering their ability to form tetramers in at least some situations, coupled with our knowledge of the related influenza neuraminidases, it seems likely that many, if not all, of the HN/H/G proteins are largely tetrameric on the virus surface.

In the HN proteins, the neuraminidase active site has been localized to one face of the β-propeller, on top of the sixfold axis. The co-crystal structures of several HN and H protein family members with substrate/receptor analogs have been solved.[96–99] The positioning of the sialic acid moiety and the overall architecture of the active site are very similar to those of other viral and bacterial neuraminidases.[110,111] Exactly how HN can balance its hemagglutination and neuraminidase activities to ensure secure attachment during entry and efficient release during budding is unclear. As a result, the location of the receptor-binding site and the possibility of secondary binding sites are matters of some debate. Several lines of evidence support the existence of a second receptor-binding site apart from the central catalytic site. First, some neuraminidases from influenza virus possess hemagglutination activity, even in the presence of inhibitors that essentially eliminate their sialidase activity.[93,112] Second, the co-crystal structure of the Newcastle disease virus HN with receptor analogs shows a well-defined sialyl moiety bound $\sim 15\,\text{Å}$ distant from the active site, making contacts with loops from the β5 and β6 propeller blades.[97] Sialic acid was bound in a similar location in an N9 influenza neuraminidase.[92] Finally, some mutagenesis and antibody inhibition studies of neuraminidase proteins have suggested that the receptor-binding and -destroying activities can be functionally separated, which implies that they are also spatially distinct.[112–115] However, much of the work carried out on neuraminidase and the NDV HN has not been reproduced in other HN or H proteins. It is possible that the second sialic acid binding site may be a unique feature of some neuraminidases and HN proteins or an artifact altogether. In contrast, the receptor-binding and -destroying activities may both reside in the active site pocket. Several crystal structures of HN and H proteins have found receptor analogs bound only in the active site pocket (or the analogous site in H).[96,98,99] In some structures, the protomer–protomer interface is altered and key catalytic residues in some of the receptor-bound HN crystal structures are displaced away from the substrate, suggestive of two distinct states: one for receptor binding and another for sialic acid hydrolysis.[98,99] In other cases, however, there are no obvious structural changes upon ligand binding.[96] Hence further studies are required in order to confirm or eliminate definitively the general possibility of a second receptor binding site.

Unlike the HN and H proteins, G-type proteins use protein receptors. The first structures of G-type proteins in complex with their cognate host cell receptors have only just become available.[100,101] The closely related receptor binding proteins NiV-G from Nipah virus and HeV-G from Hendra virus were solved in complex with ephrin-B2 and -B3. Interestingly, the ephrins bind NiV-G and HeV-G on one face of the β-propeller, near the pseudo sixfold axis (Figure 5B). This scenario contrasts with the proposed binding site for the measles virus receptor, SLAM, on the side of the viral glycoprotein.[102] It is also interesting that the G–H loop from the ephrin receptors inserts into the central pocket of NiV-G and HeV-G that corresponds to the receptor binding/neuraminidase active site in the HN and H proteins (Figure 5D).[100,101] Despite radically different cellular receptors, there is some degree of conserved binding modes.

4   Host Range, Tissue Tropism and Transmission

The specificity of a viral RBD and the distribution of its cognate receptor play a major role in determining the host range and tissue tropism of a particular virus. As an obligate intracellular parasite, viral replication cannot commence until the genome has been translocated across the plasma membrane into the host cell. Therefore, only cells that express sufficient levels of the target receptor will be susceptible to infection. At the organism level, failure to express the appropriate receptor will most often result in complete resistance to infection by the virus. Similarly, an organism that expresses a receptor homolog with substantial sequence variation over the receptor–RBD binding interface is likely to escape infection through reduced or abolished attachment of viral particles to the cell surface. Thus, it is commonly observed that viruses using proteinaceous receptors generally have a narrower host range than viruses that use more ubiquitous receptors, such as sialic acids. Despite the expression of homologous proteins on the cell surface, there is usually significant divergence between the protein sequences of all but the most closely related species. For example, poliovirus infects humans and primates and infection of the central nervous system can lead to paralysis and death. In contrast, mice and many other higher mammals are unaffected by the virus and murine cell lines do not support multicycle virus replication.[116] Mice encode a poliovirus receptor (PVR) homolog, but the protein is unable to facilitate attachment and entry. However, ectopic expression of the human PVR in transgenic mice renders them permissive to the virus and infection leads to neurological symptoms similar to those observed in humans.[117] These data demonstrate that receptor engagement restricts the poliovirus host range, at least in mice.

The interaction of the RBD from the SARS coronavirus (SARS-CoV) presents an example that highlights how just one or two amino acid differences in the receptor or the RBD can enhance or abolish host cell binding. Between late 2002 and July 2003, the CDC estimated that 8098 people were infected by the SARS-CoV worldwide, leading to 774 deaths.[118] Through a remarkable, international effort, the causative agent was identified and its genome sequenced within a few months.[118–121] Within one year of the first reported cases, the cellular receptor, angiotensin-converting enzyme 2 (ACE2), had been identified.[28] Comparison of the genome sequences of SARS-CoV and several zoonotic coronavirus isolates from a live animal market in Shenzhen, China, revealed that the human and animal viruses were 99.8% identical.[122] Most of the differences reside on the major surface glycoprotein, S, which is responsible for receptor binding and membrane fusion. The co-crystal structure of the RBD of S in complex with the ACE2 receptor provided a glimpse at how SARS-CoV was able to cross the species barrier into humans.[123] The RBD consists of a core domain and a long loop region. The core is composed of a five-stranded, antiparallel β-sheet and three short connecting helices. Aside from two short β-strands that pair separately from the sheet of the core, the bulk of the loop segment takes the form of a random coil. The extended loop packs across one side of the core domain, creating a concave surface that cradles the ACE2 receptor. All of the contacts between the RBD and ACE2 are mediated by this loop region. Of the four amino acid differences between the human and civet RBDs, two of the residues are positioned away from the ACE2 interaction surface and were shown to have little impact on the affinity of the RBD for human ACE2 (residues 334 and 360).[123,124] In contrast, the two other residues differing between the human and civet viruses are located on the receptor binding and make direct contact with human ACE2 (residues 479 and 487).[123] Mutation of the human SARS-CoV RBD at either of these latter two positions to the amino acid present in the civet protein significantly reduces affinity for human ACE2, while increasing affinity for civet ACE2.[124] Conversely, introduction of the 'human-like' amino acids at the same positions in civet SARS-CoV isolates greatly increased the ability of the civet virus to infect cells expressing human ACE2. Hence it appears that, in this case also, only two amino acid changes were sufficient to allow an animal coronavirus to use a human receptor for attachment and entry.

Looking at the SARS RBD–ACE2 interaction from the standpoint of variation between the ACE2 receptors from different animal species, a similar story emerges. Just as only two mutations were required to increase binding of the civet RBD to human ACE2, only a few amino acid changes in the receptor itself may restrict SARS replication to particular hosts. For example, SARS-CoV binds tightly to human ACE2, more weakly to mouse ACE2, and binding to rat ACE2 is barely detectable.[125] These observations are consistent with the robust replication of the virus in humans and mice and its poor replication in rats. Of the differences in the proteins from these three species, only a few amino acid substitutions map to the surface of ACE2 that interacts with the SARS RBD and seem likely to affect binding affinity.[123] Most notably, rat ACE2 contains an N-linked glycosylation site in a position that contacts the RBD in the crystal structure with human ACE2.[124] This additional glycosylation probably prevents binding of rat ACE2 by SARS-CoV by steric interference. The shielding of the ACE2 surface by a glycosylation is reminiscent of the use of glycosylation to mask and unmask epitopes in the HIV and influenza glycoproteins to escape neutralizing antibodies. It is unclear whether the glycosylation site in the rat ACE2 was retained during evolution due to continual exposure to ACE2-tropic coronaviruses, but several examples of host adaptation to persistent viral infection have been reported in the literature.[126–128]

In contrast to viruses that bind proteinaceous receptors, the use of sialic acid receptors by many respiratory viruses seems to contribute to their rather promiscuous replication and broad host range. Perhaps most noteworthy and best studied in this regard is the influenza A virus. Due to the wide distribution of sialic acids through the mucosal surfaces of most higher animals – particularly the respiratory and digestive tracts – influenza A viruses can infect a wide range of hosts, including birds, pigs, horses, rodents and humans. Through co-evolution of influenza viruses with their preferred host, the RBD of the attachment protein hemagglutinin has a finely tuned specificity for either α(2,3) or α(2,6) sialylated glycans.[129] Some dual-tropic isolates have been shown to recognize both forms of the receptor. As α(2,3) and α(2,6) sialylated glycans are often expressed in different tissues in different organisms, the receptor specificity of the hemagglutinin from a given isolate plays a major role in determining the host range. For example, avian influenza virus isolates tend to replicate in the digestive tract where α(2,3) sialic acids dominate.[130] In contrast, human viruses primarily infect the upper respiratory tract, which is rich in α(2,6) sialic acids. As a result, the hemagglutinins from avian viruses tend to be α(2,3)-specific, whereas those from human viruses are α(2,6)-specific.[129] For the most part, the differing receptor specificity between human and avian influenza viruses is sufficient to keep avian viruses from spreading through the human population. However, the hemagglutinins from the viruses that caused the 1918, 1957 and 1968 influenza pandemics are strikingly similar to those from the avian viruses that were circulating at the same time.[66–69] Hence it seems fairly likely that two (1957, 1968) and possibly all three of these pandemics were triggered by transfer of the avian influenza virus HA gene into circulating human viruses for which the avian HA protein had acquired, or quickly thereafter acquired, the ability to use human-like α(2,6) receptors to enable human-to-human transmission.

Inhomogeneous expression of receptors at the tissue level restricts the replication of many viruses to a subsection of the host. Rabies virus is the prototypical member of the *Rhabdoviridae*. While primarily affecting mammals such as bats, dogs and raccoons, humans are occasionally infected by the bite of an infected animal. The rabies virus receptor is a ligand-gated ion channel, the acetylcholine receptor.[30] Acetylcholine receptors are only expressed on electrically excitable cell types, such as muscle fibers and sensory neurons. As a result, replication of the rabies virus occurs almost exclusively in muscle and neural tissues. The bite of an infected animal inoculates the virus into the muscle, where replication begins. The infection slowly spreads up the muscle fibers to neuromuscular junctions, where the virus can be transmitted to motor neurons. Infected neurons then traffic the virus up the axon towards the cell body in the CNS, where it can rapidly spread to other excitable cells. Thus, a clear link is defined between usage of the acetylcholine receptor, neurotropism and the neurological symptoms associated with rabies infection.

In the case of HIV-1, the expression of CD4 and CCR5 in a limited proportion of human cells restricts viral replication to only a few cell types. Both of these molecules are involved in signal transduction in the immune system and are, therefore, primarily expressed on cells, such as monocytes, macrophages and T-cells. HIV-1 replicates preferentially in these cell types, leading to the eventual decline of CD4-positive T-cells and immunological exhaustion.

Host range and tissue tropism, however, are not always wholly independent. Although many zoonotic influenza viruses can replicate in humans, they are generally not well adapted to human hosts and usually fail to spread past the index case. Despite hundreds of human infections with H5N1 avian influenza viruses, only a handful of cases have been reported where human-to-human transmission has occurred.[131] Similarly, of 89 human cases of H7N7 infection in The Netherlands, only three instances of human-to-human transmission were identified.[132] Whereas humans have primarily α(2,6) sialic acids in the upper airway and would be expected to be immune to α(2,3)-specific avian viruses, α(2,3) sialic acids predominate deeper in the lungs.[133] Consequently, at least some α(2,3)-tropic avian viruses are capable of infecting humans. Most notably, the H5N1 viruses that originated in China in the mid-1990s replicate efficiently in the lower respiratory tract in humans and cause high morbidity and mortality. Recent work indicates that the replication of α(2,3)-specific avian viruses in the lower respiratory tract may account for the low transmissibility of avian H5N1 viruses. Since human influenza viruses are transmitted *via* aerosols, replication of human influenza viruses in the upper airway is thought to be important for the generation of airborne droplets through coughing and sneezing. However, the ectopic replication of avian H5N1 viruses in the α(2,3)-rich lungs may reduce the efficient production of aerosols, despite robust viral replication and high titers in the lungs.[133] The revival of several isolates of the 1918 pandemic virus from archived samples using reverse genetics has provided strong evidence for this link between receptor specificity and transmissibility.[70] Three clones of the 1918 virus that differ by merely one and two amino acid substitutions in the hemagglutinin RBD radically alter receptor specificity and transmission. A human, $\alpha(2,6)$-specific isolate, A/South Carolina/1/1918, is efficiently transmitted in a ferret model. A second human isolate with a single amino acid substitution exhibits mixed $\alpha(2,3)/\alpha(2,6)$ binding and reduced transmissibility. Finally, one additional substitution that resulted in $\alpha(2,3)$-specific receptor usage completely abolished transmission in the ferret model. These results suggest that the $\alpha(2,3)$ receptor usage of avian H5N1 viruses is responsible for the lack of transmission of the virus among humans. Moreover, these three isolates give us an approximate idea of the potential evolutionary trajectory that the 1918 pandemic virus followed – from an avian virus, through an intermediate, to a fully fledged human virus – as it adapted to human hosts.

However, it is clear that other factors also play a role in determining host range and tissue tropism. For example, HIV-1 and simian immunodeficiency virus (SIV) are two closely related viruses that both use CD4 and CCR5 as their primary receptors. HIV-1 only infects humans and SIV only infects macaques. However, hybrid viruses that encode primarily SIV internal proteins and the HIV-1 Env glycoproteins replicate efficiently in macaques, suggesting that, in this case, replication of HIV-1 in primates may be restricted by events occurring after entry.[134] Similarly, the restricted tissue tropism of the Sabin poliovirus vaccine strain that results in reduced replication in the central nervous system seems to be a consequence of the low expression levels of an essential host factor, nPTB, in neural tissues.[135] Thus, although receptor binding is a major determinant, it is not the only factor involved in restricting host range. In influenza A viruses, one of the polymerases, PB1, is also always transferred in the reassortment process between an avian and human virus when new human pandemics arise.

5   Summary

Viruses employ a wide variety of strategies to anchor themselves to the surface of a target cell and transport their genomic material to the cellular interior. Initial contact may be mediated by non-specific interactions, followed by the engagement of the primary receptor and co-receptors. The recognition and binding to host cell receptors is carried out by specific proteins on the surface of the virus particle. In addition to stabilizing the association between the virus and the target cell, receptor binding can also trigger endocytosis of virion and/ or activate the membrane fusion machinery. How viral glycoproteins drive the fusion of cellular membrane with the viral envelope will be discussed next, in Chapter 13.

References

1. M. M. Poranen, R. Daugelavicius and D. H. Bamford, *Annu. Rev. Microbiol.*, 2002, **56**, 521.

2. J. G. Shaw, *Philos. Trans. R. Soc. London B*, 1999, **354**, 603.
3. M. J. Rust, M. Lakadamyali, F. Zhang and X. Zhuang, *Nat. Struct. Mol. Biol.*, 2004, **11**, 567.
4. L. G. Wu, *Trends Neurosci.*, 2004, **27**, 548.
5. B. M. Curtis, S. Scharnowske and A. J. Watson, *Proc. Natl. Acad. Sci. USA*, 1992, **89**, 8356.
6. T. B. Geijtenbeek *et al.*, *Cell*, 2000, **100**, 587.
7. G. Campadelli-Fiume, F. Cocchi, L. Menotti and M. Lopez, *Rev. Med. Virol.*, 2000, **10**, 305.
8. J. Liu and S. C. Thorp, *Med. Res. Rev.*, 2002, **22**, 1.
9. E. Biziagos, J. Passagot, J. M. Crance and R. Deloince, *Appl. Environ. Microbiol.*, 1988, **54**, 2705.
10. M. El-Far *et al.*, *Curr. HIV/AIDS Rep.*, 2008, **5**, 13.
11. D. C. Hooper, *J. Neurovirol.*, 2005, **11**, 88.
12. M. Lafon, *Curr. Top. Microbiol. Immunol.*, 2005, **289**, 239.
13. M. Lafon, *Dev. Biol. (Basel)*, 2008, **131**, 413.
14. S. B. Sieczkarski and G. R. Whittaker, *Curr. Top. Microbiol. Immunol.*, 2005, **285**, 1.
15. E. E. Fry *et al.*, *EMBO J.*, 1999, **18**, 543.
16. T. J. Pritchett, R. Brossmer, U. Rose and J. C. Paulson, *Virology*, 1987, **160**, 502.
17. S. Kelm *et al.*, *Eur. J. Biochem.*, 1992, **205**, 147.
18. N. K. Sauter *et al.*, *Biochemistry*, 1989, **28**, 8388.
19. N. K. Sauter *et al.*, *Biochemistry*, 1992, **31**, 9609.
20. E. Lycke, M. Johansson, B. Svennerholm and U. Lindahl, *J. Gen. Virol.*, 1991, **72**, 1131.
21. J. D. Pinon *et al.*, *J. Virol.*, 2003, **77**, 9922.
22. K. S. Jones, C. Petrow-Sadowski, D. C. Bertolette, Y. Huang and F. W. Ruscetti, *J. Virol.*, 2005, **79**, 12692.
23. A. de Parseval and J. H. Elder, *J. Virol.*, 2001, **75**, 4528.
24. M. T. Shieh and P. G. Spear, *J. Virol.*, 1994, **68**, 1224.
25. A. A. Lindberg, *Annu. Rev. Microbiol.*, 1973, **27**, 205.
26. T. Schirmer, T. A. Keller, Y. F. Wang and J. P. Rosenbusch, *Science*, 1995, **267**, 512.
27. L. Randall-Hazelbauer and M. Schwartz, *J. Bacteriol.*, 1973, **116**, 1436.
28. W. Li *et al.*, *Nature*, 2003, **426**, 450.
29. P. Freimuth, L. Philipson and S. D. Carson, *Curr. Top. Microbiol. Immunol.*, 2008, **323**, 67.
30. T. L. Lentz, T. G. Burrage, A. L. Smith, J. Crick and G. H. Tignor, *Science*, 1982, **215**, 182.
31. A. G. Dalgleish *et al.*, *Nature*, 1985, **312**, 763.
32. D. Klatzmann *et al.*, *Nature*, 1985, **312**, 767.
33. T. Dragic *et al.*, *Nature*, 1996, **381**, 667.
34. H. Choe *et al.*, *Cell*, 1996, **85**, 1135.
35. B. J. Doranz *et al.*, *Cell*, 1996, **85**, 1149.
36. E. Oberlin *et al.*, *Nature*, 1996, **382**, 833.

37. K. Levesque, A. Finzi, J. Binette and E. A. Cohen, *Curr. HIV Res.*, 2004, **2**, 51.
38. S. Iwashita and S. Kanegasaki, *Biochem. Biophys. Res. Commun.*, 1973, **55**, 403.
39. S. Steinbacher *et al.*, *Proc. Natl. Acad. Sci. USA*, 1996, **93**, 10584.
40. Y. He *et al.*, *Proc. Natl. Acad. Sci. USA*, 2000, **97**, 79.
41. D. M. Belnap *et al*, *Proc. Natl. Acad. Sci. USA*, 2000, **97**, 73.
42. I. A. Wilson, J. J. Skehel and D. C. Wiley, *Nature*, 1981, **289**, 366.
43. *Bull. World Health Org.*, **58**, 1980, 585.
44. V. S. Hinshaw *et al.*, *J. Virol.*, 1982, **42**, 865.
45. Y. Kawaoka, S. Yamnikova, T. M. Chambers, D. K. Lvov and R. G. Webster, *Virology*, 1990, **179**, 759.
46. C. Rohm, N. Zhou, J. Suss, J. Mackenzie and R. G. Webster, *Virology*, 1996, **217**, 508.
47. R. A. Fouchier *et al.*, *J. Virol.*, 2005, **79**, 2814.
48. M. F. Schmidt, *Virology*, 1982, **116**, 327.
49. C. W. Naeve and D. Williams, *EMBO J.*, 1990, **9**, 3857.
50. R. Wagner, A. Herwig, N. Azzouz and H. D. Klenk, *J. Virol.*, 2005, **79**, 6449.
51. C. Fischer, B. Schroth-Diez, A. Herrmann, W. Garten and H. D. Klenk, *Virology*, 1998, **248**, 284.
52. T. Sakai, R. Ohuchi and M. Ohuchi, *J. Virol.*, 2002, **76**, 4603.
53. W. Weis *et al.*, *Nature*, 1988, **333**, 426.
54. J. Stevens *et al.*, *Science*, 2006, **312**, 404.
55. J. Stevens *et al.*, *Science*, 2004, **303**, 1866.
56. W. I. Weis, A. T. Brunger, J. J. Skehel and D. C. Wiley, *J. Mol. Biol.*, 1990, **212**, 737.
57. D. Fleury *et al.*, *Nat. Struct. Biol.*, 1999, **6**, 530.
58. S. J. Gamblin *et al.*, *Science*, 2004, **303**, 1838.
59. G. N. Rogers *et al.*, *Nature*, 1983, **304**, 76.
60. R. J. Connor, Y. Kawaoka, R. G. Webster and J. C. Paulson, *Virology*, 1994, **205**, 17.
61. M. Matrosovich *et al.*, *J. Virol.*, 2000, **74**, 8502.
62. E. Nobusawa, H. Ishihara, T. Morishita, K. Sato and K. Nakajima, *Virology*, 2000, **278**, 587.
63. J. Stevens *et al.*, *J. Mol. Biol.*, 2006, **355**, 1143.
64. J. Stevens *et al.*, *J. Mol. Biol.*, 2008, **381**, 1382.
65. J. K. Taubenberger, A. H. Reid, A. E. Krafft, K. E. Bijwaard and T. G. Fanning, *Science*, 1997, **275**, 1793.
66. A. H. Reid, T. G. Fanning, J. V. Hultin and J. K. Taubenberger, *Proc. Natl. Acad. Sci. USA*, 1999, **96**, 1651.
67. Y. Kawaoka, S. Krauss and R. G. Webster, *J. Virol.*, 1989, **63**, 4603.
68. W. J. Bean *et al.*, *J. Virol.*, 1992, **66**, 1129.
69. J. R. Schafer *et al.*, *Virology*, 1993, **194**, 781.
70. T. M. Tumpey *et al.*, *Science*, 2007, **315**, 655.
71. S. Yamada *et al.*, *Nature*, 2006, **444**, 378.

72. J. Chen *et al.*, *Cell*, 1998, **95**, 409.
73. P. A. Bullough, F. M. Hughson, J. J. Skehel and D. C. Wiley, *Nature*, 1994, **371**, 37.
74. J. J. Skehel *et al.*, *Proc. Natl. Acad. Sci. USA*, 1982, **79**, 968.
75. J. W. Yewdell, W. Gerhard and T. Bachi, *J. Virol.*, 1983, **48**, 239.
76. J. Chen *et al.*, *Proc. Natl. Acad. Sci. USA*, 1995, **92**, 12205.
77. J. Chen, J. J. Skehel and D. C. Wiley, *Biochemistry*, 1998, **37**, 13643.
78. S. E. Swalley *et al.*, *Biochemistry*, 2004, **43**, 5902.
79. P. B. Rosenthal *et al.*, *Nature*, 1998, **396**, 92.
80. Q. Wang, X. Tian, X. Chen and J. Ma, *Proc. Natl. Acad. Sci. USA*, 2007, **104**, 16874.
81. Q. Zeng, M. A. Langereis, A. L. van Vliet, E. G. Huizinga and R. J. de Groot, *Proc. Natl. Acad. Sci. USA*, 2008, **105**, 9065.
82. D. C. Chan, D. Fass, J. M. Berger and P. S. Kim, *Cell*, 1997, **89**, 263.
83. C. C. Huang *et al.*, *Science*, 2007, **317**, 1930.
84. C. C. Huang *et al.*, *Proc. Natl. Acad. Sci. USA*, 2004, **101**, 2706.
85. T. Zhou *et al.*, *Nature*, 2007, **445**, 732.
86. C. C. Huang *et al.*, *Science*, 2005, **310**, 1025.
87. P. D. Kwong *et al.*, *Nature*, 1998, **393**, 648.
88. J. Liu, A. Bartesaghi, M. J. Borgnia, G. Sapiro and S. Subramaniam, *Nature*, 2008, **455**, 109.
89. G. Zanetti, J. A. Briggs, K. Grunewald, Q. J. Sattentau and S. D. Fuller, *PLoS Pathog.*, 2006, **2**, e83.
90. P. Zhu *et al.*, *Proc. Natl. Acad. Sci. USA*, 2003, **100**, 15812.
91. P. Zhu *et al.*, *Nature*, 2006, **441**, 847.
92. J. N. Varghese *et al.*, *Proc. Natl. Acad. Sci. USA*, 1997, **94**, 11808.
93. J. Hausmann, E. Kretzschmar, W. Garten and H. D. Klenk, *J. Gen. Virol.*, 1995, **76**, 1719.
94. J. N. Varghese, W. G. Laver and P. M. Colman, *Nature*, 1983, **303**, 35.
95. T. A. Bowden *et al.*, *J. Virol.*, 2008, **82**, 11628.
96. P. Yuan *et al.*, *Structure*, 2005, **13**, 803.
97. V. Zaitsev *et al.*, *J. Virol.*, 2004, **78**, 3733.
98. M. C. Lawrence *et al.*, *J. Mol. Biol.*, 2004, **335**, 1343.
99. S. Crennell, T. Takimoto, A. Portner and G. Taylor, *Nat. Struct. Biol.*, 2000, **7**, 1068.
100. K. Xu *et al.*, *Proc. Natl. Acad. Sci. USA*, 2008, **105**, 9953.
101. T. A. Bowden *et al.*, *Nat. Struct. Mol. Biol.*, 2008, **15**, 567.
102. R. W. Ruigrok and D. Gerlier, *Proc. Natl. Acad. Sci. USA*, 2007, **104**, 20639.
103. K. N. Bossart *et al.*, *J. Virol.*, 2005, **79**, 6690.
104. M. A. Markwell and C. F. Fox, *J. Virol.*, 1980, **33**, 152.
105. A. Harris *et al.*, *Proc. Natl. Acad. Sci. USA*, 2006, **103**, 19123.
106. X. Xu, X. Zhu, R. A. Dwek, J. Stevens and I. A. Wilson, *J. Virol.*, 2008, **82**, 10493.
107. T. Deroo, W. M. Jou and W. Fiers, *Vaccine*, 1996, **14**, 561.
108. J. M. Colacino *et al.*, *Virology*, 1997, **236**, 66.

109. A. Kundu, M. A. Jabbar and D. P. Nayak, *Mol. Cell. Biol.*, 1991, **11**, 2675.

110. P. M. Colman, J. N. Varghese and W. G. Laver, *Nature*, 1983, **303**, 41.

111. S. J. Crennell, E. F. Garman, W. G. Laver, E. R. Vimr and G. L. Taylor, *Proc. Natl. Acad. Sci. USA*, 1993, **90**, 9852.

112. W. G. Laver, P. M. Colman, R. G. Webster, V. S. Hinshaw and G. M. Air, *Virology*, 1984, **137**, 314.

113. R. G. Webster *et al.*, *J. Virol.*, 1987, **61**, 2910.

114. G. M. Air and W. G. Laver, *Virology*, 1995, **211**, 278.

115. J. M. Nuss and G. M. Air, *Virology*, 1991, **183**, 496.

116. J. J. Holland, L. C. McLaren and J. T. Syverton, *Proc. Soc. Exp. Biol. Med.*, 1959, **100**, 843.

117. R. B. Ren, F. Costantini, E. J. Gorgacz, J. J. Lee and V. R. Racaniello, *Cell*, 1990, **63**, 353.

118. M. A. Marra *et al.*, *Science*, 2003, **300**, 1399.

119. C. Drosten *et al.*, *N. Engl. J. Med.*, 2003, **348**, 1967.

120. T. G. Ksiazek *et al.*, *N. Engl. J. Med.*, 2003, **348**, 1953.

121. J. S. Peiris *et al.*, *Lancet*, 2003, **361**, 1319.

122. Y. Guan *et al.*, *Science*, 2003, **302**, 276.

123. F. Li, W. Li, M. Farzan and S. C. Harrison, *Science*, 2005, **309**, 1864.

124. W. Li *et al.*, *EMBO J.*, 2005, **24**, 1634.

125. W. Li *et al.*, *J. Virol.*, 2004, **78**, 11429.

126. D. M. Sayah, E. Sokolskaja, L. Berthoux and J. Luban, *Nature*, 2004, **430**, 569.

127. R. Barrangou *et al.*, *Science*, 2007, **315**, 1709.

128. B. Mangeat *et al.*, *Nature*, 2003, **424**, 99.

129. G. N. Rogers and J. C. Paulson, *Virology*, 1983, **127**, 361.

130. R. G. Webster, M. Yakhno, V. S. Hinshaw, W. J. Bean and K. G. Murti, *Virology*, 1978, **84**, 268.

131. H. Wang *et al.*, *Lancet*, 2008, **371**, 1427.

132. R. A. Fouchier *et al.*, *Proc. Natl. Acad. Sci. USA*, 2004, **101**, 1356.

133. K. Shinya *et al.*, *Nature*, 2006, **440**, 435.

134. J. Li, C. I. Lord, W. Haseltine, N. L. Letvin and J. Sodroski, *J. Acquir. Immune Defic. Syndr.*, 1992, **5**, 639.

135. S. Guest, E. Pilipenko, K. Sharma, K. Chumakov and R. P. Roos, *J. Virol.*, 2004, **78**, 11097.

CHAPTER 13

Attachment and Entry: Viral Cell Fusion

RACHEL M. SCHOWALTER,[a] EVERETT C. SMITH[b] AND REBECCA ELLIS DUTCH[b]

[a] Laboratory of Cellular Oncology, Center for Cancer Research, National Cancer Institute, National Institutes of Health, Bethesda, MD 20892, USA; [b] Department of Molecular and Cellular Biochemistry, University of Kentucky College of Medicine, Lexington, KY 40536, USA

1 Introduction

Enveloped viruses contain integral membrane proteins which facilitate entry into new host cells by mediating fusion of the viral and cellular membranes. Virus–cell membrane fusion results in the deposition of the viral genome into the host cell cytoplasm, an essential early step in infection. Most enveloped viruses have a single protein, referred to here as the fusion protein, which is primarily responsible for promoting membrane fusion. However, in some cases a second protein, or group of proteins, has been shown to be necessary to assist the fusion protein in folding, receptor binding or regulation of activity (reviewed[1-4]). All fusion proteins examined to date are capable of undergoing dramatic conformational changes and the energy released upon transition to a more stable state is harvested to drive membrane merger.[5]

A number of structural studies on fusion proteins from a variety of viruses have recently been performed, providing important insights into the mechanism of fusion protein function. High-resolution structural determinations of both the pre- and post-fusion conformations permit the modeling of the membrane fusion mechanism and have verified hypotheses based on biochemical findings. Interestingly, the high-resolution structures have revealed that proteins which

RSC Biomolecular Sciences No. 21
Structural Virology
Edited by Mavis Agbandje-McKenna and Robert McKenna
© Royal Society of Chemistry 2011
Published by the Royal Society of Chemistry, www.rsc.org

differ immensely in amino acid sequence and initial oligomerization state may cause membrane fusion through remarkably similar mechanisms. Three structural classifications of fusion proteins have been defined, with the new type III proteins containing features reminiscent of both the type I and type II proteins. However, despite many structural differences, common themes and functional similarities have emerged. Following dramatic conformational changes, a hydrophobic domain is inserted into a target membrane, forming a bridge between viral and cellular membranes. Next, the bridge folds on itself, forcing the anchoring membranes into close contact. In the post-fusion conformation, fusion proteins from all three types are trimeric and the viral membrane-spanning domain of the protein, although not visible in the X-ray structures, appears to be in close proximity to the region of the protein that inserts into target membranes.[6–11]

In addition to providing key insights into the mechanism by which fusion proteins physically promote the merger of membranes, structural data have furthered our understanding of how the function of these proteins is regulated. The majority of fusion proteins become trapped in the post-fusion conformation once it has been attained, as this conformation is considerably more stable than the pre-fusion conformation.[12,13] The pre-fusion conformation of type I and type II fusion proteins is considered to be metastable and a triggering event is needed to destabilize the metastable state leading to the extensive structural changes that promote fusion.[14,15] Triggers of fusion are generally either receptor binding, low pH or a combination of the two (reviewed[16]). Fusion proteins must be protected from premature triggering, as this would inactivate the protein, thus preventing infection.[14,17,18] This critical regulation of triggering is often achieved with the help of a second protein and/or *via* proteolytic processing.[19–21] Intriguing structural data have also shed light on the mechanisms of fusion protein triggering and regulation.

2 Promotion of Membrane Fusion by Viral Fusion Proteins

All viral fusion proteins characterized to date promote fusion by inducing the deformation and disruption of opposing membrane bilayers, leading first to hemifusion, a state in which only the two outer layers of the lipid bilayers merge.[5,22] Further changes then result in the formation of a small lipidic pore where the two individual lipid bilayers have completely merged into one. However, this small pore may collapse unless further energy is expended to allow expansion of the pore. In fact, pore expansion has been suggested to be the most energy-demanding step in membrane fusion.[5] For each of the viral fusion proteins for which detailed structural data are available, the post-fusion conformation of the protein exhibits increased stability compared with the pre-fusion conformation, as suggested by an increase in buried surface area and the formation of highly stable amino acid contacts.[6–11] Hence, the post-fusion conformation is expected to be of lower energy than the pre-fusion conformation of the protein and the energy made available through the conformational change is likely utilized to overcome the energy barriers to membrane fusion.

Structural data indicate that all three types of viral fusion proteins promote fusion as C-terminally membrane-anchored homotrimers. Each type of fusion protein also contains a hydrophobic domain, in addition to the transmembrane (TM) domain, that can interact with and partially insert into membranes. This domain is usually referred to as a 'fusion peptide' or 'fusion loop'. The bridging of membranes which results from fusion domain insertion into a target membrane sets the stage for the subsequent critical step of hairpin formation. The portion of the protein adjacent to the viral membrane is refolded or repositioned nearer to the target membrane, resulting in a folded back conformation of the protein. Since fusion proteins at this stage are trimeric and each subunit forms a hairpin, this structure is generally referred to as a 'trimer of hairpins'. This refolding event is intimately tied to membrane fusion, as it brings the two membrane-anchored domains, and by extension their associated membranes, into close proximity. While sharing these similarities in mechanism, type I and type II viral fusion proteins display no conserved structural features. Type III proteins share defining features of both type I and type II proteins in addition to displaying unique features. In this chapter, we explore these important differences and describe how the structural characteristics of different fusion protein types provide insight into the mechanism of fusion protein function.

Type I Viral Fusion Proteins

The first X-ray structure of a viral fusion protein was that of influenza HA in its pre-fusion form,[23] published in 1981. Over a decade later, the structure of HA in its post-fusion form was finally determined,[7] and the X-ray structures of a number of other fusion protein ectodomains have been solved in recent years. In addition to HA structures from a number of strains, we now also have pre- and post-fusion structures of the type I paramyxovirus F protein[8,24] and the pre-fusion structure of a large portion of the type I Ebola GP protein.[25] Unfortunately, an X-ray structure of the complete HIV Env protein ectodomain has eluded researchers, but a six-helix core, which best characterizes type I proteins in their post-fusion form (discussed below), was demonstrated by X-ray crystallography.[26,27] Indeed, the post-fusion six-helix bundle (6HB) is the primary defining feature of type I fusion proteins,[28,29] and six-helix bundle structures have been structurally demonstrated for a large number of additional type I fusion proteins, including the SARS S protein and Ebola virus GP.[30,31]

Type I fusion proteins are proteolytically processed at or near the N-terminal side of the hydrophobic fusion peptide,[16] which is a necessary event for fusion activity, as it unlocks the fusion domain and primes the protein for triggering. Structures both before and after proteolytic processing have been determined for influenza HA,[23,32] but not for other type I fusion proteins. Interestingly, HA cleavage results in very minor changes in overall ectodomain structure, with the movement of only 19 residues adjacent to the cleavage site.[33] The movement of these residues results in the repositioning of the hydrophobic fusion peptide into a nearby negatively charged cavity.

Type I fusion proteins are trimeric in both the pre- and post-fusion conformations, yet the switch in conformation requires extensive refolding of the majority of the protein (Figure 1). The pre-fusion conformations of both influenza HA and paramyxovirus F appear superficially as a globular head on a stalk,[23,24] whereas the post-fusion structures are more rod shaped.[7,8] The Ebola GP pre-fusion structure reveals only a head domain, but portions of this protein, which likely include the stalk, were disordered, preventing structural determination.[25] The stalk of the paramyxovirus F protein is composed of a domain called heptad repeat B (HRB), whereas the equivalent domain in Ebola GP is disordered. Interestingly, the fusion peptide domains of paramyxovirus F and Ebola GP both pack around the globular head where they are wedged against a neighboring subunit in the trimer. In contrast, the fusion domain of influenza HA is found near the base of protein in the stalk domain.

Type I proteins all contain two functionally essential heptad repeat domains in each monomer, denoted HRB and heptad repeat A (HRA; also called HR1 and HR2), which contain non-polar residues in 3–4 periodicity, facilitating coiled-coil formation.[21] In the post-fusion structures of all type I fusion

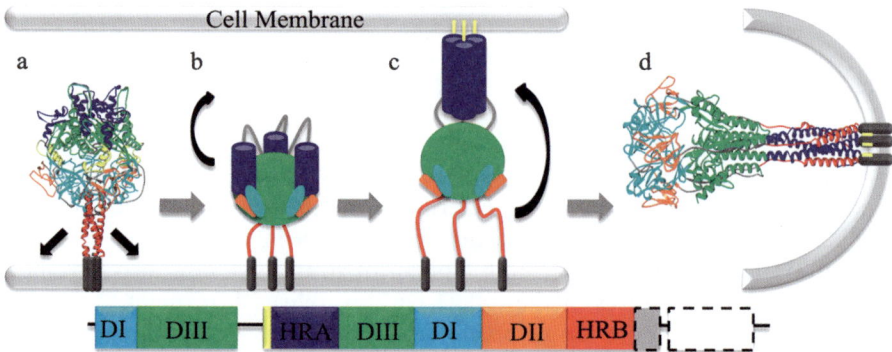

Figure 1   Model of paramyxovirus F protein-promoted fusion, an example of a class I fusion protein. Upper panel: (a) pre-fusion structure of the parainfluenza virus 5 F protein (PDB ID: 2B9B[24]) with domains colored according to the schematic below. (b) Triggering of fusion results in movement of the HRA and HRB regions. (c) HRA forms a coiled coil as the fusion peptide inserts into the target membrane. Subsequently, the protein folds to bring the HRA and HRB regions together. (d) The post-fusion structure of hPIV3 F (PDB ID: 1ZTM[8]), with the final six-helix bundle bringing the fusion peptide and TM domains into close proximity. Lower panel: schematic of the PIV5 F protein, colored as in the fusion model. HRA (129–204) = blue; HRB (446–477) = red; fusion peptide (103–128) = yellow; domain 1 (20–41; 279–369) = cyan; domain II (375–421) = orange; domain III (with the exception of the HRA region) (42–94; 205–278) = green. TM domain (not present in structure) = gray.

proteins, these domains associate in a 6HB (Figure 1), which peptide studies suggest is extremely stable.[28,34] HRA forms a three-stranded coil at the center of the bundle, whereas the HRB domain fits in an anti-parallel orientation into the grooves that form around the inner coiled coil. The extended coiled coil formed by the HRA domains is thought to act as a scaffold around which the HRB domains may zipper down on as the membranes merge. However, in the pre-fusion structures, the α-helix of the HRA domain is disrupted and folds around the head of paramyxovirus F (Figure 1) and Ebola GP proteins. In fact, the HRA domain of the paramyxovirus F protein is broken into 11 distinct segments in the pre-fusion structure,[24] all of which refold into a single extended coil in the post-fusion structure. In contrast, the influenza HA HRA domain is found primarily in the stalk of the protein in the pre-fusion structure,[23] and a large portion of it is folded into a single α-helix. The hairpin that forms during fusion lies between the heptad repeat domains, facilitating the anti-parallel interactions of the helices of HRA and HRB in the post-fusion structures. Furthermore, HRA is just C-terminal of the fusion domain whereas HRB is just N-terminal of the TM domain. Thus, 6HB formation forces the fusion domain and TM domain, and also their associated membranes, into close proximity. Moreover, insertion of the fusion domain into the target membrane is likely mediated by the refolding of the HRA domain into the coiled coil above the globular head, as the cleaved fusion domain is anchored to the end of this extended coil.

In summary, type I fusion proteins are trimeric spikes, characterized by having N-terminal fusion domains that are exposed following proteolytic processing and by the extensive refolding of their trimeric structure to form a stable 6HB.

Type II Viral Fusion Proteins

Understanding of type II fusion protein structure was considerably delayed compared with that of type I proteins, but a large volume of structural data for the fusion proteins of multiple flaviviruses (dengue virus,[11,35–37] tick-borne encephalitis virus[10,38] and West Nile virus[39]), and also the alphavirus Semliki Forest virus,[9,40,41] has flooded the field in recent years. The story of type II fusion has proven to be remarkable and different from that of type I proteins, but with several common themes. Rather than folding as homotrimers, as type I proteins do, type II protein monomers fold co-translationally with a chaperone protein, which plays an essential stabilizing role and prevents premature activation of fusion.[4] The chaperone is thought to release its grip on the fusion protein as it transitions from a dimeric state in the mature virion to the final post-fusion trimeric architecture.[35,37,42,43] During this transition, it is likely that the fusion protein must transiently exist as a monomer.

The flavivirus E protein and alphavirus E1 protein have remarkably similar topology, each containing three domains primarily composed of β-sheets (Figure 2). The pre-fusion monomeric structure is that of an elongated, bent

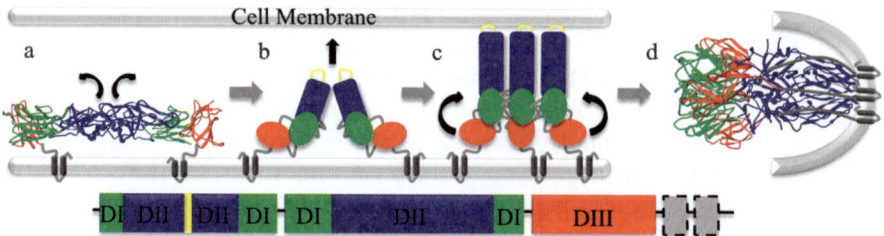

Figure 2   Model of Tick-borne encephalitis virus E protein-promoted fusion, an example of a class II fusion protein. Upper panel: (a) pre-fusion, dimeric structure of the TBE E protein (PDB ID: 1SVB[38]) with domains colored according to the schematic below. (b) Triggering by low pH results in rotation of the domains relative to each other, extending the fusion loop (yellow) towards the target membrane. (c) The fusion loops of the now trimeric E protein insert into the target, while domain III is rotated towards domain II. (d) The post-fusion structure of TBE E (PDB ID: 1URZ[10]), with the trimer of hairpins conformation bringing the TM domains and fusion loops in close proximity. Lower panel: schematic of the TBE E protein, colored as in the fusion model. Domain I (1–51; 137–189; 285–302) = green; domain II (52–136; 190–284) = blue; domain III (303–395) = red; fusion loop = yellow. TM domain (not present in structure) = gray.

rod. Domains I and II (DI and DII) lie adjacent to each other, forming a relatively straight shaft, while the bent portion of the rod is formed by domain III (DIII). DI contains the N-terminus of the protein, but forms the central of the three domains, whereas DIII leads into the C-terminal stem and anchor of the protein. During fusion, each of the domains rotate relative to each other *via* short hinges that join them. This results in further arching of DIII toward DII. The distal portion of DII contains the fusion domain, which takes the form of a loop rather than the cleavage exposed end seen in type I fusion proteins. However, as in type I proteins, large protein movements must take place for the fusion loop to come into contact with target membranes.

Interestingly, the three domains undergo very little refolding as the protein transitions through these substantial movements (Figure 2). It is the rotation and flexibility of hinge regions between the domains that facilitate the majority of the necessary conformational changes. The most apparent change between pre- and post-fusion conformations of the monomer is the movement of DIII closer to the fusion loop-containing DII. This movement brings the viral membrane in close proximity to the target membrane and results in a trimer of hairpins conformation, reminiscent of the events promoted by type I proteins. The most obvious changes that take place during fusion, however, are not at the level of the monomer, but in the oligomeric state of the protein and its placement relative to the viral surface. The pre-fusion structures of type II proteins are brick-shaped dimers, in which monomers lie anti-parallel to each other, flat on the viral surface. In this conformation, the fusion loop of the flavivirus E protein is buried in a pocket between DI and DIII on the neighboring subunit, and the alphavirus E1 protein fusion loop is shielded by the chaperone protein, E2.[44] Remarkably, post-fusion structures of type II proteins revealed a trimeric protein with subunits oriented parallel to each other and perpendicular to the viral membrane. As the fusion loops of three adjacent monomers transition towards the target membrane, the lengths of DI and DII from neighboring monomers come together at the central axis to form the trimeric core.

Thus, type II fusion proteins are defined by a dimer to trimer conformational change involving domain rearrangements rather than refolding. In addition, the β-sheet character of the domains and the internal fusion loop strongly differentiate the type II fusion proteins from the type I proteins.

Type III Viral Fusion Proteins

A third classification of viral fusion proteins was needed to accommodate the unique features of the vesicular stomatitis virus (VSV) G protein and the herpes simplex virus 1 (HSV-1) gB protein, which were discovered to have a surprisingly high level of homology in the recently solved X-ray structures.[6,45,46] HSV-1 has a very large DNA genome and fusion requires a complex of proteins,[1] whereas VSV contains a small RNA genome and its fusion glycoprotein, G, is the only membrane protein encoded by the virus.[47] hence there was no reason to suspect the glycoproteins of these viruses would be homologous.[48] Yet, although VSV G is more compact than HSV-1 gB, the individual domains of the proteins are structurally homologous and the domains are organized in a similar manner. In addition, a new type III protein was revealed when the crystal structure of the baculovirus fusion protein, gp64, was recently solved.[49] Baculoviruses infect invertebrates and are very distinct from VSV or HSV-1. Both pre- and post-fusion conformations of the VSV G protein have now been solved,[6,46] but only post-fusion structures of HSV-1 gB[45] and baculovirus gp64[49] are known. As knowledge of both pre- and post-fusion structures greatly facilitates our understanding of the fusion mechanism, our discussion will focus on the VSV G protein.

Both the pre- and post-fusion conformations of VSV G are trimeric and oriented vertically from the viral membrane (Figure 3). The pre-fusion conformation of the protein is a novel tripod shape, with the fusion domain of the protein found at the end of the tripod legs. Thus, the fusion loops of VSV G are exposed and pointed towards the viral membrane in the pre-fusion state, but kept wide apart from each other. The domain containing the fusion loop of VSV G appears superficially similar to the fusion loop-containing domain of type II proteins, yet the topologies of the strands in the domain are unrelated and they are therefore not homologous.[6] Furthermore, the fusion loop of VSV G is formed from two discontinuous segments and is thus termed a bipartite fusion loop.

The G protein can be divided into four distinct domains, which were named in order according to the original post-fusion structure. The domains shift relative to each other from the pre- to the post-fusion state and all but DII

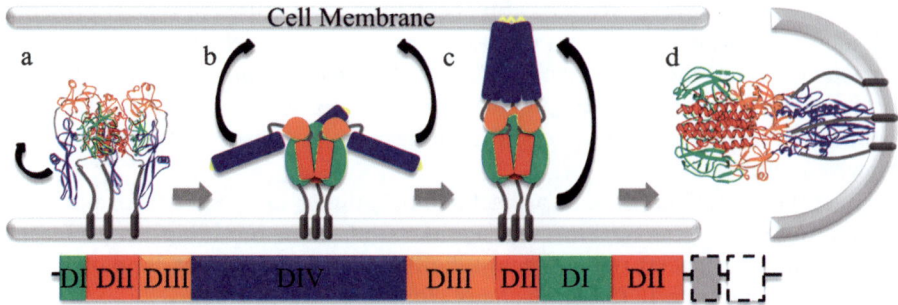

Figure 3 Model of VSV G protein-promoted fusion, an example of a class III fusion
protein. Upper panel: (a) pre-fusion structure of the VSV G protein (PDB
ID: 2J6J[46]) with domains colored according to the schematic below. (b) Low
pH triggers conformational changes which rotate domain IV and its fusion
loop (yellow) towards the target membrane. (c) Refolding of domain III
facilitates insertion of the fusion loops into the target membrane. (d)
Movement of the C-terminal portion results in the post-fusion structure
(PDB ID: 2CMZ[6]) containing a six-helix bundle. Lower panel: schematic of
the VSV G protein, colored as in the fusion model. Domain I (1–17; 310–
382) = green; domain II (18–35; 259–309; 383–405) = red; domain III (36–46;
181–258) = orange; domain IV (53–172) = blue. TM domain (not present in
structure) = gray.

retain their folds throughout this transition (Figure 3). There are no predicted
heptad repeat domains in VSV G, yet DII forms a 6HB in the post-fusion
conformation. Refolding of segments in DII results in the elongation of
the central helix in the bundle, the coils of which also become increasingly
parallel in the post-fusion structure, resulting in highly stable trimer contacts.
A different segment of DII forms a second elongated helix, which packs in
the grooves of the core trimer in anti-parallel orientation, reminiscent of the
type I 6HB.

A model for the VSV G membrane fusion mechanism can be built from
knowledge of the pre- and post-fusion structures. DIV, containing the fusion
loop, swings from its original position facing the viral membrane to a position
160 Å towards the target membrane. This movement is facilitated by rotation
around hinges between DIV and the adjacent DIII and also through the
repositioning of DIII on top of DI and DII. DI and DII remain largely stationary during this movement, serving as a central scaffold, although the
refolding of the trimerization domain (DII) also facilitates DIV movement.
Fusion loop insertion into the target membrane then bridges the viral and
target membranes and a final movement of the C-terminal anchored portion of
the protein towards the inserted fusion loop forces the two membranes together. Unfortunately, the C-terminal stem and anchor of the protein are not
visible in the X-ray structure, leaving the placement of these domains unclear.

To summarize, type III fusion proteins are characterized by an internal,
bipartite fusion loop reminiscent of type II proteins and a 6HB in the post-
fusion conformation like that of type I proteins. Unlike other fusion proteins,

however, the fusion domain is exposed in the pre-fusion state and the conformational change leading to fusion involves both refolding and rearrangement of domains.

3 Regulation of Fusion Protein Activity

The fusogenic activity of viral glycoproteins must be carefully regulated to ensure that viral entry occurs at the correct location within a suitable cell type. In addition, triggering of type I and type II fusion proteins generally results in an irreversible conformational change to the low-energy post-fusion conformation,[14,15] and therefore premature fusion protein activation results in inactivated virus particles.[14,17,18] It should be noted, however, that reversible structural changes in type II fusion proteins may take place when in the absence of specific membrane lipids.[50,51] In contrast, the conformational changes associated with fusogenic activity of the type III fusion protein VSV G appear to be completely reversible. Both the pre- and post-fusion conformations are stable and the lowest energy state is determined by the pH of the environment.[52]

Various mechanisms are utilized to protect fusion proteins from premature activation. Common themes in the control of activation include regulated proteolytic processing, association with other viral proteins and shielding of the fusion peptide or loop. Structural analysis of fusion proteins alone or in complex with accessory proteins has greatly enhanced our understanding of these regulatory mechanisms, building on a foundation laid by a large body of biochemical and biological data exploring fusion protein function and regulation. For type I and type II fusion proteins, regulation to prevent activation must be alleviated prior to the triggering of conformational changes by an additional environmental stimulus or stimuli. We will first discuss type I and type II fusion protein regulatory mechanisms, followed by a discussion of conditions which lead to triggering. VSV G is not regulated in the manner seen with type I or type II proteins, likely due to the reversibility of G protein conformational changes. However, variables controlling type III protein conformational changes will be discussed in the later paragraphs concerning protein triggering.

Regulation of Type I Fusion Protein Activity

The activity of type I fusion proteins is generally dependent on specific proteolytic cleavage of the homotrimeric protein, resulting in a trimer of heterodimers. An exception to this requirement for specific cleavage is the Ebola GP protein, for which the initial cleavage event is dispensable for fusion,[53–55] likely due to subsequent proteolytic degradation by cathepsin proteases within the endosome.[56,57] The heterodimers formed by proteolytic processing are disulfide linked in some, but not all, type I fusion proteins. The C-terminal subunit contains the TM domain, the fusion domain and the heptad repeat regions, and is therefore largely responsible for fusion activity. For many type I fusion proteins, such as influenza HA and HIV Env, the N-terminal subunit of the heterodimer forms a domain that is structurally distinct from the C-terminal subunit, and in the pre-fusion state the N-terminal subunit resides at the membrane distal end of the protein.[33,58] In these proteins, the N-terminal subunit is also the receptor-binding domain. In contrast, the paramyxovirus F protein pre-fusion structure revealed that the N-terminal cleaved subunit of this protein, F_2, is intimately folded with the C-terminal (F_1) subunit.[24] A second paramyxovirus protein, termed the attachment protein, mediates attachment of the virus to host cells and this protein is essential to the fusion activity of most paramyxoviruses (discussed below).

Proteolytic processing of type I fusion proteins results in placement of the hydrophobic fusion peptide at or near the newly created N-terminus of the TM-containing subunit.[16] This proteolytic cleavage therefore serves to unlock the fusion domain, permitting its exposure at the end of the extended HRA region following triggering. As discussed previously, structural data before and after proteolytic processing are available only for influenza HA and these structures revealed little movement of the fusion domain after proteolytic cleavage. As the pre-fusion structures of paramyxovirus F and Ebola GP show tight packing of the fusion domain against the globular head of the protein,[24,25] it is likely that this domain also does not show significant movement after proteolytic cleavage of these proteins, although further structural analysis is needed to address this important point. Thus, proteolytic processing primes type I fusion proteins for activation by unlocking the fusion domain, but the fusion domain remains stably associated with the protein until the pre-fusion conformation is destabilized.

Regulation of Type II Fusion Protein Activity

Proteolytic cleavage is also important to prime type II fusion proteins for activity, but it is a chaperone protein which must be cleaved rather than the fusion protein itself.[59,60] Chaperone protein cleavage is necessary to permit subsequent oligomeric rearrangements in the fusion protein. The chaperone protein is processed by furin during viral egress (for flaviviruses) or protein transport (for alphaviruses) through the secretory pathway. The alphavirus chaperone, E2, also plays an important role in fusion loop shielding and receptor binding.[43,44] The alphavirus E2 protein remains associated with the E1 fusion protein throughout the process of viral budding from the plasma membrane. Low pH, encountered after endosomal entry, then serves to release E2 from the fusion protein E1, a triggering event for fusion.

Flavivirus particles bud into the ER and the viral glycoproteins, E and prM, assemble into an ordered array on the viral surface with quasi-icosahedral symmetry.[61] As particles are transported through the secretory pathway, the prM protein is cleaved by furin into a membrane anchored segment with a short extracellular domain (M) and a non-covalently bound surface subunit (pr). The structure of the dengue virus prM–E heterodimer, representing the immature configuration, was recently solved by X-ray crystallography, providing valuable insight into the mechanism of chaperone protein cleavage and release.[35,37] The prM–E heterodimers are initially arranged as trimers, but the low pH of the Golgi compartment results in a conformational change which not only exposes the prM protein to furin cleavage, but also induces a trimer to dimer transition in the prM–E heterodimers. Remarkably, it seems that although the prM protein is cleaved in the Golgi, the pr subunit is not released until the virus reaches the neutral pH of the extracellular environment. The pr protein is stably associated with the E protein, covering the fusion loop, in an acidic environment, but the contacts between these proteins are weakened at neutral pH. Three prominent electrostatic patches bind the pr protein to E at low pH, but affinity of these interactions decreases at neutral pH. It is thought that the delayed release of the pr protein prevents premature triggering of the E protein by low pH in the Golgi. The virus is said to be mature once pr is released from the viral surface, since the virus is not fusion competent until this event takes place.

4   The Trigger of Fusion Protein Conformational Changes

Triggering events provide the final level of control of viral fusion proteins, dictating the start of the cascade of conformational changes leading to fusion. The classification of fusion proteins does not dictate the mechanism of triggering; however, all type II proteins examined to date are triggered by low pH. Type I and type III proteins may be triggered by low pH, receptor binding, a combination of low pH and receptor binding or through interaction with a second viral protein following receptor binding (reviewed[16]). Previously, viruses were grouped into one of two entry categories: neutral-pH entry at the plasma membrane or low-pH entry following endocytosis. However, recent research has shown that these generalities can be challenged when entry in multiple cell types is examined or when various members of the same family are analyzed. To cite examples, HSV-1 entry was long thought to take place at the plasma membrane in a pH-independent manner, but recent data indicate that entry into certain cell types occurs *via* endocytosis and necessitates low pH.[62,63] Also, whereas HIV-1 glycoprotein activation follows receptor binding and is clearly independent of pH, many other retroviruses use pH-dependant entry mechanisms.[64–66] In keeping with our emphasis on structural insights, we will focus on viral entry for which we have structural information that has provided new understanding into how the trigger stimulates the conformational changes in the fusion protein.

The Low pH Trigger of Fusion

In addition to all characterized type II proteins, type I fusion proteins, such as influenza HA, and type III proteins, such as VSV G and baculovirus gp64, are also triggered by low pH.[67,68] A low pH trigger for fusion suggests that viral entry proceeds through endocytosis of viral particles, as endocytic vesicles increase in acidity from the early endosome to the lysosome. The increased concentration of hydrogen ions results in protonation of residues with a pK_a in the physiological range. Histidine residues are the most likely to ionize within the physiological range, as the pK_a of histidine in solution is 6.0. However, the local environment of an amino acid within a folded protein affects its pK_a, and therefore the stability of the protein in areas of charged amino acids may also be affected by the pH of the environment.

Conformational changes of type I and type II fusion proteins are triggered when low-pH conditions destabilize the pre-fusion conformation, driving changes in protein conformation that lead to the final, most stable, post-fusion form. The metastability of the pre-fusion form of influenza HA was demonstrated by the ability of increased heat or urea also to trigger HA conformational changes, suggesting that the mechanism of triggering involves destabilization of the pre-fusion state.[14] Furthermore, brief exposure of influenza virus to low pH in the absence of a target cell results in a virus that is no longer capable of infecting a cell, likely because the lower energy of the post-fusion conformation traps HA in that state.[18] Interestingly, low-pH exposure induces an irreversible conformational change in type II proteins only when certain lipids are also present.[50,51] Thus, while the type II pre-fusion conformational change is also considered metastable, some structural changes that take place upon low-pH exposure are likely reversible. The need for membrane lipids to trap the post-fusion conformation suggests that lipids play a role in achieving the most stable state of the protein. Finally, the type III protein VSV G is not considered to be metastable in the pre-fusion conformation and VSV G conformational changes cannot be induced by heat or urea.[69] Instead, unfavorable interactions within the post-fusion conformation of the G protein at neutral pH favor transition back to the pre-fusion conformation.[46] Likewise, the pre-fusion conformation is not stable at low pH. Hence the VSV G protein appears to exist in a pH-sensitive equilibrium. However, the buried surface area in the protein more than doubles in the post-fusion state, potentially explaining where the energy to induce membrane fusion is generated.[6] The baculovirus gp64 protein is also thought to undergo reversible pH-triggered conformational changes,[70] and the post-fusion structure of this protein revealed a series of pH-sensitive interfaces throughout the protein, most of which involved histidine residues.[49] It is currently not known if HSV-1 gB conformational changes are reversible, but results with VSV G and baculovirus gp64 suggest that the structures of type III proteins facilitate reversible conformational changes, whereas the other fusion protein structures do not.

Electrostatic forces at work throughout the protein play an important role in low-pH triggering. Common themes involve repulsion between positive charges brought on by protonation of residues in the pre-fusion conformation and the formation of salt bridges in the post-fusion conformation. Many of the titratable residues are found at the interface of pre-fusion subunits and near the fusion peptide. X-ray structures of the HA proteins of multiple strains of influenza have been solved and a comparison of HA structures reveals that, although individual ionizable residues are not well conserved, there are multiple different ionizable residues at all regions along the chain and subunit inter-faces.[71] Furthermore, ionizable residues are buried when the cleaved fusion peptide domain moves into a charged cavity, potentially facilitating the release of the fusion peptide in low pH.[33] In addition to the electrostatic forces governing release of the chaperone protein,[35] type II fusion proteins appear to contain pH-sensing residues in the hinge regions, which undergo the major structural changes that result in domain movements.[11,72] VSV G contains a small number of conserved residues involved in stabilizing networks that differ in the pre- and post-fusion forms.[6,46] Three histidine residues, two from the fusion domain and one from the C-terminal portion of the protein, cluster together in the pre-fusion form of the protein and low pH is predicted to destabilize this region. In the post-fusion conformation, one of these histidine residues forms a salt bridge with a negatively charged residue while another histidine is involved in interactions between fusion domains. In addition, multiple acidic residues are brought into close proximity when the 6HB forms at low pH. They are protonated and stabilized by hydrogen bonds in the low-pH structure, but this arrangement would be unfavorable in a neutral environment. Indeed, these acidic residues appear to be solvent exposed in the pre-fusion structure.

Triggering of Fusion Through Receptor Binding

All viruses must bind a receptor in order to enter the target cell. For low-pH triggered viral fusion proteins, receptor binding is the key to attachment and endocytosis. In contrast, for viruses with fusion proteins which are not triggered by low pH, receptor binding is generally thought to be the key to triggering fusion. For example, HIV entry involves sequential binding by the gp120 subunit of the Env fusion protein to the CD4 receptor and then to a co-receptor, resulting in the triggering of fusion by the gp41 subunit. Entry of most paramyxoviruses and likely also HSV-1, is thought to proceed through a similar mechanism, except that a second viral protein binds to the receptor and transmits the signal to trigger fusion to the fusion protein. However, the exact mechanism by which this signal is transmitted is not understood.

In the case of HSV-1, four viral proteins are needed to promote fusion in model systems.[1] The gD protein binds receptor and undergoes a conformational change leading to the release of a portion of the protein. The liberated polypeptide may then interact with the gH/gL and gB fusion machinery to stimulate fusion. The X-ray structure of gB strongly suggests that this protein is the primary fusion protein, but it has been reported that the gH/gL heterodimer is responsible for inducing the hemifusion intermediate.[73,74] There is still much to learn in this complex system of viral fusion, including determining the importance of the low-pH-dependant fusion phenotype observed in some cell types.

All paramyxoviruses have been hypothesized to enter cells at the plasma membrane under neutral pH conditions following receptor binding by a second viral protein referred to as an attachment protein.[75] The majority of paramyxoviruses induce cell–cell fusion when their glycoproteins are expressed in tissue culture cells,[75,76] consistent with the idea that the fusion proteins (F) of these viruses are capable of promoting fusion under neutral pH conditions. Specific interactions between cell surface receptors and viral attachment proteins are hypothesized to trigger the F protein conformational changes.[3,75] For at least the *Paramyxovirinae* subfamily, the attachment protein indeed plays an essential role in membrane fusion,[75] as fusion cannot be induced with a heterotypic attachment protein, unless the two proteins are highly related.[77,78] Interestingly, it was only recently realized that multiple *Pneumovirinae* subfamily members do not strictly depend on their attachment proteins for entry into cells in culture,[79,80] suggesting the trigger of fusion for *Pneumovirinae* F proteins is different from that of *Paramyxovirinae* F proteins. Furthermore, one *Pneumovirinae* member, respiratory syncytial virus, was recently demonstrated to depend on clathrin-mediated endocytosis for entry.[81] In addition, cell–cell fusion induced by F proteins of certain strains of human metapneumovirus were shown to be stimulated by low pH,[82,83] and entry of this virus was inhibited by agents which block endocytosis or which alter endosomal pH,[84] suggesting that low pH may be involved as a trigger for some paramyxovirus F proteins.

The pre- and post-fusion X-ray structures of paramyxovirus F proteins were derived from two viruses in the *Paramyxovirinae* subfamily and the structures have shed light on the mechanism of fusion modulation by multiple mutations that had previously been examined.[24] Residues located just N-terminal of HRB, which forms the coiled stalk of the pre-fusion structure, were found to modulate fusion activity and in some cases enhanced the kinetics and extent of fusion, in addition to permitting fusion at lower temperatures.[85] This region, termed the HRB linker, formed a network of contacts with the base of the globular head that appeared to nucleate the HRB helix. Mutations in this domain could destabilize the HRB helix, which is likely a necessary and early event in the structural transition to the post-fusion state. In addition, mutations which alter interactions of HRA in the pre-fusion form[86,87] or packing of the fusion peptide[88,89] have also been shown to modulate triggering, suggesting that control of multiple regions may be important. However, the role of the attachment protein in inducing the destabilization of the pre-fusion state of paramyxovirus F proteins has yet to be determined.

5   Summary

The transmembrane domain, present in all fusion proteins, is missing from all current structural data obtained to date. We currently do not understand the role of this important domain in the fusion process and critical questions, such as the possible self-association of the TM domain in the pre-fusion form or potential association of the TM with the fusion domain in the final post-fusion state, remain to be addressed. TM domains are known to be essential for full membrane fusion. A GPI-anchored influenza HA protein was capable of forming only small pores, but the pores did not expand.[90] Hence at least some TM domains may have a role in fusion pore expansion, which has been measured to be the most energetically demanding step in fusion.[5] Deletion of the paramyxovirus TM domain resulted in a fusion protein that folded immediately or soon after synthesis into the post-fusion conformation, regardless of proteolytic processing, suggesting that the TM domain is essential for pre-fusion stability.[91] Indeed, the X-ray structure of the pre-fusion F protein was only accomplished after a helix-stabilizing domain had been appended to the truncated C-terminus.[24] Moreover, fusion proteins contain cytoplasmic domains of various length and deletion of a 20-residue paramyxovirus cytoplasmic domain resulted in a pore expansion defect.[92] Clearly, the domains in and around the TM domain are necessary to complete the final steps in fusion. It is also unknown to what extent cooperativity between fusion proteins is needed to complete fusion. For most fusion proteins, it is thought that more than one protein is needed to promote fusion efficiently. Using a combination of cryo-electron microscopy and electron crystallography, it was shown that E1 proteins of Semliki Forest virus interact cooperatively with target membranes, forming rings of five or six trimers associated through their fusion loops.[40] It will be interesting to learn if the TM and fusion domains of various fusion proteins interact and if these domains or others may also facilitate cooperation between fusion proteins.

References

1. P. G. Spear and R. Longnecker, *J. Virol.*, 2003, **77**, 10179.
2. B. Moss, *Virology*, 2006, **344**, 48.
3. R. M. Iorio and P. J. Mahon, *Trends Microbiol.*, 2008, **16**, 135.
4. M. Kielian and F. A. Rey, *Nat. Rev. Microbiol.*, 2006, **4**, 67.
5. F. S. Cohen and G. B. Melikyan, *Trends Membr. Biol.*, 2004, **199**, 1.
6. S. Roche, S. Bressanelli, F. A. Rey and Y. Gaudin, *Science*, 2006, **313**, 187.
7. P. A. Bullough, F. M. Hughson, J. J. Skehel and D. C. Wiley, *Nature*, 1994, **371**, 37.
8. H. S. Yin, R. G. Paterson, X. Wen, R. A. Lamb and T. S. Jardetzky, *Proc. Natl. Acad. Sci. USA*, 2005, **102**, 9288.
9. D. L. Gibbons, M. C. Vaney, A. Roussel, A. Vigouroux, B. Reilly, J. Lepault, M. Kielian and F. A. Rey, *Nature*, 2004, **427**, 320.
10. S. Bressanelli, K. Stiasny, S. L. Allison, E. A. Stura, S. Duquerroy, J. Lescar, F. X. Heinz and F. A. Rey, *EMBO J.*, 2004, **23**, 728.
11. Y. Modis, S. Ogata, D. Clements and S. C. Harrison, *Nature*, 2004, **427**, 313.

12. E. Zaitseva, A. Mittal, D. E. Griffin and L. V. Chernomordik, *J. Cell Biol.*, 2005, **169**, 167.
13. L. J. Earp, S. E. Delos, H. E. Park and J. M. White, *Curr. Top. Microbiol. Immunol.*, 2005, **285**, 25.
14. C. M. Carr, C. Chaudhry and P. S. Kim, *Proc. Natl. Acad. Sci. USA*, 1997, **94**, 14306.
15. F. X. Heinz, K. Stiasny and S. L. Allison, *Arch. Virol. Suppl.*, 2004, **18**, 133.
16. J. M. White, S. E. Delos, M. Brecher and K. Schornberg, *Crit. Rev. Biochem. Mol. Biol.*, 2008, **43**, 189.
17. S. A. Connolly and R. A. Lamb, *Virology*, 2006, **355**, 203.
18. F. Boulay, R. W. Doms, I. Wilson and A. Helenius, *EMBO J.*, 1987, **6**, 2643.
19. A. Salminen, J. M. Wahlberg, M. Lobigs, P. Liljestrom and H. Garoff, *J. Cell Biol.*, 1992, **116**, 349.
20. G. Wengler and G. Wengler, *J. Virol.*, 1989, **63**, 2521.
21. R. E. Dutch, T. S. Jardetzky and R. A. Lamb, *Biosci. Rep.*, 2000, **20**, 597.
22. S. C. Harrison, *Nat. Struct. Mol. Biol.*, 2008, **15**, 690.
23. I. A. Wilson, J. J. Skehel and D. C. Wiley, *Nature*, 1981, **289**, 366.
24. H. S. Yin, X. Wen, R. G. Paterson, R. A. Lamb and T. S. Jardetzky, *Nature*, 2006, **439**, 38.
25. J. E. Lee, M. L. Fusco, A. J. Hessell, W. B. Oswald, D. R. Burton and E. O. Saphire, *Nature*, 2008, **454**, 177.
26. D. C. Chan, D. Fass, J. M. Berger and P. S. Kim, *Cell*, 1997, **89**, 263.
27. W. Weissenhorn, A. Dessen, S. C. Harrison, J. J. Skehel and D. C. Wiley, *Nature*, 1997, **387**, 426.
28. S. B. Joshi, R. E. Dutch and R. A. Lamb, *Virology*, 1998, **248**, 20.
29. R. A. Lamb and T. S. Jardetzky, *Curr. Opin. Struct. Biol.*, 2007, **17**, 427.
30. V. N. Malashkevich, B. J. Schneider, M. L. McNally, M. A. Milhollen, J. X. Pang and P. S. Kim, *Proc. Natl. Acad. Sci. USA*, 1999, **96**, 2662.
31. Y. Xu, Z. Lou, Y. Liu, H. Pang, P. Tien, G. F. Gao and Z. Rao, *J. Biol. Chem.*, 2004, **279**, 49414.
32. J. Chen, K. H. Lee, D. A. Steinhauer, D. J. Stevens, J. J. Skehel and D. C. Wiley, *Cell*, 1998, **95**, 409.
33. J. J. Skehel and D. C. Wiley, *Annu. Rev. Biochem.*, 2000, **69**, 531.
34. J. J. Skehel and D. C. Wiley, *Cell*, 1998, **95**, 871.
35. L. Li, S. M. Lok, I. M. Yu, Y. Zhang, R. J. Kuhn, J. Chen and M. G. Rossmann, *Science*, 2008, **319**, 1830.
36. Y. Modis, S. Ogata, D. Clements and S. C. Harrison, *Proc. Natl. Acad. Sci. USA*, 2003, **100**, 6986.
37. I. M. Yu, W. Zhang, H. A. Holdaway, L. Li, V. A. Kostyuchenko, P. R. Chipman, R. J. Kuhn, M. G. Rossmann and J. Chen, *Science*, 2008, **319**, 1834.
38. F. A. Rey, F. X. Heinz, C. Mandl, C. Kunz and S. C. Harrison, *Nature*, 1995, **375**, 291.

39. R. Kanai, K. Kar, K. Anthony, L. H. Gould, M. Ledizet, E. Fikrig, W. A. Marasco, R. A. Koski and Y. Modis, *J. Virol.*, 2006, **80**, 11000.
40. D. L. Gibbons, I. Erk, B. Reilly, J. Navaza, M. Kielian, F. A. Rey and J. Lepault, *Cell*, 2003, **114**, 573.
41. A. Roussel, J. Lescar, M. C. Vaney, G. Wengler, G. Wengler and F. A. Rey, *Structure*, 2006, **14**, 75.
42. S. Mukhopadhyay, W. Zhang, S. Gabler, P. R. Chipman, E. G. Strauss, J. H. Strauss, T. S. Baker, R. J. Kuhn and M. G. Rossmann, *Structure*, 2006, **14**, 63.
43. S. R. Wu, L. Haag, L. Hammar, B. Wu, H. Garoff, L. Xing, K. Murata and R. H. Cheng, *J. Biol. Chem.*, 2007, **282**, 6752.
44. J. Lescar, A. Roussel, M. W. Wien, J. Navaza, S. D. Fuller, G. Wengler, G. Wengler and F. A. Rey, *Cell*, 2001, **105**, 137.
45. E. E. Heldwein, H. Lou, F. C. Bender, G. H. Cohen, R. J. Eisenberg and S. C. Harrison, *Science*, 2006, **313**, 217.
46. S. Roche, F. A. Rey, Y. Gaudin and S. Bressanelli, *Science*, 2007, **315**, 843.
47. A. T. Da Poian, F. A. Carneiro and F. Stauffer, *Braz. J. Med. Biol. Res.*, 2005, **38**, 813.
48. A. C. Steven and P. G. Spear, *Science*, 2006, **313**, 177.
49. J. Kadlec, S. Loureiro, N. G. Abrescia, D. I. Stuart and I. M. Jones, *Nat. Struct. Mol. Biol.*, 2008, **15**, 1024.
50. K. Stiasny, S. L. Allison, J. Schalich and F. X. Heinz, *J. Virol.*, 2002, **76**, 3784.
51. M. R. Klimjack, S. Jeffrey and M. Kielian, *J. Virol.*, 1994, **68**, 6940.
52. S. Roche and Y. Gaudin, *Virology*, 2002, **297**, 128.
53. G. Neumann, H. Feldmann, S. Watanabe, I. Lukashevich and Y. Kawaoka, *J. Virol.*, 2002, **76**, 406.
54. R. J. Wool-Lewis and P. Bates, *J. Virol.*, 1999, **73**, 1419.
55. G. Neumann, T. W. Geisbert, H. Ebihara, J. B. Geisbert, K. M. Daddario-DiCaprio, H. Feldmann and Y. Kawaoka, *J. Virol.*, 2007, **81**, 2995.
56. K. Chandran, N. J. Sullivan, U. Felbor, S. P. Whelan and J. M. Cunningham, *Science*, 2005, **308**, 1643.
57. K. Schornberg, S. Matsuyama, K. Kabsch, S. Delos, A. Bouton and J. White, *J. Virol.*, 2006, **80**, 4174.
58. J. Liu, A. Bartesaghi, M. J. Borgnia, G. Sapiro and S. Subramaniam, *Nature*, 2008, **455**, 109.
59. K. Stadler, S. L. Allison, J. Schalich and F. X. Heinz, *J. Virol.*, 1997, **71**, 8475.
60. M. Lobigs and H. Garoff, *J. Virol.*, 1990, **64**, 1233.
61. R. Perera, M. Khaliq and R. J. Kuhn, *Antiviral Res.*, 2008, **80**, 11.
62. A. V. Nicola, A. M. McEvoy and S. E. Straus, *J. Virol.*, 2003, **77**, 5324.
63. A. V. Nicola, J. Hou, E. O. Major and S. E. Straus, *J. Virol.*, 2005, **79**, 7609.
64. M. A. Brindley and W. Maury, *J. Virol.*, 2005, **79**, 14482.

65. W. Mothes, A. L. Boerger, S. Narayan, J. M. Cunningham and J. A. Young, *Cell*, 2000, **103**, 679.
66. P. Bertrand, M. Cote, Y. M. Zheng, L. M. Albritton and S. L. Liu, *J. Virol.*, 2008, **82**, 2555.
67. R. Blumenthal, A. Bali-Puri, A. Walter, D. Covell and O. Eidelman, *J. Biol. Chem.*, 1987, **262**, 13614.
68. J. White, A. Helenius and M. J. Gething, *Nature*, 1982, **300**, 658.
69. Y. Yao, K. Ghosh, R. F. Epand, R. M. Epand and H. P. Ghosh, *Virology*, 2003, **310**, 319.
70. J. Zhou and G. W. Blissard, *Virology*, 2006, **352**, 427.
71. Q. Wang, F. Cheng, M. Lu, X. Tian and J. Ma, *J. Virol.*, 2008, **82**, 3011.
72. R. Fritz, K. Stiasny and F. X. Heinz, *J. Cell Biol.*, 2008, **183**, 353.
73. D. Atanasiu, J. C. Whitbeck, T. M. Cairns, B. Reilly, G. H. Cohen and R. J. Eisenberg, *Proc. Natl. Acad. Sci. USA*, 2007, **104**, 18718.
74. R. P. Subramanian and R. J. Geraghty, *Proc. Natl. Acad. Sci. USA*, 2007, **104**, 2903.
75. R. A. Lamb, *Virology*, 1993, **197**, 1.
76. M. L. Bissonnette, S. A. Connolly, D. F. Young, R. E. Randall, R. G. Paterson and R. A. Lamb, *J. Virol.*, 2006, **80**, 3071.
77. C. M. Horvath, R. G. Paterson, M. A. Shaughnessy, R. Wood and R. A. Lamb, *J. Virol.*, 1992, **66**, 4564.
78. X. L. Hu, R. Ray and R. W. Compans, *J. Virol.*, 1992, **66**, 1528.
79. S. Biacchesi, M. H. Skiadopoulos, L. Yang, E. W. Lamirande, K. C. Tran, B. R. Murphy, P. L. Collins and U. J. Buchholz, *J. Virol.*, 2004, **78**, 12877.
80. S. Techaarpornkul, N. Barretto and M. E. Peeples, *J. Virol.*, 2001, **75**, 6825.
81. A. A. Kolokoltsov, D. Deniger, E. H. Fleming, N. J. Roberts Jr, J. M. Karpilow and R. A. Davey, *J. Virol.*, 2007, **81**, 7786.
82. S. Herfst, V. Mas, L. S. Ver, R. J. Wierda, A. D. Osterhaus, R. A. Fouchier and J. A. Melero, *J. Virol.*, 2008, **82**, 8891.
83. R. M. Schowalter, S. E. Smith and R. E. Dutch, *J. Virol.*, 2006, **80**, 10931.
84. R. M. Schowalter, A. Chang, J. G. Robach, U. J. Buchholz and R. E. Dutch, *J. Virol.*, 2008, **83**, 1511.
85. R. G. Paterson, C. J. Russell and R. A. Lamb, *Virology*, 2000, **270**, 17.
86. A. E. Gardner and R. E. Dutch, *J. Virol.*, 2007, **81**, 8303.
87. L. E. Luque and C. J. Russell, *J. Virol.*, 2007, **81**, 3130.
88. A. E. Gardner, K. L. Martin and R. E. Dutch, *Biochemistry*, 2007, **46**, 5094.
89. J. Rawling, B. Garcia-Barreno and J. A. Melero, *J. Virol.*, 2008, **82**, 5986.
90. R. M. Markosyan, F. S. Cohen and G. B. Melikyan, *Mol. Biol. Cell*, 2000, **11**, 1143.
91. S. A. Connolly, G. P. Leser, H. S. Yin, T. S. Jardetzky and R. A. Lamb, *Proc. Natl. Acad. Sci. USA*, 2006, **103**, 17903.
92. R. E. Dutch and R. A. Lamb, *J. Virol.*, 2001, **75**, 5363.

CHAPTER 14

Structural Studies on Antibody–Virus Complexes

THOMAS J. SMITH

Donald Danforth Plant Science Center, 975 North Warson Road, St. Louis, MO 63132, USA

1 Introduction

Over the past 20 years, there have been numerous advances in the study of antibody interactions with viruses. In particular, advances in technology and software have had a tremendous impact on both X-ray crystallography and cryo-electron microscopy (cryo-EM) and image reconstruction. The kinds of analysis that were rare 20 years ago are now nearly routine.

This chapter discusses many of the structural studies on antibody–virus interactions in approximately chronological order with the intent to summarize the key points. Leading off that discussion is a brief review of antibody structure and B cell development and a general overview of the commonly discussed mechanisms of antibody-mediated neutralization.

2 Antibody Structure and Diversity

Since the main goal of this chapter is to discuss the process of antibody recognition of viral epitopes, it is important first to review the basic structure of antibodies and how they are formed during the adaptive immune response. Humans express five kinds (isotypes) of antibodies: IgG, IgM, IgD, IgA and IgE. Their basic architecture is that of a 'Y' where two heavy chains (of molecular weight ~ 50–70 kDa) are connected to each other *via* one or more disulfide bonds and one light chain (~ 25 kDa) is attached to each heavy chain

RSC Biomolecular Sciences No. 21
Structural Virology
Edited by Mavis Agbandje-McKenna and Robert McKenna
© Royal Society of Chemistry 2011
Published by the Royal Society of Chemistry, www.rsc.org

via a disulfide bond at the C-terminus. Each type of chain is made up of immunoglobulin domains that are comprised of a β-barrel. The immunoglobulin domain at the tip of the two arms of the Y is called the variable region and the very tip of this domain lies the portion that contacts the antigen (the hypervariable region) that is composed of three loops from the light chain and three from the heavy chain. In terms of the amino acid sequence, the loops are found around residues 30, 60 and 90. The first two immunoglobulin domains of the heavy and light chains form the Fab (*f*ragment that is the *a*ntigen *b*inding portion of the antibody) that represent the two arms of the 'Y'. In the case of IgG, IgD and IgA antibodies, this Fab arm is connected to the remainder of the antibody, the Fc (*f*ragment that *c*rystallized in early studies) portion *via* an extremely flexible loop called the hinge. Similar flexibility is found in IgM and IgE antibodies due to flexible connections between Fc regions.

Each of the main types of antibodies has a particular function that driven by its structure. IgM antibodies are five 'Y' constructs connected together at the base of the 'Y' *via* an additional domain at the carboxyl end of the heavy chain and a linker protein called the J chain. IgM and IgD are co-expressed on B cells *via* differential RNA processing of the same mRNA. This processing is developmentally regulated with immature B cells making mainly the μ transcript (for IgM) and mature B cells mainly making the δ transcript (for IgD) along with some μ transcript. The exact role of the IgD is unclear since mice lacking the μ exon appear to have normal immune systems. IgG antibodies have archetypal 'Y' shape, can activate the complement cascade, are the major type of serum antibodies, can cross the placental barrier and bind to phagocytic cells such as macrophages. IgA antibodies are found as either monomers or as dimers, are secreted to the mucosal surface and can also bind to phagocytic cells. IgE antibodies are more specialized in that they play an important role in attacking parasites but, because they bind with high affinity to mast cells and basophils, are also involved in allergenic responses to antigens.

It is estimated that an individual has the potential to produce more than 10^{11} different antibodies through a series of recombination and mutational events. During B cell development, the heavy chain variable region is the first to form from recombination of 65 variable (V), 27 diversity (D) and six joining (J) regions. Diversity is created by making various combinations of the VDJ regions and the splicing event itself introduces variations at the junction regions. After the developing B cell creates a viable heavy chain exon, then one of the 40 light chain variable region combines with one of the 5 J regions. There are two possible loci, κ and λ, for the creation of the light chain exon. Prior to exposure to antigen, these naïve B cells link the heavy chain VDJ region to either the δ or μ exons to make IgD or IgM antibodies, respectively. Upon stimulation by antigen and T helper cells, these B cells undergo further changes through somatic hypermutation with the driving force being selection of those B cells that, through mutations, have higher affinity for the antigen on their cell surface antibodies than other B cells in the pool. Therefore, the various combinations of VDJ (heavy chain) and VJ (light chain) cassettes, combined with the two possible light chains, leads to ∼3.5×10^6 different antibodies. This is increased to $\sim 10^{11}$ antibodies when junctional diversity is included and there are even more combinations created during somatic hypermutation in the presence of antigen.

The driving force in stimulating the propagation and affinity maturation of B cells is the interaction with antigen and not the subsequent ability of those antibodies to neutralize the target pathogen. This strongly infers that, by far, the major role of antibodies is to bind tightly to antigen. Upon binding *in vivo*, the antibodies cause a number of responses as they interact with other components of the immune system such as phagocytic cells and complement proteins. Indeed, some antibodies may either neutralize weakly or not at all *in vitro* but, combined with other components of the immune system, can protect the individual from the pathogen. It is also clear that the incredible diversity in the antigen-binding (hypervariable) region offers nearly unlimited potential in recognizing antigens. Indeed, it was initially surprising when it was found that antibodies could recognize man-made compounds – hence there was not the possibility of evolving that particular specificity. Therefore, any observed limitation in antibody recognition of viruses is unlikely to be due any lack of genetic potential. The architecture is also well suited for the maturation process and for binding to multivalent surfaces. Naïve B cells have not yet undergone affinity maturation and therefore their antibodies (IgM) have a relatively weak affinity. To compensate for this, the IgMs have 10 Fab arms that increase the apparent affinity (avidity) by essentially requiring all arms to disassociate from the antigen for the antibody to return to the bulk solvent. As the B cells are stimulated by antigens, the μ exon is replaced by other more stable and soluble heavy chain constant domains (*e.g.* γ for an IgG antibody) and somatic mutations optimize the hypervariable region to compensate for this loss in binding valency. Nevertheless, the two Fab arms of these other antibody types also have the ability to cross-link antigens or bind bivalently to the antigenic surface. This simple structural feature can increase the apparent antibody affinity by as much as 1000-fold.[1]

3 Mechanisms of *In vitro* Antibody-mediated Neutralization of Viruses

This section summarizes several of the major proposed mechanisms of antibody-mediated neutralization. As a caveat, however, it must noted that how the *in vitro* assays are performed can greatly impact the measured neutralization efficacy of the antibodies. This has been shown most strikingly in the case of human rhinovirus 14 (HRV14). When the antibody is mixed with the virus, added to the target cells and then the unattached antibody–virus complex is washed away, some monoclonal antibodies are far more efficacious than others. However, if the excess antibody is not removed after the attachment phase, then most of these differences in antibody efficacy disappear. Therefore, the extent to which the results of *in vitro* studies can be extrapolated to vaccine development may be limited in some viral systems.

Aggregation

It was suggested that aggregation and neutralization occur concomitantly and that antibody:virus ratios *in vivo* favor aggregation.[2–4] However, studies on HRV14 have shown that antibodies that bind bivalently to virions and do not aggregate the virions are strong neutralizers.[5] Other HRV14 antibodies that strongly aggregate HRV14 were shown to neutralize virus even at antibody:virus ratios that do not favor aggregation. Antibodies that tend to aggregate often have optimal neutralization activity at ratios where immunoprecipitation is greatest. This enhanced neutralization may come from a decrease in the number of independent infectious particles or from avidity effects caused by antibodies bound to neighboring particles in the large immuno-complexes. *In vivo*, aggregation may be important for innate immunological responses such as opsonization.

Virion Stabilization

Antibodies have also been thought to neutralize virions by stabilizing the capsid that in turn might prevent uncoating. In the case of HRV14, both aggregating and non-aggregating NIm-IA antibodies stabilize virions to varying extents against low pH.[5] Antibodies to the other sites did not cause similar stabilization, nor did stabilization correlate with neutralization efficacy or binding valency. Furthermore, antibodies to all four NIm sites prevent cellular attachment,[6] thereby blocking infectivity prior to possible stabilization effects. Notably, no escape mutation has yet been observed that prevents neutralization without affecting antibody binding. It seems that if capsid stabilization/destabilization was a major determinant of neutralization, certainly some escape mutations would have been observed that could abrogate these effects without directly affecting antibody binding.

Induction of Conformational Changes

Antibodies and Fab fragments cause an apparent decrease in the pI of the poliovirus and rhinovirus capsid concomitant to neutralization,[6,7] and this observation was used to suggest that antibodies neutralize by distorting the capsid structure. The crystal structure of the Fab17–HRV14 complex clearly demonstrated that efficacious neutralization does not require large conformational changes in the capsid.[8] In the case of HRV, all antibodies to the four different antigenic sites caused apparent changes in the pI of the capsid,[6] but it seems unlikely that all antibodies binding to all over the capsid cause the same change in the capsid. As discussed in the following subsections, there may be some examples of antibodies binding to particular conformations being sampled by normal capsid dynamics. However, from the discussion above about B cell maturation, it is also unlikely that B cells could specifically create antibodies that would necessarily cause conformational changes in the pathogen.

Abrogation of Cellular Attachment

Antibodies to all four HRV14 antigenic sites clearly block cellular attachment.[6] However, only NIm-IA antibodies were shown to overlap directly the binding site footprint of ICAM-1 as determined by cryo-TEM.[9] As reviewed below, there are now many examples of antibodies directly contacting the viral receptor binding area but abrogation of cell attachment could be more indirect and by virtue of the large size ($\sim 150\,\text{Å}$) of antibodies. There are also examples reviewed below where antibodies may not have large, direct affects on receptor binding but may affect subsequent release of genomic material into the cell (Figure 1).

Other *In situ* Effects

There is evidence that some antibodies neutralize in a manner not easily understood. For example, it has been shown that antibodies to Sindbis

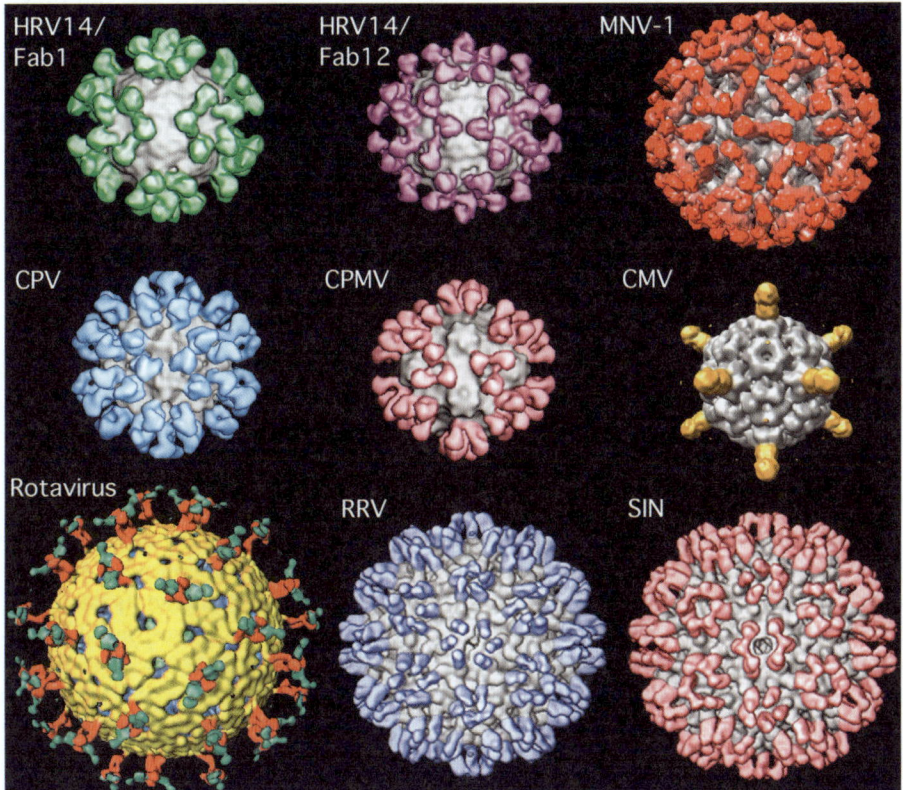

Figure 1   Some of the Fab–virus complexes examined using cryo-TEM methods. Except for the rotavirus complex, the bound Fab fragments are highlighted by color whereas the capsid protein is presented in gray. In the case of the rotavirus reconstruction, the bound antibodies are colored green.

virus[10,11] and poliovirus[12,13] can eliminate infection or progression of infection even when added to cells hours after infection. In the case of Sindbis virus, the exact mechanism of this viral clearance is unknown, but appears to be related to antibody cross-linking.[11] Similarly, the post-adsorption neutralization properties of at least some of the antibodies to poliovirus appeared to be related to binding valency as well.[13] Therefore, antibodies, or antibodies interacting with viral components, may be triggering some unknown defensive mechanism within the infected cell.

Significance of *In vitro* Neutralization Mechanisms *In vivo*

It is important to note, however, that these *in vitro* mechanisms may not represent the primary mode by which antibodies protect animals from viral infections. For example, antibodies against Sindbis virus[14] and FMDV[15] that are not efficacious *in vitro* still protect animals from viral challenge. Also, non-neutralizing antibodies against the neuraminidase spike of influenza can block disease progression *in vivo*.[16] Therefore, the primary role of antibodies *in vivo* may be to act synergistically with other components of the immune system. The challenge for vaccine design, therefore, is to create an antigen that best represents the authentic antigen and can effectively induce high-affinity antibodies to the pathogen.

Antibody–Virus Complexes

Over the past few decades, numerous structural studies have examined antibody–virus complexes in order to understand better how antibodies neutralize viruses. Some of these studies are reviewed below to see if common mechanisms of antibody neutralization emerge.

Influenzavirus

Influenzaviruses are enveloped viruses that belong to the *Orthomyxoviridae* family. These viruses have a pleomorphic or spherical morphology with a diameter of 800–1200 Å. The viral genome consists of 7–8 segments of linear negative-sense single-stranded RNA. There are two types of spikes on the outer envelope; the major protein is hemagglutinin (HA) and the less prominent neuraminidase (NA) (Figure 2).

When the high-resolution structure of influenza virus N9 NA was determined, it was noted that the conserved residues involved in sialic acid binding were located in a crevasse.[17] Analogous to most enzymes, a cavity or pocket-like structural feature may have evolved to facilitate contact with receptor. Since residues within this deep depression are conserved while the residues about the rim vary with serotype, it was suggested that conserved residues are hidden from antibody recognition.[17]

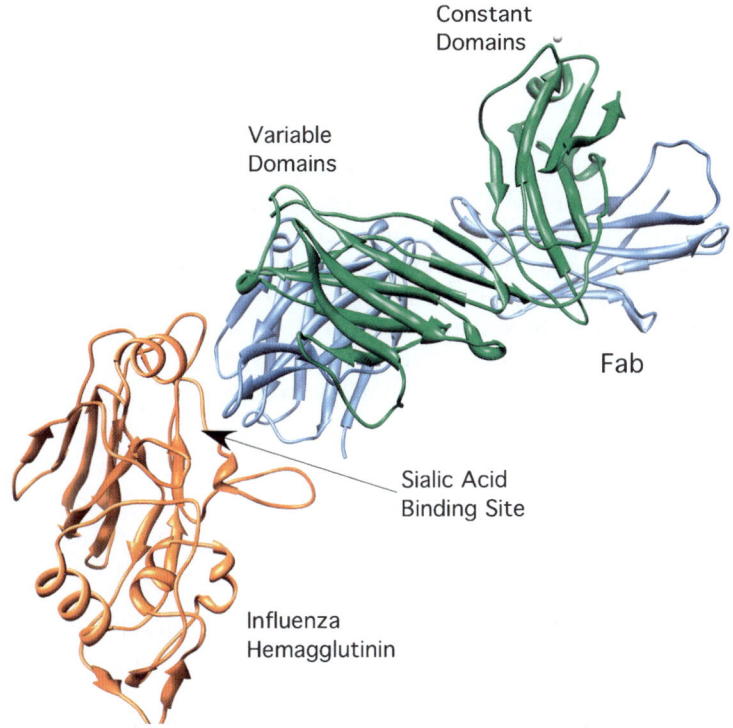

Constant Domains

Variable Domains

Fab

Sialic Acid Binding Site

Influenza Hemagglutinin

Figure 2   Structure of a portion of the influenza hemagglutinin spike complexed with a neutralizing antibody. The antibody is represented by blue and green ribbon diagrams and the hemagglutinin is in orange. The general location of the sialic acid binding pocket is indicated by the arrow.

More recent studies demonstrated that about one-third of the conserved binding region in this depression is contacted by a neutralizing antibody.[18] To explain how viruses might evade antibody attack while leaving conserved residues immunologically exposed, Colman proposed that this capability may reflect the potential for different proteins to recognize identical protein surfaces.[19] In this way, receptors and antibodies can bind to overlapping areas of the viral surface, but can exhibit differing sensitivities to mutations at these contact surfaces.

Rotavirus

Rotaviruses are a member of the family *Reoviridae*. Although immunity after infection is incomplete, repeat infections tend to be less severe than the original infection. The complete rotavirus particle has three shells; an outer capsid, inner capsid and core. These viruses have an 11-segmented genome of double-stranded RNA that encodes six structural and five nonstructural proteins.

The outer shell is comprised of two of the structural proteins, VP7 (the glycoprotein or G protein) and VP4 (the protease cleaved or P protein). These proteins define the serotype of the virus and are the major antigens involved in virus neutralization.[20]

The structure of rotavirus complexed with Fab fragments of a neutralizing monoclonal antibody raised against VP4 was the first Fab–virus structure determined using cyro-EM.[21] VP4 has been implicated in several important functions such as cell penetration, hemagglutination, neutralization and virulence. These results also demonstrated that the surface spikes on rotavirus particles are comprised of VP4. The antigenic sites were found to be located near the distal ends of the spikes and two Fab fragments bound to each of the 60 spikes. These studies also showed that the elbow region (the junction between the $V_H V_L$ and the $C_{H1} C_L$ domains) is highly flexible in solution.

Cowpea Mosaic Virus

Cowpea mosaic virus (CPMV) is the type species of the comovirus genus of the *Comoviridae* family. The CPMV coat contains 60 copies of two viral proteins: the large subunit (molecular weight 42 000) and the small subunit (24 000). The large subunit contains two of the canonical viral eight-stranded β-barrels that are arranged about the icosahedral threefold axes, whereas the small subunit is adjacent to the fivefold axes.[22]

Unlike subsequent studies with animal viruses, the antibody recognition site was unknown due to the lack of natural escape mutation analysis. The cryo-TEM structure of CPMV complexed with Fab fragments[23] were therefore important to define the most antigenic region of the virion surface to aide in the design of CPMV-based vaccines.[24] While the dogma had been that antibodies would predominantly recognize the large protruding domains of viral capsids, this antibody actually recognized the flattened surface between the protruding pentameric towers located at the icosahedral fivefold axes. Furthermore, this was the first structural evidence that antibodies need not cause gross conformational changes in the virions upon binding. Subsequent reconstructions with IgGs from 5B2[25] demonstrated that the intact antibody binds in a monodentate fashion with only one Fab arm attached to the virus surface. Because of the marked flexibility of the hinge region, the unbound Fc and Fab arms were disordered and formed islands above the viral surface.

Human Rhinovirus 14

Picornaviruses are among the largest of animal virus families and include polio-, rhino-, foot-and-mouth disease, Coxsackie and hepatitis A viruses. The rhinoviruses, of which there are more than 100 serotypes, are major causative agents of the common cold in humans.[26] The virus is non-enveloped and has an ~300 Å diameter protein shell that encapsidates a single-stranded, plus-sense RNA genome of about 7200 bases. The human rhinovirus 14 (HRV14) capsid exhibits a pseudo T = 3 (P = 3) icosahedral symmetry and consists of 60 copies each of four viral proteins, VP1, VP2, VP3 and VP4. VP4 is smaller, has an extended structure and lies at the RNA–capsid interface.[27] An ~20 Å deep canyon surrounds each of the 12 icosahedral fivefold vertices. The canyon regions of HRV14 and HRV16, both major receptor group rhinoviruses, were shown to contain the binding site of the cellular receptor, intercellular adhesion molecule 1 (ICAM-1).[28–30] Four major neutralizing immunogenic (NIm) sites, NIm-IA, NIm-IB, NIm-II and NIm-III, were identified by studies of neutralization-escape mutants with monoclonal antibodies[31,32] and then mapped to four protruding regions on the viral surface[27] (Figure 3).

Neutralizing monoclonal antibodies against HRV14 have been divided into three groups: strong, intermediate and weak.[33,34] All strongly neutralizing antibodies bind to the NIm-IA site, defined by natural escape mutations at residues D91 and E95 of VP1. Because strongly neutralizing antibodies form stable, monomeric antibody–virus complexes with a maximum stoichiometry of 30 antibodies per virion, it was concluded that they bind bivalently to the virions.[33,34] Weakly neutralizing antibodies form unstable, monomeric complexes with HRV14 and bind with a stoichiometry of ~60 antibodies per virion.[34,35] The remaining antibodies, all of which precipitate the virions, are classified as intermediate neutralizers.[33,34]

The structures of three different Fab–HRV14 (Fab17, Fab12 and Fab1) complexes and of one mAb–HRV14 (mAb17) complex have been determined.

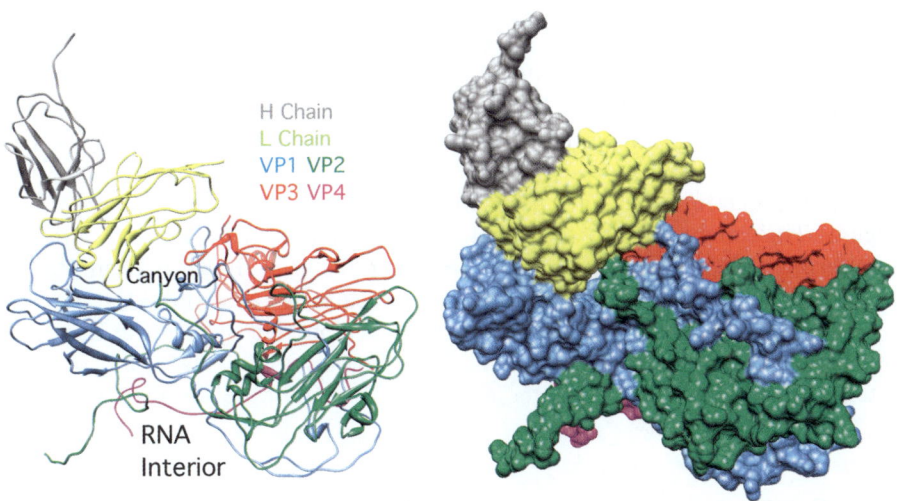

Figure 3   Crystal structure of human rhinovirus 14 complexed with a neutralizing Fab fragment. On the left is a ribbon diagram demonstrating that the antibody binds down into the canyon that serves as a receptor binding region for the major group of HRV serotypes. On the right, the surface rendering of the same structure demonstrates how the antibody fills and overlaps the entire receptor binding region.

Although all bind to the same antigenic site, mAb17 and mAb12 are strongly neutralizing antibodies, whereas mAb1 is a weakly neutralizing antibody. Fab17 and Fab12 both bind to the NIm-IA site at a somewhat tangential orientation that placed the constant domains ($C_{H1}C_V$) of twofold related Fabs in close proximity to each other that could facilitate bidentate binding across icosahedral twofold axes. In contrast, Fab1 binds almost vertically to the virion surface with a 'twist' that made it seem unlikely that these antibodies could bind bivalently. Bivalent binding was subsequently visualized by the structure of the mAb17–HRV14 complex.[36]

The atomic structures of Fab1, Fab17 and HRV14 were used to construct pseudo-atomic models and then tested using site-directed mutagenesis.[5,35] In the Fab17–HRV14 complex, the loop of the NIm-IA site on HRV14 is clamped in the cleft between the heavy and light chain hypervariable regions and forms complementary electrostatic interactions with corresponding side chains of Fab17. In addition, a cluster of lysines on HRV14 VP1 (K236, K97, K85) interacts with two acidic residues, Asp45H and Asp54H, in the CDR2 region of the Fab heavy chain.[8] Using site-directed mutagenesis, it was found that even though K1236, K1097 and K1085 were not identified as sites of naturally occurring escape mutations, they do affect antibody binding. Therefore, electrostatic interactions can dominate paratope–epitope interactions and naturally occurring escape mutations represent only a small subset of residues crucial for antibody binding.

These studies, at resolutions of 20–30 Å, consistently demonstrated that antibodies do not induce conformational changes in the virion upon binding. This was further tested at higher resolution with the crystal structure of the Fab17–HRV14 complex.[8] In this structure, the only observable changes in the virus were that the side-chains of VP1 D91 and E95 that were found to rotate slightly in order to form salt bridges with the basic residues in the paratope cleft. There were not, however, any significant changes in the NIm-IA loop or the rest of the capsid.

It was apparent that Fab17 penetrates into the receptor-binding canyon region. This might lead to the conclusion that Fab17 neutralizes by directly interfering with ICAM-1 binding. However, it has been shown that antibodies to all four antigenic sites can abrogate cell attachment,[6] even though some sites (*e.g.* NIm-III) are fairly distal to the canyon region. The simplest explanation for this is that, owing to the large bulk of antibodies, it would take only a few to interfere with virus–receptor interactions at the membrane surface.

Parvovirus

Members of the parvovirus genus cause a number of diseases in mammals, including enteritis[37,38] and childhood fifth disease. These non-enveloped viruses have a capsid diameter of ~ 255 Å that encases a double-stranded DNA genome and only infect cells that are in the S phase.[39] In canine parvovirus (CPV), the T = 1 capsid is composed mainly of 60 copies of viral protein 2 (VP2).

To understand better the mechanism of antibody-mediated neutralization of CPV, the cryo-TEM structure of the CPV–Fab complex was determined.[40] For these studies, Fab fragments from the antibody A3B10 that recognizes epitope B were used. Since Fab fragments from this antibody were nearly as efficacious as the mAb, this complex represents a neutralized state. Since the Fab molecules bound perpendicularly to the virion surface, it was clear that these antibodies were unlikely to bind bivalently to the surface of the virion. From these modeling exercises, it was also apparent that this epitope region was not nearly as hydrophilic as the NIm-IA site of HRV14. As was the case with the HRV14 work, this antibody does not recognize the viral protein Western blots, nor does it bind to the peptide representing this epitope loop. Therefore, it was concluded that the antibody is probably recognizing both the epitope loop and the context in which it is being presented.

From these results, several mechanisms of neutralization were eliminated. Since the Fab fragments are as efficacious as the intact IgG, neutralization must be independent of aggregation and bivalent attachment. No changes were observed upon antibody binding, making it also unlikely that the antibody neutralizes by inducing gross conformational changes in the virion. It was suggested that the proximity of the B site to icosahedral symmetry axes might allow for Fabs to stabilize the capsids. However, it is also possible that the antibodies block infectivity by virtue of their bulk.

Poliovirus

Poliovirus is a member of the enterovirus genus that belongs to the *Picornaviridae* family. The structure of poliovirus is remarkably similar to that of human rhinovirus 14.[27,41] There is a direct link between the receptor binding site and the serotypic determinants in poliovirus. Upper, exposed regions of the poliovirus canyon are crucial for receptor interactions, but residues at the bottom of the poliovirus canyon are not. In fact, changes at the top of the canyon that affect antibody-neutralizing sites also alter receptor–virus interactions.[42] Yielding similar conclusions, other studies showed that mutations at the north and south walls of the canyon overcome deleterious defects in the poliovirus receptor. These mutations, which are fairly distal to the canyon floor, lie very close to the antigenic sites and appear to represent destabilizing mutations.[43]

Studies on polioviruses were the first to show that picornaviruses undergo a dynamic transition or 'breathing' that is part of their natural infection cycle. Infectious poliovirus is thought to have at least two conformations, A and B. The A form is preferred at and above neutral pH whereas the B form has been reported to have a p*I* below 4.5. The main difference between the poliovirus B particles and eclipsed (non-infectious, 80S) particles is that B particles still have VP4 and can revert back to the A form. Roivainen *et al.*[44] demonstrated that antibodies to the buried N-terminus of poliovirus VP1 could immunoprecipitate intact poliovirus with preference for 80S particles over intact native virions.

Indeed, based on peptide scanning results, the first 17 residues of VP1 are immunodominant. These authors were the first to suggest that the capsid might be undergoing 'breathing' in solution. This was followed up by similar studies that demonstrated antibodies to VP1 and VP4 could react in a temperature dependent manner and further supported the idea of picornavirus 'breathing'.[45]

In subsequent studies, Fab fragments from a neutralizing antibody, C3, were used.[46] This antibody was originally raised against heat inactivated virus particles and strongly neutralized the Mahoney strain of poliovirus type 1. The Fab was complexed with a peptide corresponding to the viral epitope and the structure was determined to a resolution of 3.0 Å. The carboxyl end of the peptide was found to interact extensively with the paratope and adopted a conformation that differed significantly from the structure of the corresponding residues in the virus. This apparent difference between the bound peptide and the authentic antigenic loop suggested that this antibody might induce structural changes important for neutralization.

Antibodies can actually facilitate viral infection only when the virions are in the 135S state. Poliovirus binding to its receptor (PVR) on the cell surface induces a conformational transition that generates an altered particle with a sedimentation value of 135S *versus* the 160S of the native virion.[47] These altered 135S particles are much less infectious than native virions. In earlier studies, it was found that neutralizing antibodies to the native virion block attachment to target cells. When cells were made to express Fc receptors, the antibody–virus complex was again able to enter the cells but was still non-infectious.[48] Subsequently, it was shown that a poliovirus receptor–IgG2a (Fc portion) hybrid molecule permitted poliovirus to enter and infect *via* this Fc receptor.[49] This was followed by studies showing that when antibodies specific for 135S particles are added to the 135S particles, infectivity increases by 2–3 orders of magnitude.[50] This suggests that the lack of infectivity in 135S particles is due to the loss of cell binding. These results further imply that one function of neutralizing antibodies is to prevent these viruses from interacting with receptor and becoming 'primed' for uncoating.

Alphavirus

The alphaviruses are a group of 26 icosahedral, positive-sense RNA viruses that are primarily transmitted by mosquitoes.[51] These ∼ 700 Å diameter viruses are some of the simplest of the membrane-enveloped viruses and members of this group cause serious tropical diseases.[52] The viral RNA genome and 240 copies of the capsid protein form the nucleocapsid core[53–58] and the E1 and E2 glycoproteins form heterodimers that associate as 80 trimeric spikes on the viral surface. E1 has a putative fusion domain that may facilitate host membrane penetration.[59,60] E2 contains most of the neutralizing epitopes and is also probably involved in host cell recognition.[61–63]

To examine the mechanism of antibody neutralization and to identify the portion of E2 involved in receptor recognition, two antibodies were examined,

SV209 and T10C9.[64] Anti-idiotypic antibodies to SV209 compete with SIN for its cellular receptor and block viral attachment by $\sim 50\%$.[62] This implies that the original SV209 antibody is recognizing at least a portion of the spike involved in cellular recognition. The naturally occurring mutation in Ross River virus that facilitates escape from the T10C9 antibody maps to residue T216 of E2.[65] This residue is presumably near the cell receptor-binding site since residue N218 was found to vary as the virus adapted to growth in chicken cells[66] and residue T219 mutates to ALA during the course of an epidemic in humans.[67]

In both virus–Fab reconstructions, the Fab fragments were observed to bind to the outermost tips of the trimeric spikes.[64] When compared with reconstructions of the virus alone, the binding of the antibody did not appear to cause conformational changes in the virion. While the two antibodies bound to their respective viruses with markedly different orientations, their binding footprints were nearly identical on the highly exposed tip of the spike. Further, the results with the anti-idiotypic antibodies suggest that there is a direct overlap between the antibody and receptor contact areas.

Human Rhinovirus 2

HRV2 is a member of the minor group of human rhinoviruses. The receptor for the minor group is low-density lipoprotein receptor[68–72] and binds to the BC and HI loops of VP1 that lies on the surface of the virus near the fivefold axis.[73] Whereas all of the work on HRV14 has focused on the NIm-IA site (site A in HRV2), the structural work on HRV2 has focused on the B site (NIm-II in HRV14). Both of the antibodies used in these studies, mAb-8F5 and mAb-3B10, are weakly neutralizing antibodies[74,75] and neither antibody grossly impairs viral attachment to cells. Since mAb-8F5, but not mAb-3B10, binds with a stoichiometry of 30 and does not cause apparent immunoprecipitation, it was proposed that only mAb-3F5 binds bivalently to the viral surface.

The structures of the Fab[76] and the Fab–antigenic loop complex[77] were used in modeling of the Fab structures into the cryo-TEM envelopes.[74,75] The mAb-8F5 antibody binds nearly perpendicular to the surface and it was suggested that it binds bivalently to the virion surface across the nearest twofold axis. This orientation is not what was predicted using the HRV1A structure since this antigenic loop bends towards the twofold axis when the antibody binds. In contrast, mAb-3B10 binds at an angle of $\sim 45°$ away from the icosahedral twofold axes, making it impossible to model a bivalently bound antibody.

Calicivirus

Rabbit hemorrhagic disease virus (RHDV) and murine norovirus (MNV) are members of the *Caliciviridae* family. The positive-sense ssRNA genome is encapsidated by a $T = 3$ icosahedral capsid with an architecture is similar to

TBSV with dimers of the capsid protein forming 90 arch-like capsomers on the viral surface.

The first cryo-TEM study on an antibody–calicivirus complex was performed using the neutralizing antibody mAb-E3 and virus-like particles of RHDV.[78] It was suggested that this antibody might bind bivalently to the virion surface since there was relatively little aggregation upon complex formation. From the image reconstruction, it was clear that the antibody was binding to the top of the dimeric bridges characteristic of the caliciviruses. As with all of the other reconstructions that used intact antibodies, the Fc region was not visible, presumably due to hinge flexibility.[79,80] Due to spatial overlap, only one antibody could bind to these dimeric arches at a time and therefore the antibody density was relatively weak compared with the capsid, with the density of constant domains of the Fab arms being weaker than the variable domains. It was suggested that the antibodies may be cross-linking these arch dimers about the icosahedral threefold axes. This would also explain the diffuse density observed above the threefold axes. In terms of antibody neutralization, it is clear that neither aggregation nor gross conformational changes are responsible for neutralization. This antibody blocks attachment of the virus to human group O red blood cells and therefore steric abrogation of cellular attachment is one possible mechanism of neutralization.

Murine norovirus (MNV) is the only norovirus of the calicivirus family that can be grown in tissue culture, studied in an animal model, reverse engineered *via* an infectious clone and to which neutralizing antibodies have been isolated.[81] Both the intact and Fab fragments of the antibody used in these studies were shown to be capable of neutralizing MNV. The cryo-TEM structures of the MNV T = 3 authentic virion were determined in the presence and absence of neutralizing Fab fragments. Unlike all of the previous calicivirus structures, the P domains of MNV were found to twist and rise up off the surface of the shell domains by ~ 16 Å. In this new orientation, they form an outer shell of P domains that could represent a form of maturation akin to other viral systems.[82–84] Although this conformation of the P domains is unusual compared with some of the other known calicivirus structures, it is interesting that the image reconstruction of RHDV showed a similar and large gap between the P and shell domains (see Figures 2 and 4 in Thouvenin *et al.*[78]). Importantly, the structure of MNV was unchanged by the bound antibodies. These results suggest that antibody-mediated neutralization is independent of conformational changes and bivalent attachment.

Herpes Simplex Virus

Herpes simplex virus type 1 (HSV-1) is a member of the *Herpesviridae* family and is a large, complex virus with a diameter of ~ 1250 Å. The shell is made up of ~ 3000 polypeptides and the virion has a total mass of ~ 200 MDa.[85,86] The majority of the capsid is made up of 960 copies of VP5 (149 kDa molecular weight) that forms the 150 hexons and 12 pentons. At 320 sites of threefold symmetry are triplexes. These triplexes are heterotrimers and are proposed to have to have an $\alpha_2\beta$ stoichiometry[87] of VP23 and VP19c. These three proteins are essential for capsid assembly and co-assemble with pre-VP22a that acts as an internal scaffolding protein during procapsid formation.[88,89] During maturation, pre-VP22a is cleaved and expelled from the capsid. The fourth abundant protein in the capsid is called VP26 and has a molecular weight of 12 kDa.[90,91] VP26 is dispensable for assembly but incorporates in an equimolar ratio with VP5.[87]

Other than their differences in symmetry, the pentons and hexons have similar structures.[87,92–95] The hexons are cylindrical projections with a diameter of 170 Å and a height of 110 Å. A channel with a diameter of 50 Å runs through the center of the hexons. Each of the VP5 subunits making up the hexons has three domains; a diamond-shaped upper domain, a stem-like central domain and a base domain. The lower domains for both hexons and pentons form the 30–40 Å capsid floor. In spite of these similarities, cryo-TEM studies have demonstrated that VP26 associates with the tips of hexons but not the pentons.

To augment better the structural differences between the hexons and pentons that are being discerned by VP5, the virus was decorated with the antibody 6F10.[96] The residues being recognized by this antibody were determined using a combination of limited proteolysis, immunoblotting with GST–peptide fusions and reactivity to synthetic peptides. From these results, this antibody mostly likely binds to peptide region ~862–880. In the image reconstructions, the antibody binds on the outer surface of the capsid just inside the opening of the channel that runs through the capsomers. Because these antibodies are binding near a symmetry axis, they have more of a 'turret' shape than the well-defined structures observed in some of the other antibody–virus complexes. Nevertheless, it is clear that the antibodies do not induce gross conformational changes in the virion.

Adenovirus

Adenoviruses are non-enveloped viruses that are a significant cause of respiratory, ocular and gastrointestinal infections in humans.[97,98] Adenoviruses have icosahedral shells with diameters ranging from 80 to 110 Å with 240 hexons formed by three copies of viral protein II. Twelve copies of protein IX are found between nine hexons in the center of each icosahedral facet. Several copies of VI form a ring under the peripentonal hexons. The pentons at each of the 12 fivefold axes are composed of five copies of protein III and are tightly associated with one or two fibers each composed of three copies of protein IV. The 22 Å fibers are a shaft with a knob at the very tip.

The pentons play an important role in viral entry. As expected, since the co-receptors are integrins, both antibodies to the functional domains of integrin and RGD peptides can block viral entry.[99] From cryo-TEM studies, it was proposed that the RGD loop was located on the outermost tip of these fibers.[100] Although both $\alpha_v\beta3$ and $\alpha_v\beta5$ integrins can serve as co-receptors, only

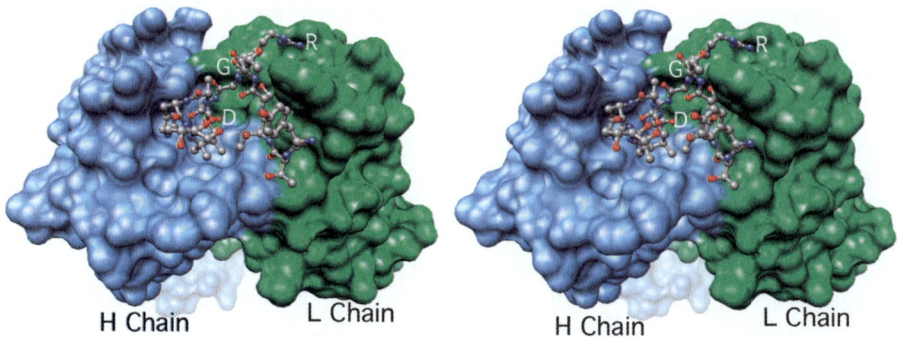

Figure 4 Interactions between the receptor binding RGD loop of FMDV and a
neutralizing antibody. In this stereo diagram, the paratope of the antibody is
represented by the blue and green surfaces and the viral RGD loop is shown
as a ball and stick model. Note how the antibody makes extensive contact
with the canonical Arg–Gly–Asp residues.

$\alpha_v \beta 5$ enhances membrane permeablization.[101] Upon treatment of the capsid
with pH 5.0 buffers, the pentons become highly hydrophobic and are therefore
thought to interact with the endosomal membranes.[102] In image reconstructions of the virus alone, weak density is observed ~ 24 Å above the penton base
protein – suggestive of a mobile loop decorating the penton base protein
(Figure 4).[103]

To understand better antibody neutralization of these complex virions, the
cryo-TEM structure of a Fab–adenovirus complex was determined.[103] For
these studies, an unusual antibody, DAV-1, was chosen. Using peptide-scan-
ning techniques, it was found that this antibody recognizes a nonapeptide with
the sequence of **IRGDTFATR**. This is consistent with the fact that this antibody recognizes several different adenovirus serotypes that have similar
sequences flanking the RGD motif but poorly recognizes Epstein–Barr virus
that has dissimilar flanking residues.

In contrast to all other antibodies discussed in this review, the Fab fragments
of DAV-1 were better able to neutralize viral infectivity than the intact mAb
fragments. This is in spite of the fact that both the mAb and Fab were both
potent inhibitors of penton base–cell interactions. Indeed, the IgG had about a
fourfold higher affinity for the pentons than did the Fab yet were ineffective at
viral neutralization. A possible reason for this difference came from the fact
that the Fab fragments bind with a stoichiometry of 5/penton whereas the
mAbs bind with a stoichiometry of 2.8/penton. Therefore, although the affinities of the Fab and IgG were similar, their binding stoichiometries were not.

The image reconstruction of the Fab–virus complex clearly showed that the
Fab is binding to a flexible portion of the penton base, but this did not result in
any observable structural change in the virus. It was proposed that IgGs had
lower neutralization efficacy than Fabs because they tend to occlude themselves
about the penton base. Although it was suggested that mobility in this RGD

loop might allow virus to escape antibody neutralization, it should be noted that antibody binding itself is sufficient for antibody-mediated opsonization *in vivo*. Furthermore, the fact that the Fabs neutralize better than IgGs, presumably due to their higher binding stoichiometry, supports the hypothesis that antibody neutralization is primarily due to steric interference between the virus and its receptor.

Hepatitis B virus

Hepatitis B virus (HBV) is a member of the *Hepadnaviridae* family and of the genus orthohepadnavirus.[20] HBV causes chronic, acute and fulminate hepatitis and is still a major health issue, with hundreds of millions of individuals infected despite the development of a number of efficacious vaccines.[104] There are two sizes of HBV composed of 90 or 120 capsid protein dimers in a T = 3 or T = 4 icosahedral arrangement, respectively.[105,106]

HBV has three major clinical antigens. One of these antigens is the viral glycoprotein or 'surface antigen'. The other two antigens, the 'core antigen' (HBcAg) and the 'e-antigen' (HBeAg) determinants, are both found on the capsid protein. The difference between these two antigens is that HBcAg is the capsid protein assembled into icosahedral particles and appears early in the infection whereas HBeAg appears late in the infection,[107] correlates with disease progression and is the capsid protein in a non-capsid form. Antibodies to these two capsids are not cross-reactive. A major goal of structural studies was to ascertain how the same protein can result in two different antigens.[108]

The Fab fragments were found to bind directly on the top of the 4-helical spikes that are formed by the interactions of the A/B and C/D subunits within the T = 4 asymmetric unit. The proximity of symmetry-related epitopes limited binding to 30–40% of the possible epitopes. The densities at the two different spikes were roughly equal, suggesting that the all four quasi-equivalent antigenic sites are immunologically indistinguishable. As observed with most of the other antibody–virus complexes, no significant conformational changes were observed upon antibody binding.

The question remaining is why these two forms of capsid protein are so different antigenically. One possibility is that HBeAg is 29 residues shorter at the C-terminus compared with HBcAg. It is also possible that these antigenic differences are due to masking of e2 determinants in the capsid and/or conformational changes in the dimer as it is assembled in the capsid. In the case of the latter, differences in antigenic structures between quasi-equivalent subunits have been observed in herpes simplex virus[109] and cucumber mosaic virus.[110]

Foot and Mouth Disease Virus

Foot and mouth disease (FMDV) is a highly contagious member of the picornavirus family that infects cloven-hoofed animals. FMDV differs from rhinovirus in several important ways.[111,112] Unlike the convoluted surfaces of rhinovirus and poliovirus, FMDV has a relatively smooth surface. Protruding up from the shell is a long flexible loop connecting the βG–βH strands of VP1. At the tip of this loop is a conserved RGD sequence that is recognized by the viral receptor, integrin αvβ3. While there are four or five immunogenic sites on VP1, -2 and -3 in the seven serotypes of FMDV, most flank the RGD motif residues. This led these to the suggestion that FMDV can use 'camouflage' to hide crucial residues 'in plain sight'.[111] According to this hypothesis, crucial residues might be exposed to antibodies but simply do not change in response to antibodies since doing so will be lethal to the virus while residues adjacent to the RGD sequence can change and thwart antibody binding (Figure 5).

A direct test of the 'camouflage' hypothesis came from the structures of the Fab–peptide[113–115] and FMDV–heparin complexes.[116] For these studies, two strongly neutralizing antibodies were used, SD6 and 4C4. In the case of the

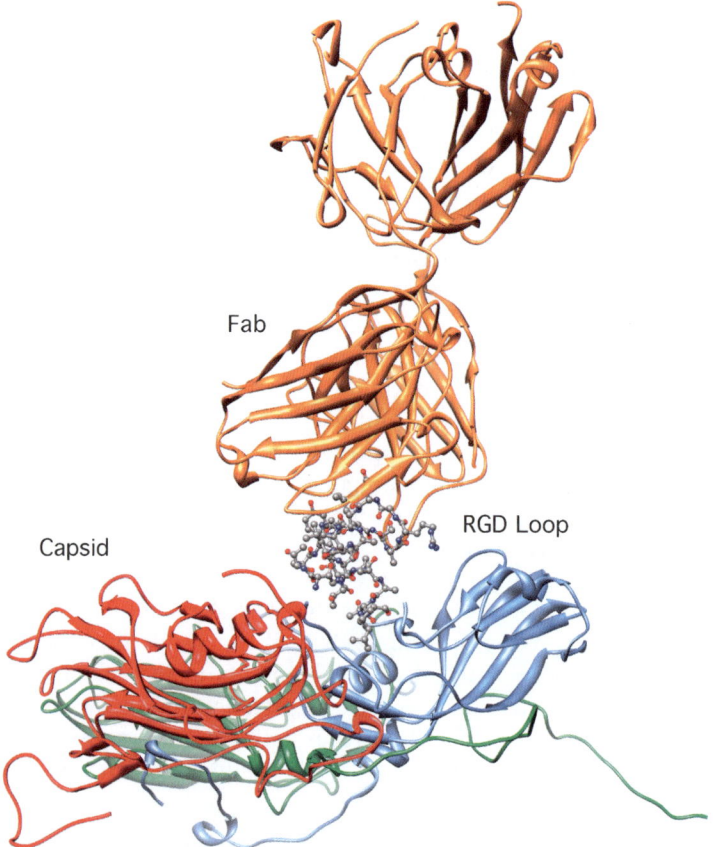

Figure 5   Pseudo atomic model of FMDV complexed with neutralizing Fab fragments. Using the atomic structures of FMDV and the Fab–peptide complexes along with the cryo-TEM electron density, the structure of the antibody–virus complex was assembled.

SD6–peptide complex, there was a great deal of rearrangement of the CDR3 loop of the heavy chain and a number of residues in the peptide also. This altered conformation was observed at a low occupancy in the Fab structure, suggesting that the induced fit conformation is somewhat 'natural' to the paratope and therefore antigen binding does not come at a high energy cost.

A secondary receptor binds to a different region than the RGD loop. Heparin sulfate has been suggested to be involved in a two-step attachment process where low-affinity interaction with heparin sulfate at one site is followed by high-affinity binding to an integrin receptor *via* the RGD sequence.[117] The interactions between heparin and FMDV determined in the structure of FMDV complexed with a heparin sulfate.[116] Like the RGD sequence,[118] this oligosaccharide binding site is not only exposed but is also part of one of the antigenic sites.[116] Again, these results clearly demonstrate that viruses do not necessarily hide receptor-binding regions in convolutions on the virion surface.

To examine further the interactions between FMDV and neutralizing antibodies, the atomic structures of peptide–Fab complexes and the cryo-EM structures of the Fab–virus complexes were determined.[119] In the image reconstructions, SD6 has a well-defined orientation on the virion surface, whereas the density for 4C4 is extremely diffuse. One possible reason for this difference is that, while the RGD loop is in an extended conformation in the SD6 complex, a hinge rotation at the base of the loop may bring it closer to the capsid surface and stabilize the orientation.

Papillomavirus

Papillomavirus infections usually cause benign epithelial papillomas, but a subset of human papillomaviruses is associated with cervical cancer.[120] Papillomavirus has a 600 Å diameter shell that encases a histone-bound 8 kb double-stranded, covalently closed circular genome.[20] This icosahedral shell is composed of a major (L1) and minor (L2) capsid protein in a ratio of ~ 30:1.[121] Image reconstructions of papillomavirus show that the capsid is composed of pentameric, star-shaped capsomers arranged in a $T = 7$ icosahedron.[122] Studies on BPV to 9 Å have suggested that L2 may be located at the fivefold vertices, in the center of the pentavalent capsomers.[123]

Virus-like particles can be generated from viral protein L1 alone and these assemblies retain the antigenic determinants of the authentic virion.[124,125] Antibody-mediated neutralization can occur by more than one mechanism. One set of antibodies (*e.g.* mAb #9 to BPV1 L1) prevents virions from binding to cell surfaces, presumably by abrogating interactions with the cell surface receptor (possibly α6 integrin).[126–128] A second group, (*e.g.* 5B6 to BPV1 L1) neutralizes but does not significantly block cell attachment.

To determine the structural basis for these apparent differences in neutralization mechanism, the cryo-TEM image reconstructions of both the mAb #9/BPV and mAb 5B6/BPV complexes were determined.[129] The epitope for mAb #9 is between the protrusions of density at the very tip of the capsomers.

Antibody was observed bound to each of the pentavalent capsomers, but only three of the five L1 molecules in the hexavalent capsomers had antibody bound. It was suggested that steric hindrance prevents antibody binding to the two adjacent L1 molecules of the hexavalent capsomers. mAb 5B6 bound in a different manner to mAb #9. No antibody was observed to bind to the fivefold pentamer and the epitope is on the side of the capsomere about 25 Å above the capsid floor. 5B6 adopts a more linear conformation and binds deep into the cleft between the hexavalent capsomers and fills all of the space between capsomers around the epitope. In both cases, antibodies did not induce conformational changes in the virion. Although mAb #9 tended to aggregate the virions, the fact that this antibody can neutralize the virus post-attachment makes it unlikely that such behavior is responsible for neutralization. mAb #9 covers much of the viral surface and may account for its abrogation of cell attachment. Since mAb 5B6 binds close to the putative inter-capsomer linkages, this antibody may neutralize by stabilizing the virion.

Reovirus

Reoviruses are non-enveloped virions consisting of two concentric protein shells, (outer capsid and core[20]) that encapsidate a genome of 10 double-stranded RNA gene segments. Reovirus strain type 1 Lang (T1L) has a diameter of 850 Å and has 600 projections comprised of the σ3 protein.[84] σ3 interdigitates with a more internal layer composed of 600 copies of μ1 protein that form the outer capsid. At each icosahedral fivefold axis, pentamers of λ2 protein form turrets. At the center of each fivefold axis is the viral attachment protein σ1. In virions, the σ1 has a retracted conformation[84,130] where it may interact with σ3.[131] During cellular entry *via* the endosomes, the σ3 protein is removed from virions by acid-dependent proteolysis[132,133] that is hypothesized to facilitate a conformational change in σ1 to a more extended form.[134] Monoclonal antibodies to each of the reovirus outer-capsid proteins have been isolated and characterized.[131,135] σ1-specific mAbs are serotype specific[131,135] and some of these mAbs are effective at neutralizing infectivity *in vitro*.[131,135,136]

mAb 4F2, which is specific for outer-capsid protein σ3,[131] blocks the binding of σ1 protein to sialic acid and inhibits reovirus-induced hemagglutination (HA). The structure of the 4F2–T3D complex was determined to ascertain whether mAb 4F2 inhibits HA by altering σ1–σ3 interactions or by steric hindrance.[137] In this case, the intact 4F2 was > 16-fold better than the corresponding Fab fragments at inhibiting T3D-induced HA. However, the affinity of the Fab fragments was only about threefold weaker than that of the mAbs.

From comparing these fitting results of the virion and antibody–virion complexes,[137] it was proposed that the antibody binding induced a small change in σ3 orientation. A small spur of density was also observed at a radius of ~385 Å and was thought to be indicative of an antibody-induced rearrangement of the μ1 protein. As was expected, the hypervariable region of the bound antibody contacted residue 116 that was previously shown to be important for 4F2 binding. The constant domains for the mAb needed to be adjusted by an elbow rotation compared with the Fab reconstruction in order to optimize the fit, suggesting that elbow motion is required for bivalent binding.

From these results, the most likely mode of antibody-mediated abrogation of virally induced HA appears to be steric hindrance of sialic acid binding. Both Fabs and mAbs cause identical apparent changes in the outer capsid. The problem with this model is that the isolated σ1 in the extended conformation has a length of ~ 480 Å while an mAb has an extended length of ~ 150 Å. It was proposed that the increased length and bulk of the intact antibody is able to block sialic acid binding to σ1, but the shorter Fab fragments that lack the Fc portion cannot.

Cucumber Mosaic Virus

Cucumber mosaic virus (CMV), the type member of the genus cucumovirus (family *Bromoviridae*), infects over 800 plant species and causes economically important diseases of many crops worldwide.[138] CMV is transmitted by aphids in a non-persistent manner; it does not circulate or replicate in the aphid.

The X-ray crystal structure of CMV revealed an exposed βH–βI loop,[139] the sequence of which is highly conserved among strains of CMV and other cucumoviruses.[140] Mutations in several of the loop residues (D191, D192, L194 and E195) had no significant affect on virion formation or stability but they did reduce or eliminate aphid transmission.[140] To understand better the molecular basis for virus transmission by insects, antibodies were developed against this loop and the structure of this antibody–virus complex was determined.[110]

The cryo-TEM and modeling results clearly demonstrated that this antibody binds immediately adjacent to an axis of icosahedral symmetry and only one Fab binds per penton. Indeed, each antibody bound to several antigenic loops at the same time. Although antibodies are known to be able to bind simultaneously to multiple viral subunits,[25,32] this was the first example in which an antibody bridges the same regions of two or more identical and adjacent capsid subunits. Since the βH–βI loop that is thought to interact with the receptor molecule in the aphid, it may be that the aphid receptor, similar to antibody 3A8-5C10, also exhibits quasi-equivalent specificity and may interact only with pentons or only with hexons.

West Nile and Dengue Viruses

West Nile virus (WNV) and dengue virus (DENV) are members of the flavivirus family, of which yellow fever is also a member. These ~ 500 Å diameter viruses are membrane enveloped with an external icosahedral scaffold of 180 glycoproteins that recognize the host receptor and are involved in a pH-mediated membrane fusion event during endocytosis. On comparing different monoclonal antibodies with WNV, it was found that antibody E16 blocked attachment by ~3.5-fold (using an RT-PCR assay to quantify the bound virus), whereas antibodies that bind outside of this epitope are approximately 10-fold weaker at viral neutralization and block attachment by ~10-fold. In addition, E16 is effective at preventing macrophage infection (representing the process of antibody dependent enhancement of viral infection) and can inhibit replication when added to the virus after it is allowed to initially attach to the cells at 4 °C. It was proposed that antibodies like E16 might prevent conformational changes required for host cell fusion.[141,142] It will be interesting to see whether this model can be confirmed by future studies that directly monitor E glycoprotein rearrangement and the effect of these monoclonal antibodies on that conformational change. In the case of DENV, an antibody (1A1D-2) was found to bind to the virus at 37 °C but not at room temperature.[143] This implies that the higher temperatures induce 'breathing' in DENV that expose the antibody epitope. Indeed, large changes were observed in the organization of the envelope protein in the antibody cryo-TEM image reconstruction.[143] It could be argued that at the higher temperatures the epitopes are reversibly exposed (breathing) and, as antibodies bind, the conformational equilibrium shifts towards the antibody-bound altered state. However, since similar changes were not observed when antibody E16 bound to a similar location in WNV at room temperature, it was suggested that initial binding by antibody might cause a cascade of changes in the icosahedral arrangement of the enveloped protein.[143] Delineation between these models will require further analysis of dengue virus structure at 37 °C to ascertain whether this altered virus conformation is part of the normal viral structure ensemble at these elevated temperatures. For example, it may even be that DENV breathing at 37 °C, in the absence of antibody, occurs in a cooperative manner so as to maintain icosahedral contacts.

4 Summary

From this wealth of structural information about antibody–virus complexes, what conclusions can be drawn? With the ability of the individual theoretically to create $> 10^{11}$ different types of antibodies, it is not surprising to see that antibodies can bind to all kinds of viral surfaces; from canyons to protrusions and from proteins to carbohydrates. It is also clear that the driving force in selecting for the appropriate antibody response to a particular pathogen is affinity during B cell maturation. In many cases, this can result in a spectrum of antibodies with *in vitro* neutralization activity ranging from strong to weak. Further, it is also clear that antibodies can neutralize *in vitro* by a combination of various mechanisms. Perhaps the most common mechanism is the interference with the manner that viruses interact with the host cell by blocking attachment or by affecting the processes that are required for the transfer of genetic material across the membrane into the cell. However, in most cases it is not clear what is the proportional relevance of this antiviral action compared with the overall protection afforded by antibodies acting in concert with other components of the immune system. Further, from the manner by which B cells are clonally expanded, it seems highly improbable that antibodies with particular modes of action are made in response to a pathogen. Indeed, all escape mutants to date have been localized to the antibody contact areas and not to distal sites that might compensatory mutants that counteract antibody effects (*e.g.* stabilization or destabilization).

The next generation of vaccines will most certainly include insights from all of these structural studies. For a particular target, is there an epitope that is more highly conserved than a more immunodominant one? Does cross-reactivity in an immune response come at a cost of affinity to a particular serotype(s)? During 'breathing' or when the virus interacts with receptor, are conserved portions of the capsid exposed and does the virus expose these epitopes sufficiently for efficacious neutralization? Finally, is the adaptive humoral response sufficient to block pathogenesis or are other specific components of the immune system necessary for protection? Such approaches are already being applied to a number of vaccine systems and it will be interesting to see how well the immune system can be augmented.

References

1. C. L. Hornick and F. Karush, *Immunochemistry*, 1972, **9**, 325.
2. P. Brioen, D. Dekegel and A. Boeyé, *Virology*, 1983, **127**, 463.
3. P. Brioen, A. A. M. Thomas and A. Boeyé, *J. Gen. Virol.*, 1985, **66**, 609.
4. A. A. Thomas, R. Vrijsen and A. Boeyé, *J. Virol.*, 1986, **59**, 479.
5. Z. Che, N. H. Olson, D. Leippe, W.-M. Lee, A. Mosser, R. R. Rueckert, T. S. Baker and T. J. Smith, *J. Virol.*, 1998, **72**, 4610.
6. R. J. Colonno, P. L. Callahan, D. M. Leippe and R. R. Rueckert, *J. Virol.*, 1989, **63**, 36.
7. B. Mandel, *Virology*, 1967, **31**, 247.
8. T. J. Smith, E. S. Chase, T. J. Schmidt, N. H. Olson and T. S. Baker, *Nature*, 1996, **383**, 350.
9. N. H. Olson, T. J. Smith, P. R. Kolatkar, M. A. Oliveira, R. R. Rueckert, J. M. Greve, M. G. Rossmann and T. S. Baker, *Proc. Electron Microsc. Soc. Am.*, 1992, **50**, 524.
10. B. Levine, J. M. Hardwick, B. D. Trapp, T. O. Crawford, R. C. Bollinger and D. E. Griffen, *Nature*, 1991, **254**, 856.
11. S. Ubol, B. Levine, S. H. Lee, N. S. Greenspan and D. E. Griffin, *J. Virol.*, 1995, **69**, 1990.
12. E. A. Tolskaya, T. A. Ivannikova, M. S. Kolesnikova, S. G. Drozdov and V. I. Agol, *J. Virol.*, 1992, **66**, 5152.
13. R. Vrijsen, A. Mosser and A. Boeye, *J. Virol.*, 1993, **67**, 3126.
14. A. L. Schmaljohn, E. D. Johnson, J. M. Dalrymple and G. A. Cole, *Nature*, 1982, **297**, 70.

15. K. C. McCullough, F. De Simone, E. Brocchi, L. Capucci, J. R. Crowther and U. Kihm, *J. Virol.*, 1992, **66**, 1835.
16. J. L. Schulman, in *The Influenza Viruses and Influenza*, ed. E. D. Kilbourne, Academic Press, New York, 1975, p. 373.
17. P. M. Colman, J. N. Varghese and W. G. Laver, *Nature*, 1983, **303**, 41.
18. T. Bizebard, B. Gigant, P. Rigolet, B. Rasmussen, O. Diat, P. Bosecke, S. A. Wharton, J. J. Skehel and M. Knossow, *Nature*, 1995, **376**, 92.
19. P. M. Colman, *Structure*, 1997, **5**, 591.
20. F. A. Murphy, C. M. Fauquet, D. H. L. Bishop, S. A. Ghabrial, A. W. Jarvis, G. P. Martelli and M. A. Mayo, *Virus Taxonomy. Sixth Report of the International Committee on Taxonomy of Viruses*, 6th edn, Springer, Vienna, 1995.
21. B. V. V. Prasad, J. W. Burns, E. Marietta, M. K. Estes and W. Chiu, *Nature*, 1990, **343**, 476.
22. Z. Chen, C. Stauffacher, Y. Li, T. Schmidt, W. Bomu, G. Kamer, M. Shanks, G. Lomonossoff and J. Johnson, *Nature*, 1989, **245**, 154.
23. P. Wang, C. Porta, Z. Chen, T. S. Baker and J. E. Johnson, *Nature*, 1992, **355**, 275.
24. C. Porta, V. Spall, J. Loveland, J. Johnson, P. Barker and G. Lomonossoff, *Virology*, 1994, **202**, 949.
25. C. Porta, R. H. Cheng, Z. Chen, T. S. Baker and J. E. Johnson, *Virology*, 1994, **204**, 777.
26. R. R. Rueckert, *Picornaviridae and Their Replication*, Raven Press, New York, 1996.
27. M. G. Rossmann, E. Arnold, J. W. Erickson, E. A. Frankenberger, J. P. Griffith, H. J. Hecht, J. E. Johnson, G. Kamer, M. Luo, A. G. Mosser, R. R. Rueckert, B. Sherry and G. Vriend, *Nature*, 1985, **317**, 145.
28. R. J. Colonno, J. H. Condra, S. Mizutani, P. L. Callahan, M. E. Davies and M. A. Murcko, *Proc. Natl. Acad. Sci. USA*, 1988, **85**, 5449.
29. N. H. Olson, P. R. Kolatkar, M. A. Oliveira, R. H. Cheng, J. M. Greve, A. McClelland, T. S. Baker and M. G. Rossmann, *Proc. Natl. Acad. Sci. USA*, 1993, **90**, 507.
30. P. R. Kolatkar, J. Bella, N. H. Olson, C. M. Bator, T. S. Baker and M. G. Rossmann, *EMBO J.*, 1999, **18**, 6249.
31. B. Sherry and R. R. Rueckert, *J. Virol.*, 1985, **53**, 137.
32. B. Sherry, A. G. Mosser, R. J. Colonno and R. R. Rueckert, *J. Virol.*, 1986, **57**, 246.
33. A. G. Mosser, D. M. Leippe and R. R. Rueckert, in *Molecular Aspects of Picornavirus Infection and Detection*, ed. B. L. Semler and E. Ehrenfeld, American Society for Microbiology, Washington, DC, 1989, pp. 155–167.
34. D. M. Leippe, PhD. Thesis, University of Wisconsin, 1991.
35. T. J. Smith, N. H. Olson, R. H. Cheng, H. Liu, E. Chase, W. M. Lee, D. M. Leippe, A. G. Mosser, R. R. Ruekert and T. S. Baker, *J. Virol.*, 1993, **67**, 1148.
36. T. J. Smith, N. H. Olson, R. H. Cheng, E. S. Chase and T. S. Baker, *Proc. Natl. Acad. Sci. USA*, 1993, **90**, 7015.

37. M. J. Studdert, in *CRC Handbook of Parvoviruses*, Vol. II, ed. P. Tijssen, CRC Press, Boca Raton, FL, 1990, p. 3.
38. C. R. Parrish, *Adv. Virus Res.*, 1990, **38**, 403.
39. G. Siegl, in *The Parvoviruses*, ed. K. I. Berns, Plenum Press, New York, 1984, pp. 297–362.
40. W. R. Wikoff, G. Wang, C. R. Parrish, R. H. Cheng, M. L. Strassheim, T. S. Baker and M. G. Rossmann, *Structure*, 1994, **2**, 595.
41. J. M. Hogle, M. Chow and D. J. Filman, *Nature*, 1985, **229**, 1358.
42. J. Harber, G. Bernhardt, H. H. Lu, J. Y. Sgro and E. Wimmer, *Virology*, 1995, **214**, 559.
43. M. W. Wien, S. Curry, D. J. Filman and J. M. Hogle, *Nat. Struct. Biol.*, 1997, **4**, 666.
44. M. Roivainen and L. Piirainen et al., *Virology*, 1993, **195**, 762.
45. Q. Li, A. G. Yafal, Y. M. H. Lee, J. Hogle and M. Chow, *J. Virol.*, 1994, **68**, 3965.
46. M. W. Wien, D. J. Filman, E. A. Stura, S. Guillot, F. Delpeyroux, R. Crainic and J. M. Hogle, *Nat. Struct. Biol.*, 1995, **2**, 232.
47. C. E. Fricks and J. M. Hogle, *J. Virol.*, 1990, **64**, 1934.
48. P. W. Mason, B. Baxt, F. Brown, J. Harber, A. Murdin and E. Wimmer, *Virology*, 1993, **192**, 568.
49. M. Arita, H. Horie, M. Arita and A. Nomoto, *J. Virol.*, 1999, **73**, 1066.
50. Y. Huang, J. M. Hogle and M. Chow, *J. Virol.*, 2000, **74**, 8757.
51. C. H. Calisher, N. Karabatsos, J. S. Lazuick, T. Monath and K. L. Wolff, *Am. J. Trop. Med. Hyg.*, 1988, **38**, 447.
52. B. H. Kay and J. G. Aaskov, in *The Arboviruses: Epidemiology and Ecology*, Vol. IV, ed. T. P. Monath, CRC Press, Boca Raton, FL, 1989, p. 93.
53. K. Coombs and D. T. Brown, *Virus Res.*, 1987, **7**, 131.
54. H.-K. Choi, L. Tong, W. Minor, P. Dumas, U. Boege, M. G. Rossmann and G. Wengler, *Nature*, 1991, **354**, 37.
55. A. M. Paredes, M. Simon and D. T. Brown, *Virology*, 1993, **187**, 324.
56. A. M. Paredes, D. T. Brown, R. Rothnagel, W. Chiu, R. E. Johnston and B. V. V. Prasad, *Proc. Natl. Acad. Sci. USA*, 1993, **90**, 9095.
57. L. Tong, G. Wengler and M. G. Rossmann, *J. Mol. Biol.*, 1993, **230**, 228.
58. R. H. Cheng, R. J. Kuhn, N. H. Olson, M. G. Rossmann, H. Choi, T. J. Smith and T. S. Baker, *Cell*, 1995, **80**, 1.
59. H. Garoff, A. M. Frischauf, K. Simons, H. Lehrach and H. Delius, *Proc. Natl. Acad. Sci. USA*, 1980, **77**, 6376.
60. C. M. Rice and J. H. Strauss, *Proc. Natl. Acad. Sci. USA*, 1981, **78**, 2062.
61. E. G. Strauss, D. S. Stec, A. L. Schmaljohn and J. H. Strauss, *J. Virol.*, 1991, **65**, 4654.
62. S. Ubol and D. E. Griffin, *J. Virol.*, 1991, **65**, 6913.
63. K.-S. Wang, A. L. Schmaljohn, R. J. Kuhn and J. H. Strauss, *Virology*, 1991, **181**, 694.

64. T. J. Smith, R. H. Cheng, N. H. Olson, P. Peterson, E. Chase, R. J. Kuhn and T. S. Baker, *Proc. Natl. Acad. Sci. USA*, 1995, **92**, 10648.
65. S. Vrati, C. A. Fernon, L. Dalgarno and R. C. Weir, *Virology*, 1988, **162**, 346.
66. P. J. Kerr, R. C. Weir and L. Dalgarno, *Virology*, 1993, **193**, 446.
67. A. T. Burness, I. Pardoe, S. G. Faragher, S. Vrati and L. Dalgarno, *Virology*, 1988, **167**, 639.
68. B. Ronacher, T. C. Marlovits, R. Moser and D. Blaas, *Virology*, 2000, **278**, 541.
69. T. C. Marlovits, C. Abrahamsberg and D. Blaas, *J. Biol. Chem.*, 1998, **273**, 33835.
70. T. C. Marlovits, T. Zechmeister, M. Gruenberger, B. Ronacher, H. Schwihla and D. Blaas, *FASEB J.*, 1998, **12**, 695.
71. H. Mischak, C. Neubauer, E. Kuechler and D. Blaas, *Virology*, 1988, **163**, 19.
72. V. Okun, R. Moser, B. Ronacher, E. Kenndler and D. Blaas, *J. Biol. Chem.*, 2001, **276**, 1057.
73. E. A. Hewat, E. Neumann, J. Conway, R. Moser, B. Ronacher, T. C. Marlovits and D. Blaas, *EMBO J.*, 2000, **19**, 6317.
74. E. A. Hewat and D. Blaas, *EMBO J.*, 1996, **15**, 1515.
75. E. A. Hewat, T. C. Marlovits and D. Blaas, *J. Virol.*, 1998, **72**, 4396.
76. J. Tormo, E. Stadler, T. Skern, H. Auer, O. Kanzler, C. Betzel, D. Blaas and I. Fita, *Protein Sci.*, 1992, **1**, 1154.
77. J. Tormo, D. Blaas, N. R. Parry, D. Rowlands, D. Stuart and I. Fita, *EMBO J.*, 1994, **13**, 2247.
78. E. Thouvenin, S. Laurent, M.-F. Madelaine, D. Rasschaert, J.-F. Vautherot and E. A. Hewat, *J. Mol. Biol.*, 1997, **270**, 238.
79. K. R. Roux, *Eur. J. Immunol.*, 1984, **14**, 459.
80. R. H. Wade, J. C. Taveau and J. N. Lamy, *J. Mol. Biol.*, 1989, **206**, 349.
81. C. E. Wobus, S. M. Karst, L. B. Thackray, K.-O. Chang, S. V. Sosnovtsev, G. Belliot, A. Krug, J. M. Mackensie, K. Y. Green and H. W. I. Virgin, *PLoS Biol.*, 2004, **2**, e432.
82. W. R. Wikoff, J. F. Conway, J. Tang, K. K. Lee, L. Gan, N. Cheng, R. L. Duda, R. W. Hendrix, A. C. Steven and J. E. Johnson, *J. Struct. Biol.*, 2006, **153**, 300.
83. M. A. Canady, M. Tihova, T. N. Hanzlik, J. E. Johnson and M. Yeager, *J. Mol. Biol.*, 2000, **299**, 573.
84. K. A. Dryden, G. Wang, M. Yeager, M. L. Nibert, K. M. Coombs, D. B. Furlong, B. N. Fields and T. S. Baker, *J. Cell Biol.*, 1993, **122**, 1023.
85. F. J. Rixon, *Semin. Virol.*, 1993, **4**, 135.
86. A. C. Steven and P. G. Spear, in *Structural Biology of Viruses*, ed. W. Chiu, R. M. Burnett and R. L. Garcea, Oxford University Press, New York, 1997, pp. 312–351.
87. W. W. Newcomb, B. L. Trus, F. P. Booy, A. C. Steven, J. S. Wall and J. C. Brown, *J. Mol. Biol.*, 1993, **232**, 499.

88. W. W. Newcomb, F. L. Homa, F. P. Booy, D. R. Thomsen, B. L. Trus, A. C. Steven, J. V. Spencer and J. C. Brown, *J. Mol. Biol.*, 1996, **263**, 432.
89. B. L. Trus, F. P. Booy, W. W. Newcomb, J. C. Brown, F. L. Homa, D. R. Thomsen and A. C. Steven, *J. Mol. Biol.*, 1996, **263**, 447.
90. G. H. Cohen, M. Ponce de Leon, H. Diggelmann, W. C. Lawrence, S. K. Vernon and R. J. Eisenberg, *J. Virol.*, 1980, **34**, 521.
91. C. J. Heilman, M. Zweig, J. R. Stephenson and B. Hampar, *J. Virol.*, 1979, **29**, 34.
92. F. P. Booy, B. L. Trus, W. W. Newcomb, J. C. Brown, J. J. Conway and A. C. Steven, *Proc. Natl. Acad. Sci. USA*, 1994, **91**, 5652.
93. B. L. Trus, F. L. Homa, F. P. Booy, W. W. Newcomb, D. R. Thomsen, N. Cheng, J. C. Brown and A. C. Steven, *J. Virol.*, 1995, **69**, 7362.
94. Z. H. Zhou, J. He, J. Jakana, J. D. Tatman, F. J. Rixon and W. Chiu, *Nat. Struct. Biol.*, 1995, **2**, 1026.
95. Z. H. Zhou, B. V. V. Prasad, J. Jakana, F. J. Rixon and W. Chiu, *J. Mol. Biol.*, 1994, **242**, 456.
96. J. V. Spencer, B. L. Trus, F. P. Booy, A. C. Steven, W. W. Newcomb and J. C. Brown, *Virology*, 1997, **228**, 229.
97. C. D. Brandt, H. W. Kim, A. J. Vargosko, B. C. Jeffries, J. O. Arrobio, B. Rindge, R. H. Parrott and R. M. Chanock, *Am. J. Epidemiol.*, 1969, **90**, 484.
98. M. S. Horwitz, in *Virology*, ed. B. N. Fields and N. M. Knipe, Raven Press, New York, 1990, pp. 1679–1721.
99. T. J. Wickham, P. Mathias, D. A. Cheresh and G. R. Nemerow, *Cell*, 1993, **73**, 309.
100. P. I. Stewart, S. D. Fuller and R. M. Burnett, *EMBO J.*, 1993, **12**, 2589.
101. T. J. Wickham, E. J. Filardo, D. A. Cheresh and G. R. Nemerow, *J. Cell Biol.*, 1994, **127**, 257.
102. P. Seth, M. C. Wilingham and I. Pastan, *J. Biol. Chem.*, 1985, **260**, 14431.
103. P. L. Stewart, C. Y. Chiu, S. Huang, T. Muir, Y. Zhao, B. Chait, P. Mathias and G. R. Nemerow, *EMBO J.*, 1997, **16**, 1189.
104. B. S. Blumberg, *Proc. Natl. Acad. Sci. USA*, 1997, **94**, 7121.
105. R. A. Crowther, N. A. Kiselev, B. Bottcher, J. A. Berriman, G. P. Borisova, V. Ose and P. Pumpens, *Cell*, 1994, **77**, 943.
106. P. T. Wingfield, S. J. Stahl, R. W. Williams and A. C. Steven, *Biochemistry*, 1995, **34**, 4919.
107. F. B. Hollinger, in *Fields Virology*, ed. B. N. Fields, D. M. Knipe and P. M. Powley, Lippincott-Raven, Philadelphia, PA, 1996, pp. 2738–2808.
108. J. F. Conway, N. Cheng, A. Zlotnick, S. J. Stahl, P. T. Wingfield, D. M. Belnap, U. Kanngiesser, M. Noah and A. C. Steven, *J. Mol. Biol.*, 1998, **279**, 1111.
109. P. T. Wingfield, S. J. Stahl, D. R. Thomsen, F. L. Homa, F. P. Booy, B. L. Trus and A. C. Steven, *J. Virol.*, 1997, **71**, 8955.
110. V. D. Bowman, E. S. Chase, A. W. E. Franz, P. R. Chipman, X. Zhang, K. L. Perry, T. S. Baker and T. J. Smith, *J. Virol.*, 2002, **76**, 12250.

111. R. Acharya, E. Fry, E. Stuart, G. Fox, E. Rowlands and F. Brown, *Nature*, 1989, **327**, 709.

112. R. Acharya, E. Fry, D. Stuart, G. Fox, D. Rowlands and F. Brown, *Vet. Microbiol.*, 1990, **23**, 21.

113. E. Domingo, N. Verdaguer, W. F. Ochoa, C. M. Ruiz-Jarabo, N. Sevilla, E. Baranowski, M. G. Mateu and I. Fita, *Virus Res.*, 1999, **62**, 169.

114. E. A. Hewat, N. Verdaguer, I. Fita, W. Blakemore, S. Brookes, A. King, J. Newman, E. Domingo, M. G. Mateau and D. I. Stuart, *EMBO J.*, 1997, **16**, 1492.

115. W. F. Ochoa, S. G. Kalko, M. G. Mateu, P. Gomes, D. Andreu, E. Domingo, I. Fita and N. Verdaguer, *J. Gen. Virol.*, 2000, **81**, 1495.

116. E. E. Fry, S. M. Lea, T. Jackson, J. W. I. Newman, F. M. Ellard, W. E. Blakemore, R. Abu-Ghazaleh, A. Samuel, A. M. Q. King and D. I. Stuart, *EMBO J.*, 1999, **18**, 543.

117. T. Jackson, F. M. Ellard, R. Abu Ghazaleh, S. M. Brookes, W. E. Blakemore, A. H. Corteyn, D. I. Stuart, J. W. I. Newman and A. M. Q. King, *J. Virol.*, 1996, **70**, 5282.

118. N. Verdaguer, M. G. Mateu, D. Andreu, E. Giralt, E. Domingo and I. Fita, *EMBO J.*, 1995, **14**, 1690.

119. N. Verdaguer, G. Schoehn, W. F. Ochoa, I. Fita, S. Brookes, A. King, E. Domingo, M. G. Mateu, D. Stuart and E. A. Hewat, *Virology*, 1999, **255**, 260.

120. D. R. Lowy, R. Kirnbauer and J. T. Schiller, *Proc. Natl. Acad. Sci. USA*, 1994, **91**, 2436.

121. H. Pfister, in *Papovaviridae: Volume 2. The Papillomaviruses*, ed. N. P. Salzman and P. M. Howley, Plenum Press, New York, 1987, p. 1.

122. T. S. Baker, W. W. Newcomb, N. H. Olson, L. M. Cowsert, C. Olson and J. C. Brown, *Cell*, 1991, **60**, 1007.

123. B. L. Trus, R. B. S. Roden, H. L. Greenstone, M. Vrhel, J. T. Schiller and F. P. Booy, *Nat. Struct. Biol.*, 1997, **4**, 413.

124. M. Hagensee and D. Galloway, *Papillomavirus Rep.*, 1993, **4**, 121.

125. R. Kirnbauer, F. Booy, N. Cheng, D. R. Lowy and J. T. Schiller, *Proc. Natl. Acad. Sci. USA*, 1992, **89**, 12180.

126. R. B. S. Roden, N. L. Hubbert, R. Kirnbauer, F. Breitburd, D. R. Lowy and J. T. Schiller, *J. Virol.*, 1995, **69**, 5147.

127. R. B. S. Roden, E. M. Weissinger, D. W. Henderson, F. Booy, R. Kirnbauer, J. F. Mushinski, D. R. Lowy and J. T. Schiller, *J. Virol.*, 1994, **68**, 7570.

128. M. Evander, I. H. Frazer, E. Payne, Y. M. Qi, K. Hengst and N. A. J. McMillan, *J. Virol.*, 1997, **71**, 2449.

129. F. P. Booy, R. B. S. Roden, H. L. Greenstone, J. T. Schiller and B. L. Trus, *J. Mol. Biol.*, 1998, **281**, 95.

130. D. B. Furlong, M. L. Nibert and B. N. Fields, *J. Virol.*, 1988, **62**, 246.

131. H. W. I. Virgin, M. A. Mann, B. N. Fields and K. L. Tyler, *J. Virol.*, 1991, **65**, 6772.

132. G. S. Baer and T. S. Dermody, *J. Virol.*, 1997, **71**, 4921.

133. L. J. Sturzenbecker, M. Nibert, D. Furlong and B. N. Fields, *J. Virol.*, 1987, **61**, 2351.
134. M. L. Nibert, J. D. Chappell and T. S. Dermody, *J. Virol.*, 1995, **69**, 5057.
135. S. J. Burstin, D. R. Spriggs and B. N. Fields, *Virology*, 1982, **117**, 146.
136. D. R. Spriggs and B. N. Fields, *Nature*, 1982, **297**, 68.
137. E. L. Nason, D. Wetzel, S. K. Mukherjee, E. S. Barton, B. V. V. Prasad and T. S. Dermody, *J. Virol.*, 2001, **75**, 6625.
138. P. Palukaitis, M. J. Roossinck, R. G. Dietzgen and R. I. B. Francki, *Adv. Virus Res.*, 1992, **41**, 281.
139. T. J. Smith, E. Chase, T. J. Schmidt and K. Perry, *J. Virol.*, 2000, **74**, 7578.
140. S. Liu, X. He, G. Park, C. Josefsson and K. L. Perry, *J. Virol.*, 2002, **76**, 9756.
141. G. E. Nybakken, T. Oliphant, S. Johnson, S. Burke and M. S. Diamond, *Nature*, 2005, **437**, 764.
142. B. Kaufmann, G. E. Nybakken, P. R. Chipman, W. Zhang, M. S. Diamond, D. H. Fremont, R. J. Kuhn and M. G. Rossmann, *Proc. Natl. Acad. Sci. USA*, 2006, **103**, 12400.
143. S.-M. Lok, V. Kostyuchenko, G. E. Nybakken, H. A. Holdaway, A. J. Battisti, S. Sukupolvi-Petty, D. Sedkal, D. H. Fremont, P. R. Chipman, J. T. Roehrig, M. S. Diamond, R. J. Kuhn and M. G. Rossmann, *Nat. Struct. Mol. Biol.*, 2008, **15**, 312.

Section 3

CHAPTER 15
Development of Anti-HIV Drugs

ROXANA M. COMAN AND ROBERT MCKENNA

Department of Biochemistry and Molecular Biology, Center for Structural Biology, The McKnight Brain Institute, University of Florida, Gainesville, FL 32610, USA

1 Introduction: World Human Immunodeficiency Virus (HIV) Epidemic Status

Acquired immunodeficiency syndrome (AIDS) continues to spread largely unchecked since its first documentation in 1983.[1–3] Even though promising developments have been seen in recent years in the global effort to control the AIDS epidemic, the number of people living with HIV continues to grow, as does the number of deaths due to AIDS. A total of 33.2 million people were living with HIV at the end of 2008, but this number may be as high as 36.1 million, according to figures released by the Joint United Nations Program on HIV/AIDS (UNAIDS) and the World Health Organization (WHO).[4] The rate of new HIV infections has fallen in several countries, but globally these favorable trends are at least partially offset by increases in new infections in other countries. One of the developing regions most devastated by HIV epidemic is sub-Saharan Africa, accounting for two-thirds of all adults and children living with HIV and three-quarters of all adult and child deaths due to AIDS in 2007. Overall, sub-Saharan Africa is home to an estimated 25 million adults and children infected with HIV. However, some of the most worrisome increases in new HIV infections are now occurring in populous countries in other regions, such as Indonesia, China, Russia and various high-income countries.[4]

RSC Biomolecular Sciences No. 21
Structural Virology
Edited by Mavis Agbandje-McKenna and Robert McKenna
© Royal Society of Chemistry 2011
Published by the Royal Society of Chemistry, www.rsc.org

Another important shift in the trend from 2006 that continues with the 2008 AIDS epidemic figures is that globally, and in every region, more adult women (15 years or older) than ever before are now living with HIV.[4]

Therefore, HIV remains a global health problem of unprecedented dimensions. Unknown 27 years ago, HIV has already caused an estimated 25 million deaths worldwide and has generated profound demographic changes in the most heavily affected countries.

2   HIV Genome and Structure

HIV is part of a family or group of viruses called lentiviruses and it is now generally accepted that HIV is a descendant of simian (monkey) immunodeficiency virus (SIV).[5] The HIV family of viruses can be subdivided into two types, HIV-1 and HIV-2, with HIV-1 being responsible for the majority of infections worldwide. HIV-1 is divided in three groups: major (M), outlier (O) and non-M, non-O (N). Group M is further classified in nine subtypes (A–D, F–H, J and K), subsubtypes and a myriad of circulating and unique recombinant forms (CRFs and URFs) (Figure 1A). HIV is a retrovirus that packages two copies of positive-sense RNA strands in its genome. Its genome is about 9 kb in length and is flanked by two long terminal repeats (LTRs) that are involved in integration and regulation of the viral genome. The genome can be read in three frames and there are several overlaps of viral genes in different reading frames, allowing for the encoding of many proteins in a small genome. The viral genes encode for structural (gag and env), enzymatic (pol), accessory (vif, vpr, vpu, nef) and regulatory (rev, tat) proteins (Figure 1B).[6]

HIV has a diameter of 100–120 nm. The outer shell of the virus, known as the viral envelope, consists of a lipid bilayer that is acquired as the virus buds from the cell surface. Embedded in the viral envelope is a complex protein known as gp120/gp41 (envelope, ENV), with an important role in viral entry. Inside the viral ENV is a protein called p17 (matrix, MA) and within this is the viral core, which is made of another viral protein called p24 (core antigen, capsid, CA). The major elements contained within the viral core, besides the viral RNA, are protein p7 (nucleocapsid, NC) that associates with the RNA molecules and three enzymatic proteins, p66/p51 (reverse transcriptase, RT), p11 (protease, PR) and p32 (integrase, IN). Some other regulatory proteins, such as nef, vpr and vif, are also packaged in the virion (Figures 2 and 3).

Current antiretroviral (ARV) drugs target various components in the HIV structure, disrupting essential steps in the viral lifecycle. There are several classes of anti-HIV drugs: nucleoside RT inhibitors (NRTIs) and non-nucleoside RT inhibitors (NNRTIs), PR inhibitors (PIs), IN inhibitors, CD4 + fusion inhibitors, and, most recently, chemokine receptor 5 (CCR5) antagonists. The next paragraphs will briefly describe biochemical and structural characteristics of these viral proteins (Figure 4).

ENV protein consists of an outer protruding cap that is a glycoprotein named gp120 and a stem glycoprotein called gp41. Gp120 units form trimers

A

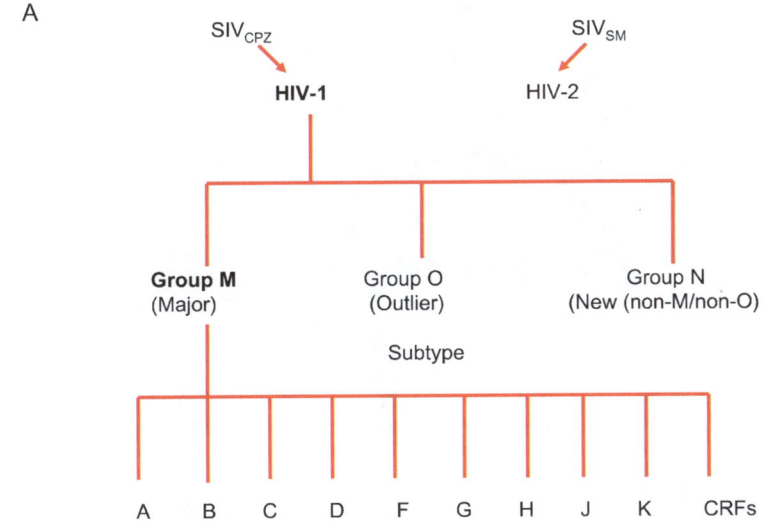

B

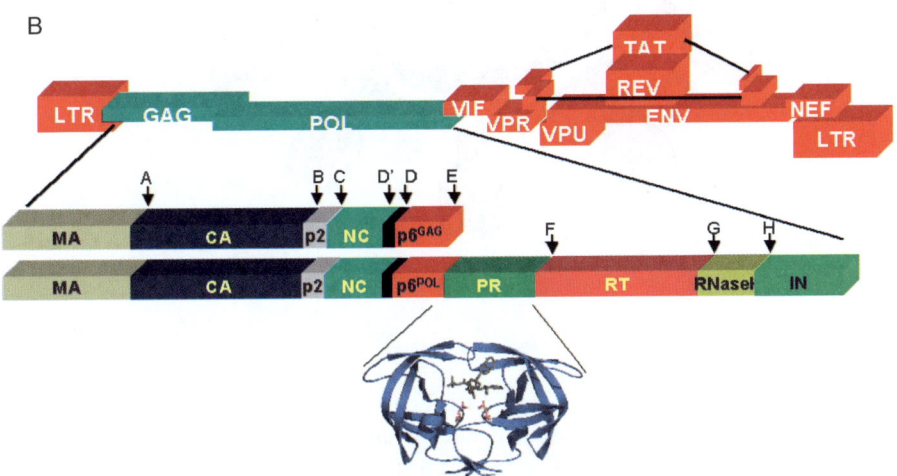

Figure 1 (A) Genetic epidemiology of HIV. Classification of HIV in types, groups, subtypes, subsubtypes. HIV-1 recombinants are categorized in two classes: circulating recombinant forms (CRFs) and unique recombinant forms (URFs). Adapted from Takeb *et al.*[151] (B) HIV gag and gag/pol polyproteins. The gag precursor contains the structural proteins of the viral core: matrix (MA), capsid (CA) and nucleocapsid (NC) in addition to regulatory proteins (p1, p2, p6[GAG]). The gag domain of gag/pol encodes for approximately the same protein as gag precursors and the pol domain of gag/pol contains PR, reverse transcriptase (RT) and integrase (IN). The HIV PR is represented as a blue ribbon.

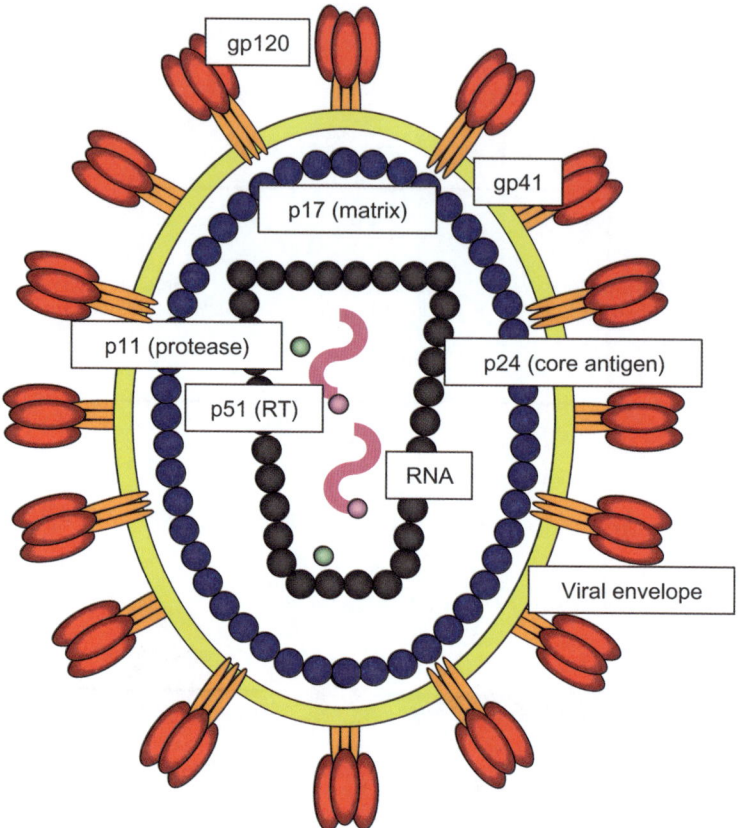

Figure 2 The structure of the mature HIV.

that are non-covalently associated with gp41 trimers. Five well-exposed variable regions (V1–V5) are located on the surface of gp120 and are interspersed by five conserved regions (C1–C5) that form a central core; all of these are heavily glycosylated.[7–9] The gp120 core is composed of an inner and an outer domain and a β-sheet (the 'bridging sheet').[8–10] The inner envelope protein, gp41, interacts with the outer membrane gp120 *via* the C1 and C5 domains of gp120. Gp41 is a transmembrane protein that anchors itself and gp120 to the viral membrane. Gp41 interacts with the viral NC *via* its cytoplasmic tail. Cell–virus binding is enabled *via* the interaction of the N-terminus of the immunoglobulin-like (Ig-CDR) of the CD4 with a cavity in gp120, located at the junction of both the inner and outer domains as well as the bridging sheet (Figure 4A).[8,9,11]

MA is a structural protein located at the N-terminus of gag polyprotein. MA is a relatively small protein of 133 amino acids that folds in a compact core with mainly α-helical structure and loose N- and C-terminal regions. One of its roles

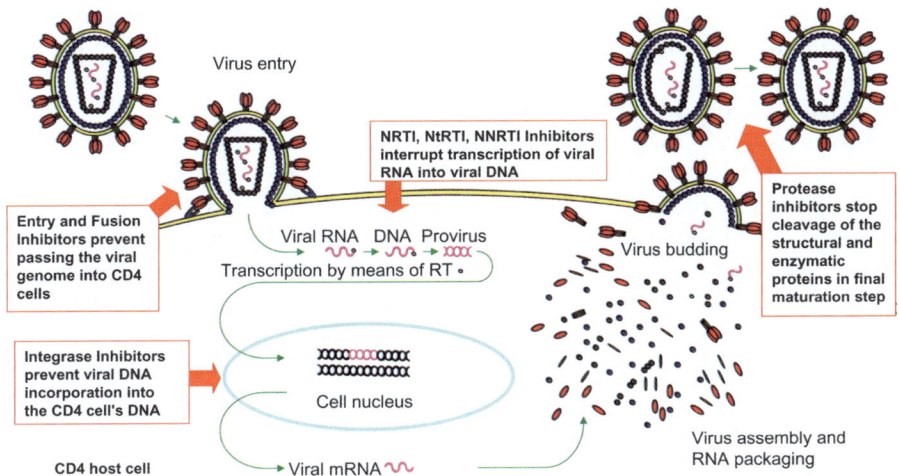

Figure 3 HIV lifecycle and drug targets.

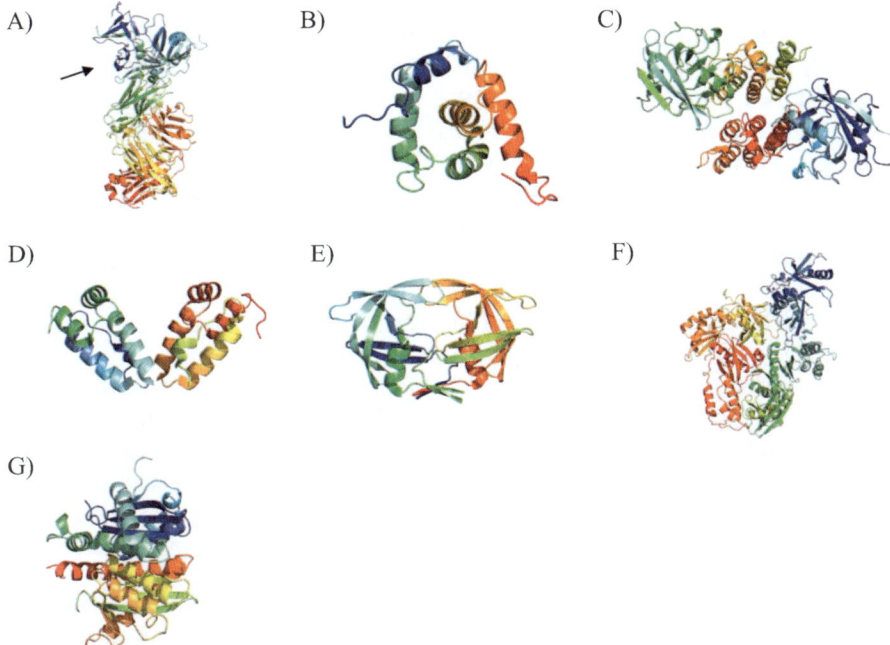

Figure 4 Various structures of HIV proteins. Obtained from the Protein DataBase. (A) ENV (PDB code, 2NYZ, arrow points towards ENV); (B) MA (PDB code, 1TAM); (C) CA [PDB code, 1AK4 (dimer)]; (D) NC [PDB code, 3DPH (dimer)]; (E) PR [PDB code, 1HPS (dimer)]; (F) RT (PDB code, 3HVT); (G) IN 1BIS). All proteins are depicted as ribbon drawings (colored red–blue).

is to target gag and gag/pol polyproteins to the plasma membrane and participate in viral assembly. This is mediated by both a cluster of conserved basic residues located near the N-terminal region[12,13] (Figure 4B) and a myristyl group that is added to the N-terminus (not shown).[14]

Similarly, CA is also a structural protein that plays an important role in virus assembly. It is located between MA and spacer protein 1 (SP1) within the gag and gag/pol polyproteins. CA is predominantly α-helical and composed of two domains: an N-terminal 'core' domain (residues 1–145), which functions in virion maturation and incorporation of the cellular protein cyclophilin A (CypA), and a C-terminal 'dimerization' domain (residues 151–231), which is necessary for particle assembly and also viral core formation. These two domains are connected by a flexible linker region.[15–20] The core domain is highly helical, being composed of seven helices, two β-hairpins and an exposed loop.[15] The dimerization domain has also a high helical content and contains the major homology region (MHR), a stretch of 20 amino acids which is required for both viral particle assembly and the correct assembly of the viral core (Figure 4C).[21]

NC is another important structural protein that is synthesized as part of gag and gag/pol polyproteins. In virions, NC is found in the core, tightly associated with the viral RNA. A distinguishing feature of NC protein, both structurally and functionally, is the presence of two Cys–X2–Cys–X4–His–X4–Cys (CCHC) domains reminiscent of the so-called zinc-finger motifs found in many cellular DNA binding proteins. NMR analysis of NC indicates that the zinc finger domains are located in a central globular domain, while the N and C termini of the protein are relatively disordered.[22,23] Each of the two zinc fingers of NC (like those of other retroviruses) coordinates a zinc ion.[23–25] The existence of zinc-finger motifs in NC protein is highly conserved among retroviruses; one or two such domains are found in all retroviruses except the spumaviruses (Figure 4D).[26]

PR cleaves the gag and gag/pol polyproteins into their mature components. This enzyme functions as an obligatory homodimer that consists of two identical 99-residue subunits.[27] The dimer is stabilized by a β-sheet formed between the N- and C-termini of each subunit.[28,29] The active site of the enzyme is located directly above the dimer interface and contains two aspartic residues, one Asp residue being provided by each monomer.[30] The active site is covered by two β-hairpins, one from each monomer, called flaps. The flaps are thought to undergo a large conformational change to open up and allow access to the active site (Figure 4E).[31]

RT is composed of an extended asymmetric heterodimer of two related subunits: a 51 kDa subunit (p51) of 440 amino acids and a larger 66 kDa subunit (p66) of 560 amino acids. The subunit p66 is composed of two spatially distinct domains, polymerase and RNase H. The polymerase domain performs the catalytic function, having a readily accessible active site that accommodates the double-stranded DNA for catalysis. It is composed of four subdomains: fingers (residues 1–85 and 118–155), palm (residues 86–117 and 156–236, which also contains the catalytic residues Asp110, Asp185 and Asp186), thumb

Table 1 HIV protein structures and PDB references.

HIV protein	Alternative names	Example PDB references
Matrix	MA, p17	1hiw, 2hmx, 1uph
Capsid	CA, p24	1afv, 1ak4
Nucleocapsid	NC, p7	3dph
Protease	PR, p11	2uxz, 1g35
Reverse transcriptase	RT, p66/p51	3hvt, 1hni, 1lw2, 2iaj, 2vg5
Integrase	IN, p32	1bis, 1bi4, 1qs4, 2b4j
Envelope	ENV, gp120/gp41	2ny7

(residues 237–318) and connection (residues 319–426).[32–34] The RNase H (catalytic residues: Asp553, Glu478, Asp498 and Asp549) subunit has the role of degrading the template RNA that is incorporated in the RNA–DNA duplex. Apparently, the smaller p51 subunit plays mainly a structural role rather than an enzymatic function. It folds into the same four subdomains as p66 (fingers, palm, thumb and connection), but the relative arrangement of the subdomains is different in the two subunits (Figure 4F).

IN is the enzyme that catalyzed the multi-step process of integration of provirus into the host genome. The full enzyme has 288 amino acids and three domains: the catalytic core (residues 50–212) and the C- and N-terminal domains. The structure of the core domain has a five-strand β-sheet (at the center) and six helices. HIV IN, like other DNA processing enzymes, possesses a DDE motif, which is a catalytic triad of D64, D116 and E152. The active site region is identified by the position of two of the conserved carboxylate residues (Asp64 and Asp116) which are essential for catalysis. The N-terminal region is characterized by an HHCC 'zinc finger'-like sequence and the C-terminal domain appears to be important for binding to the HIV LTR DNA region (Figure 4G and Table 1).[35,36]

3 Viral Life and Replication Cycles

The HIV-1 lifecycle is a complex, multistage process involving interactions between HIV proteins and host macromolecules. The early phase of the virus lifecycle comprises the infection of host cell and integration of viral genome. The late phase comprises the regulation of the expression of the viral gene products and the production of viral particles.[37] Infection begins when an HIV particle encounters a host cell with a surface receptor called CD4. The cells mainly targeted by HIV are T-helper lymphocytes, macrophages and dendritic cells. The virus particle uses gp120 to attach itself to the CD4 receptor and this is sufficient for binding, but co-receptors are necessary for viral entry.[38–40] Several proteins have been identified as possible co-receptors, but HIV, generally, uses mainly two co-receptors to enter a target cell, either CCR5 or CXCR4, depending on the strain of virus.[41–44] The strains of HIV most commonly seen early in HIV disease, known as macrophage-tropic (M-tropic) viruses, use CCR5 for cell entry. The importance of CCR5 for the virus lifecycle has been demonstrated by the identification of polymorphisms within the CCR5 gene which affect transmission and/or disease progression.[43]

The binding of viral gp120 ENV protein to the CD4 receptor causes a structural change in gp120 exposing the binding site for the co-receptors. Once the co-receptor is bound, further structural rearrangements occur, mostly in gp41 transmembrane protein, which lead to virus entry. Once within the cell, the virus particle releases its RNA and the enzyme RT makes a DNA copy of the viral RNA. Of importance in the process of retroviral DNA synthesis, for the purpose of developing adequate therapies, are the high degree of genome recombination and the high error rate of RT, which provide means for a high level of viral genome variability. Recombination is facilitated by the packaging into the virus particle of two copies of viral RNA and the ability of RT to jump from one copy of the genome to another. The high error rate is due to lack of proofreading ability of RT. It is estimated that HIV-1 RT has an average error rate of 3×10^{-5} errors per base per replication cycle.[44,45] This value is an average of all types of point mutations, but deletions, insertions and frameshift mutations are also commonly observed.

The template viral genomic RNA is degraded by RNase H and the new HIV dsDNA then moves into the nucleus of the cell where, with the help of the enzyme IN, is inserted into the host cell's DNA. This is the last step of the early phase of the HIV lifecycle. Formation of the preintegration complex must occur before the integration can take place. The preintegration complex carries sequences that interact with the cellular system and signal the transport of the viral proteins and nucleic acids into the nucleus. Proteins such as MA, vpr and IN have been postulated to be involved in the integration of the viral DNA into the host genomic DNA. This process involves not only the viral IN enzyme but also the host repair system. Initially it was believed that the site of integration was random. However, there are several reports that sites of HIV integration in the human genome are not randomly distributed but instead are enriched in active genes and regional hotspots.[46] Global analysis of cellular transcription indicated that active genes were preferential integration targets, particularly genes that were activated in cells after infection by HIV-1. Regional hotspots for integration were also found, including a 2.4 kb region containing 1% of integration sites. These data document unexpectedly strong biases in integration site selection and suggest how selective targeting promotes aggressive HIV replication.[46]

Once located in the genome of the cell, HIV DNA is called a provirus. The HIV provirus is replicated by transcription into viral RNA, some of which becomes new viral genomic material and some of which is needed to direct the synthesis of viral polyproteins env, gag and gag/pol. After transcription, the viral mRNA, as any other cellular RNA, is modified by addition of a polyA tail. Also, in order to produce proteins such as rev and tat, the mRNA must be properly spliced. However, genomic viral RNA has to be transported out of the nucleus without further splicing. To circumvent the strong aversion of the cellular machinery for transporting 'improperly' spliced RNA molecules, the retroviruses code for a constitutive transport element. This nucleotide sequence allows for the transport of the un-spliced mRNA and, in the case of HIV, this is aided by the rev protein.[47] This creates a point of regulation where regulatory genes that are coded as nascent sequences and arise due to RNA splicing are transcribed and exported out of the nucleus for translation first. One of the these genes is *rev* that encodes the regulatory protein rev. Once Rev reaches high enough concentrations, it promotes the export of the intact mRNA for translation of *gag*, *pol* and *env* genes.[48]

The final step in virus replication cycle is budding and maturation. The association of viral RNA, gag and gag/pol polyproteins just underneath the cell membrane precedes budding. Aggregation of gag and gag/pol polyproteins is mediated by both protein–protein interactions and protein–RNA interactions. Specifically, gag–gag interaction is mediated by NC, MA and CA interactions.[17,49–51] The viral genomic RNA is recruited through interactions with the NC protein within gag.

The accumulation of env, gag and gag/pol proteins within the lipid rafts in the cell membrane induces changes that promote budding (Figure 3).[52–55]

4 Antiretroviral (ARV) Therapy and Drug Resistance

Access to treatment and care has increased greatly in the recent years. Even though the coverage is still low in many HIV/AIDS-stricken regions, the benefits are dramatic. Through the expanded provision of ARV treatment, an estimated two million life-years were gained from 2002 to 2006 in low- and middle-income countries.[56]

The evolution of ARV therapy is an ongoing process aimed at discovering potent and tolerable drug regimens. The first ARV drug approved by the US Food and Drug Administration (FDA) in 1987 was zidovudine (AZT), a previously known potential anticancer agent. Shortly after, rapid progress in the understanding of the structure and lifecycle of the virus led to the unprecedented development of other drugs targeted to a variety of viral proteins. The retroviral enzymes – RT, IN and PR – were the obvious targets for drug discovery (Figure 3). The first drugs to be identified were inhibitors of RT,[57] which were discovered and developed long before the structure of RT itself was solved.[34] Newer RT-targeted drugs – NNRTIs and PIs – have been developed bearing the enzyme structure in mind. The structure-assisted drug design and discovery process utilizes structural biochemical methods, such as protein crystallography (Chapter 7), NMR spectroscopy (Chapter 8) and computational biochemistry, to guide the synthesis of potential drugs. This information can, in turn, be used to help explain the basis of their activity and to improve the potency and specificity of new lead compounds. Crystallography plays a particularly important role in this process. Recent years have seen a virtual explosion of crystallographic studies aimed at the characterization of the structures of HIV enzymes and of HIV enzyme–inhibitor complexes on an atomic level.

Initially, the ARV drugs were given as monotherapy or dual therapy. However, this approach often led to treatment failures due to the development of resistant viruses. Patients who were receiving monotherapy or dual therapy initially experienced decreases in viral load, increases in CD4 count and improvement in quality of life; then they experienced viral rebound and decreased CD4 counts further into therapy. The concept of highly active antiretroviral therapy (HAART) was introduced once the PIs were developed. The first PI approved in 1995 by the FDA to be administered to HIV-positive patients was saquinavir (SQV). To date, 23 individual ARV compounds within four classes have been approved for the treatment of HIV-1 infection: seven nucleoside (NRTIs) and one nucleotide analog RT inhibitors (NtRTI), three non-nucleoside RT inhibitors (NNRTIs), nine PR inhibitors (PIs), one IN inhibitor and two fusion and entry inhibitors (Table 1).

5   Inhibitors

Reverse Transcriptase Inhibitors (RTIs)

The mode of action of NRTIs and NtRTIs is essentially the same; they are analogs of the naturally occurring deoxynucleotides needed to synthesize the viral DNA and they compete with the natural deoxynucleotides for incorporation into the growing viral DNA chain. The difference between these two subclasses of drugs is that NRTIs, in order to be incorporated into the viral DNA, must be activated in the cell by the addition of three phosphate groups to their deoxyribose moiety. This phosphorylation step, carried out by the host cellular kinase enzymes, converts NRTIs into NtRTIs. Taking NtRTIs directly allows conversion steps to be skipped, causing less toxicity. Unlike the natural deoxynucleotides substrates, NRTIs and NtRTIs lack a 3'-hydroxyl group on the deoxyribose moiety. As a result, following incorporation of an NRTI or an NtRTI, the next incoming deoxynucleotide cannot form the next 5'–3' phosphodiester bond needed to extend the DNA chain. Thus, when an NRTI or NtRTI is incorporated, viral DNA synthesis is halted, a process known as chain termination. All NRTIs and NtRTIs are classified as competitive substrate inhibitors.

In contrast, NNRTIs have a completely different mode of action. NNRTIs block RT by binding at a different site on the enzyme when compared with NRTIs and NtRTIs. NNRTIs are not incorporated into the viral DNA but instead inhibit the movement of protein domains of RT that are needed to carry out the process of DNA synthesis. NNRTIs are therefore classified as noncompetitive inhibitors of RT.

The RTIs are also available as combination pills with the advantage of a reduced burden pill: zidovudine–lamivudine (Combivir), zidovudine–lamivudine–abacavir (Trizivir), lamivudine–abacavir (Epzicom), tenofovir–emtricitabine (Truvada) and efavirenz–tenofovir–emtricitabine (Atripla).

The most common side-effects of RTIs as a class are digestive problems (diarrhea, nausea, vomiting, abdominal pain) and constitutional problems (fatigue, fever) (Table 2).

Protease Inhibitors (PIs)

PIs are relatively small, rigid and highly hydrophobic molecules. They bind in the active site of the PR and work by inhibiting the proteolytic cleavage of the structural and enzymatic proteins, which prevents the virus from maturing into an infectious virion. The addition of PIs to the ARV therapeutic regimens significantly improved the life expectancy of HIV-infected patients.

Unfortunately, the older PIs came with multiple scheduling requirements and the need to take 10–16 capsules per day. Ritonavir (RTV) boosting is a relatively new concept and today is one of the mainstays of therapy. Boosting reduces the frequency of dosing and the number of required forms per day.[58,59] The concept of boosting involves the use of subtherapeutic doses of RTV but enough to inhibit the cytochrome P450 (CYP) 3A4 enzyme, resulting in pharmacokinetic enhancement or 'boosting' of PI serum levels and prolongation of their half-life. Metabolized mostly through the 2C19 enzyme of the cytochrome P450, nelfinavir (NFV) is the only PI not markedly boosted by RTV.[60]

The adverse effects of PIs are listed in Table 2. It is important to mention that atazanavir (ATV), one of the newest and most potent PIs, is more tolerable that other PIs. However, it has some notable drug interactions: (1) its levels are decreased by 25% when combined with tenofovir and (2) it has significant interactions with H2 blockers and proton pump inhibitors.[61]

Integrase (IN) Inhibitors

Raltegravir is the first approved IN inhibitor; it targets the strand transfer step of viral integration. It binds to the HIV preintegration complex and it dissociates at a rate slower than the half-life of the complex itself, which makes it binding essentially irreversible. It is currently approved by the FDA in combination with other ARV agents in treatment-experienced adult patients with ongoing viral replication and multidrug-resistant viral strains.[62–66]

Entry and Fusion Inhibitors

This new class of ARV targets prevention of the fusion of HIV and CD4 cell and prevents entry of the viral genome to the CD4 cell. The only FDA-approved fusion inhibitor, enfuvirtide, is a structural analog of HR2 domain of gp41 and binds to the HR1 region, more specifically to the N-terminal heptad repeat (NHR) region, preventing the change in conformation that allows the viral entry in to the cell.[67] Enfuvirtide is a synthetic peptide recommended for

Table 2   FDA-approved individual anti-HIV drugs.

Generic name	Alternative names	Brand name	Comments
Reverse transcriptase inhibitors			
Nucleoside analog reverse transcriptase inhibitors			
Zidovudine	AZT, ZDV, azidothymidine	Retrovir	First ARV drug approved by the FDA for the treatment of HIV
Didanosine	ddI	Videx	Second FDA-approved ARV drug
Zalcitabine	ddC, dideoxycytidine	Hivid	Due to lower potency and serious side-effects, it is now rarely used for the treatment of HIV
Stavudine	d4T	Zerit	
Lamivudine	3TC	Epivir	It is also used for the treatment of chronic hepatitis B
Abacavir	ABC	Ziagen	The most powerful NRTI to treat HIV
Emtricitabine	FTC	Emtriva	It is the newest NRTI and is very similar to 3TC
Nucleotide analog reverse transcriptase inhibitors			
Tenofovir	Tenofovir disoproxil fumarate, PMPA, TDF	Viread	It is also tested for treatment of hepatitis B
Non-nucleoside reverse transcriptase inhibitors			
Nevirapine	NVP	Viramune	The first NNRTI approved by the FDA
Delavirdine	DVD	Rescriptor	It is now rarely used to its lower potency and complex drug interactions
Efavirenz	EFV	Sustiva/Stocrin	It is always given in combination with other ARV drugs
Etravirine	TMC125	Intelence	The latest ARV drug approved by the FDA
Protease inhibitors			
Saquinavir	SQV	Invirase	The first PI approved by the FDA
Ritonavir	RTV	Norvir	It is widely used as a booster for other PIs
Indinavir	IDV	Crixivan	It requires very precise dosing schedule
Nelfinavir	Nelfinavir mesylate, AG1343, NFV	Viracept	The only PI approved to treat HIV in pregnant women
Amprenavir	APV		Marketed as the prodrug fosamprenavir (Lexiva)
Lopinavir	ABT-378, LPV		Due to its insufficient bioavailability, it is marketed only as a co-formulation with ritonavir (Kaletra)
Atazanavir	ATV	Reyataz	The first PI approved for once-daily dosing

Table 2 (*Continued*)

Generic name	Alternative names	Brand name	Comments
Tipranavir	TPV	Aptivus	The first non-peptidic PI FDA-approved for HIV treatment
Darunavir	DNV	Prezista	The latest PI drug on the market
Integrase inhibitors			
Raltegravir	RAL, MK-0518	Isentress	Recommended for use, in combination with other ARV, in treatment-experienced adults
Fusion and entry inhibitors			
Enfuvirtide	T-20	Fuzeon	Due to its high cost ($25 000) per year per patient and inconvenient dosing schedule, it is used for 'salvage' therapy only
Maraviroc	MCV	Selzentry	CCR5 co-receptor antagonist. It has many drug–drug interactions. Also, it is recommended for use, in combination with other ARV, in treatment-experienced adults

use in patients who are treatment experienced or who have had multiple treatment failure regimens and in those with ongoing viral replication or viral load greater than 400 copies mL^{-1}.[68]

Maraviroc is the first HIV-1 CCR5 co-receptor antagonist that works by blocking entry of HIV-1. It is a specific, slowly reversible, non-competitive, small-molecule antagonist of the chemokine receptor CCR5, targeting a host protein rather than a viral component. This recently approved drug has demonstrated clinically significant decreases in plasma concentrations of HIV-1 RNA and increases in CD4+ cell counts; however, it is indicated only for use as salvage therapy. The main mechanism of resistance to maraviroc appears to be the ability of the virus to use maraviroc-bound (inhibitor-bound) CCR5 co-receptors as a result of selection of multiple mutations in the V3 loop of gp120.[69] Also, given its specific activity against HIV that is exclusively CCR5 tropic, determination of tropism is necessary before initiation of therapy with the drug. Interestingly, there is no cross-resistance with the fusion inhibitor enfuvirtide, which selects for mutations in the gp41 region of the viral envelope complex.[70]

Highly Active Antiretroviral Therapy (HAART)

HAART are a combination of three or more drugs from two drug classes and has improved significantly the prognosis of HIV-infected individuals.[71,72] Since the advent of HAART, the following improvements have been noted: better patient quality of life, increased survival, slowed disease progression, decreased opportunistic infections, decreased viral loads and increased CD4 counts. Many studies have shown that HAART is the most powerful and essential combination to treat HIV infection effectively and decrease the likelihood of emergence of resistant viruses. However, even with this complex therapeutic approach, the development of drug resistance poses the greatest challenge for treating HIV infection and the margin of success for achieving and maintaining virus suppression is narrow. The evolution of HIV drug resistance within an individual depends on three main variables: different immunological status and response to therapy, HIV genetic variability that allows for rapid development of drug resistance mutations and, very important, the adherence to therapy Table 3.

Adherence to a complex regimen is often a significant barrier to treatment success. Several years ago, HAART was expensive, required the patient to ingest a large number of pills under a complex dosing schedule and specific food requirements.[73] The development of new anti-HIV drugs allowed simplification of the treatment regimens to several pills daily, with minimal food requirements, if any. However, providing anti-retroviral treatment is still costly and resource intensive and the majority of the world's infected individuals cannot access treatment services. HAART is a life-long treatment, associated with various and severe side-effects. Also, resistance develops rapidly if patients miss doses.

The occurrence of the drug-induced resistance mutations results from the inability of HAART to eradicate viral replication totally. With the extremely fast replication rate of HIV, which can reach 10^{10} viral particles per day, the high error rate of RT that incorporates, on average, one error per 10 000 bases and the capacity for genomic recombination, the virus can quickly develop genomic variations that translate into protein structural changes. These mutations tend to decrease the affinity of the drug for the target enzyme promoting therapeutic failure.[74] As of May 2006, the HIV guidelines recommend resistance testing in patients with acute or chronic HIV infection before therapy is started.[75] This may help detect a virus that is resistant to initial treatment regimens and regimens can be adjusted based on sensitivity to ARV agents.

6   HIV Diversity

Groups and Subtypes

The large genomic diversity of viral subtypes in different geographic regions is the consequence of the high mismatch error rate of the HIV RT enzyme (between 1×10^{-4} and 5×10^{-5}) coupled with the absence of an exonuclease proofreading activity.[45,76] Other factors that contribute to the rapid pace of genetic diversification include the replicative rate of each viral subtype, the number of mutations arising in each replicative cycle, the viral propensity for

Table 3 Summary of class and agent-specific side-effects of antiretroviral agents.

Antiretroviral agent	Adverse effects

Reverse transcriptase inhibitors

Nucleoside analog reverse transcriptase inhibitors
Class effects: lactic acidosis, hepatic steatosis, pancreatitis, bone marrow toxicity, rash

Zidovudine	Severe headache, nausea, hepatotoxicity
Didanosine	Peripheral neuropathy
Zalcitabine	Peripheral neuropathy, stomatitis
Stavudine	Peripheral neuropathy, lipodystrophy, hyperlipidemia
Lamivudine	Minimal toxicity
Abacavir	Severe hypersensitivity, nausea, diarrhea
Emtricitabine	Minimal toxicity, palmar discoloration

Nucleotide analog reverse transcriptase inhibitors

Tenofovir	Headache, diarrhea, nausea, vomiting, renal insufficiency

Non-nucleoside reverse transcriptase inhibitors
Class effects: rash, elevated transaminase levels, nausea, abdominal pain, fatigue

Nevirapine	Steven–Johnson syndrome, hepatitis
Delavirdine	Diarrhea
Efavirenz	Insomnia, abnormal dreams, confusion, impaired concentration
Etravirine	Peripheral neuropathy, headache, hypertension

Protease inhibitors
Class effects: nausea, vomiting, diarrhea, hyperlipidemia (except atazanavir), fat maldistribution, hyperglycemia, possible increased bleeding in patients with hemophilia, elevated transaminase levels

Saquinavir	Headache
Ritonavir	Abdominal pain, peripheral and peri-oral parasthesias
Indinavir	Nephrolithiasis, indirect hyperbilirubinemia, metallic taste, alopecia
Nelfinavir	Severe diarrhea
Amprenavir	Skin rash
Lopinavir	Pancreatic toxicity
Atazanavir	Indirect hyperbilirubinemia, prolongation of the PR interval on ECG
Tipranavir	Increased risk of intracranial hemorrhage
Darunavir	Cold-like symptoms

Integrase inhibitors

Raltegravir	Nausea, diarrhea, headache, dizziness, abnormal dreams, pyrexia, rash, CPK elevation

Fusion and entry inhibitors

Enfuvirtide	Skin reactions at the injection site, severe allergic reaction, renal toxicity, paralysis, neutropenia
Maraviroc	Diarrhea, nausea, hepatotoxicity/hepatitis, fatigue, dizziness, headache

genomic recombination and viral fitness. In addition, high rates of genomic evolution may result from host, environment and/or therapeutic selection pressure.[77]

HIV is characterized by significant genetic diversity among distinct types, groups and subtypes,[78-80] and this variability has implications in prevention, diagnostic tests, therapy response and vaccine development.[80-85] This variability of HIV has led to the development of various distinct types, groups and subtypes of HIV.[86] There are two types of HIV: 1 and 2. HIV type 1 (HIV-1) is responsible for more than 99% of HIV infections worldwide,[87-89] whereas the majority of HIV-2 infections are limited to West Africa. HIV-1 is further divided in three groups: M (major), O (outlier) and N (non-M, non-O). HIV-1 group M, which accounts for most infections worldwide,[88] is classified into nine distinct subtypes (A–D, F–H, J and K) and many subsubtypes, circulating (CRFs) and unique (URFs) recombinant forms.[90] These subtypes and intersubtype recombinants differ from one another by 25–30% in the *env* gene and by 10–15% in the *pol* gene (Figure 1A).[91-93]

HIV-1 Non-B Subtypes

While HIV-1 subtype B has been the most widely studied, subtypes A, C and D predominate worldwide. Among the approximately 40 million people living with HIV/AIDS in 2003, more than 80% are infected with HIV-1 non-B subtypes.[87,88]

In the absence of any drug exposure, RT and PR sequences from B and non-B HIV-1 are polymorphic among about 40% of the first 240 RT amino acids and 30% of the 99 PR amino acids.[94,95] These differences/substitutions occur at high rates in certain non-subtype B viruses and are designated as naturally occurring or baseline polymorphisms. Based on the observation of differences between sequences from untreated and treated persons, there are mutations at PR positions 10, 20, 36, 63, 71, 77, 93 that are characterized as secondary resistance mutations in subtype B PR but occur in the absence of drug exposure in non-B subtypes.[92,95] Although in treatment-naïve patients many of these baseline polymorphisms do not confer resistance to drugs *per se* among different subtypes, they may facilitate the development of drug resistance.[94,96]

With the expanding access to ARV therapy for patients living with HIV/AIDS in developing countries and with the spread of HIV-1 non-B variants in developed countries,[97-99] an important issue of concern is the behavior of non-B subtypes and their recombinant forms under the selective pressure of ARV regimens. It has been argued that these polymorphisms might play different roles such as (1) increasing the catalytic activity of non-B subtype PRs and enhancing virus viability;[100,101] (2) resulting in development of diverse mutational pathways during ARV treatment;[102,103] (3) influencing the speed of acquiring PI-related resistance mutations;[104,105] and (4) contributing to resistance and/or maintenance of viral fitness once primary resistance mutations occur.[106,107]

It has also been hypothesized that the HIV genetic variability might modulate the transmissibility rates and modes of different viral subtypes or CRFs. A study presented in 2007 found that Kenyan women infected with subtype D had more than twice the risk of death over 6 years than those infected with subtype A.[108] An earlier study of sex workers in Senegal, published in 1999, found that women infected with subtype C, D or G were more likely to develop AIDS within 5 years of infection than those infected with subtype A.[109]

It has been observed that certain subtypes/CRFs are predominantly associated with specific modes of transmission. It was initially believed that subtype B spread mostly by homosexual contact and intravenous drug use (essentially *via* blood), while subtype C and CRF A/E tend to fuel heterosexual epidemics (via a mucosal route).[110] However, this theory, based on observations made at the beginnings of HIV epidemic, has not been conclusively proven.[111]

More recent studies have looked for variations within subtypes in rates of mother-to-child transmission. One of these found that such transmission is more common with subtype D than subtype A.[112] Another reached the opposite conclusion (A more commonly transmitted than D) and also found that subtype C was more often transmitted than subtype D.[113] A third study concluded that subtype C is more transmissible than either D or A.[114] Other researchers have found no association between subtype and rates of mother-to-child transmission.[115,116]

Several recent studies argue that there are no substantial differences regarding known resistance-associated mutations and the newly emergent substitutions between non-B and B subtype strains.[117,118] Other studies showed that non-B isolates were statistically associated with a more rapid progression to resistance after ARV therapy and they had different mutational patterns when compared with those of B isolates.[119,120] Santos *et al.* studied the differences in the impact of several PI-selected mutations on subtypes B and G in 500 patients and showed that the mutation L90M confers different levels of resistance in subtypes B and G.[121] Another recent study indicated that even though, in general, drug selective pressure and resistance pathways are relatively similar between subtypes B and G, some differences do occur, leading to subtype-dependent substitutions.[122] With respect to the development of drug resistance mutations (DRMs), a recent study showed that non-B subtypes tend to select for the same DRMs as described in subtype B in PR and RT, although at distinct proportions.[123,124] Also, an increasing body of evidence suggests that certain non-B subtypes are often more susceptible to specific ARV drugs. For example, the circulating recombinant form CRF02_AG PR presents a higher susceptibility to nelfinavir and ritonavir than subtypes C, F and G, whereas subtype G isolates are more susceptible to tipranavir.[125] Unexpectedly, natural hypersusceptibility to lopinavir was also detected in subtypes C and G.[126,127] In an extensive review, Martinez-Cajas *et al.* searched 11 databases and retrieved 3486 citations on all aspects of non-B subtype-related resistance research.[128] They concluded that the genetic diversity of HIV-1 could affect the type of resistance mutations, degree of resistance and timing of emergence of anti-retroviral resistance. These findings led to the conclusion that the combined effects of naturally existing polymorphisms and drug resistant mutations might have important consequences on the feasibility of continuing to use current HIV-1 PIs for non-subtype B infections.

All these observed differences in resistance pathways may impact cross-resistance and the selection of second-line regimens with PIs. However, the knowledge of the NOPs in non-B subtypes and their clinical relevance is limited or controversial. Thus, the main concern remains how these various subtypes will respond to the current therapeutic strategies. Most current HIV-1 ARV drug regimens were designed for use against subtype B and so hypothetically might not be equally effective in Africa or Asia where other strains are more common. At present, there is no compelling evidence that subtypes differ in their sensitivity to ARV drugs.[129–134] However, some subtypes may occasionally be more likely to develop resistance to certain drugs. In some situations, the types of mutations associated with resistance may vary and the drug resistance threshold might be lower due to pre-existing polymorphisms that could act as secondary resistant mutations, increasing the drug resistance while conserving the catalytic activity and fitness of the PR.[100,101,106,135] To date there have been no long-term studies analyzing the speed of acquiring drug resistance in HIV-1 non-B subtypes *versus* B subtype. This is an important subject for future research.

The effectiveness of HIV-1 treatment is monitored using viral load tests. It has been demonstrated that some such tests are sensitive only to subtype B and can produce a significant underestimate of viral load if used to process other strains.[136,137] The latest tests do claim to produce accurate results for most group M subtypes, but the false-positive results and the need for expensive PCR instruments limit the implementation of these test in resource-limited settings.[138–140] It is important that health workers and patients are aware of the subtype/CRF they are testing for and of the limitations of the test they are applying.

It is important to mention that HIV-2, even though it accounts for a lower percentage of HIV-infected individuals, has several specific traits. HIV-2 is common in certain regions in West Africa but very rare in other parts of the world. HIV-2 appears to have a milder disease course than HIV-1, with a longer time to the development of AIDS.[141,142]

Given the low prevalence of HIV-2 in developed countries, the clinical course and optimal treatment strategies are not well known.[143] Not all of the drugs used to treat HIV-1 infection are as effective against HIV-2. In particular, HIV-2 has a natural resistance to NNRTI ARV drugs and they are therefore not recommended for treating individuals harboring this HIV type.[144,145] Also, protease inhibitors seem to have varying efficacy against HIV-2 and their use should be guided by genotype/phenotype testing.[146–149] As yet there is no FDA-licensed viral load test for HIV-2 and those designed for HIV-1 are not reliable for monitoring the other type.[150] Instead, response to treatment may be monitored by following CD4 + T-cell counts and indicators of immune system deterioration. More research and clinical experience are needed to determine the most effective treatment for HIV-2.

References

1. M. S. Gottlieb, J. E. Groopman, W. M. Weinstein, J. L. Fahey and R. Detels, *Ann. Intern. Med.*, 1983, **99**, 208.
2. J. E. Groopman and M. S. Gottlieb, *Nature*, 1983, **303**, 575.
3. A. Karpas, *Biol. Rev. Camb. Philos. Soc.*, 2004, **79**, 911.
4. UNAIDS, *Joint United Nations Program on HIV/AIDS*, UNAIDS, Geneva, 2008.
5. V. M. Hirsch, R. A. Olmsted, M. Murphey-Corb, R. H. Purcell and P. R. Johnson, *Nature*, 1989, **339**, 389.
6. S. Osmanov and J. Esparza, Development and evaluation of preventive HIV-1 vaccines. In *Human Immunodefficiency Viruses: Biology, Immunology and Molecular Biology*, N. Saksena, Editor, Medical Systems, Genoa, 1998, pp. 501–505.
7. L. Guilhaudis, A. Jacobs and M. Caffrey, *Eur. J. Biochem.*, 2002, **269**, 4860.
8. P. D. Kwong, R. Wyatt, J. Robinson, R. W. Sweet, J. Sodroski and W. A. Hendrickson, *Nature*, 1998, **393**, 648.
9. R. Wyatt, P. D. Kwong, E. Desjardins, R. W. Sweet, J. Robinson, W. A. Hendrickson and J. G. Sodroski, *Nature*, 1998, **393**, 705.
10. C. D. Rizzuto, R. Wyatt, N. Hernandez-Ramos, Y. Sun, P. D. Kwong, W. A. Hendrickson and J. Sodroski, *Science*, 1998, **280**, 1949.
11. S. Bour, R. Geleziunas and M. A. Wainberg, *Microbiol. Rev.*, 1995, **59**, 63.
12. X. Yuan, X. Yu, T. H. Lee and M. Essex, *J. Virol.*, 1993, **67**, 6387.
13. W. Zhou, L. J. Parent, J. W. Wills and M. D. Resh, *J. Virol.*, 1994, **68**, 2556.
14. P. P. Lee and M. L. Linial, *J. Virol.*, 1994, **68**, 6644.
15. T. R. Gamble, F. F. Vajdos, S. Yoo, D. K. Worthylake, M. Houseweart, W. I. Sundquist and C. P. Hill, *Cell*, 1996, **87**, 1285.
16. R. K. Gitti, B. M. Lee, J. Walker, M. F. Summers, S. Yoo and W. I. Sundquist, *Science*, 1996, **273**, 231.
17. A. Borsetti, A. Ohagen and H. G. Gottlinger, *J. Virol.*, 1998, **72**, 9313.
18. T. Dorfman, A. Bukovsky, A. Ohagen, S. Hoglund and H. G. Gottlinger, *J. Virol.*, 1994, **68**, 8180.
19. J. McDermott, L. Farrell, R. Ross and E. Barklis, *J. Virol.*, 1996, **70**, 5106.
20. A. S. Reicin, A. Ohagen, L. Yin, S. Hoglund and S. P. Goff, *J. Virol.*, 1996, **70**, 8645.
21. F. Mammano, A. Ohagen, S. Hoglund and H. G. Gottlinger, *J. Virol.*, 1994, **68**, 4927.
22. N. Morellet, N. Jullian, H. De Rocquigny, B. Maigret, J. L. Darlix and B. P. Roques, *EMBO J.*, 1992, **11**, 3059.
23. T. L. South and M. F. Summers, *Protein Sci.*, 1993, **2**, 3.
24. M. F. Summers, L. E. Henderson, M. R. Chance, J. W. Bess, T. L. South, P. R. Blake, I. Sagi, G. Perez-Alvarado, R. C. Sowder and D. R. Hare *et al.*, *Protein Sci.*, 1992, **1**, 563.

25. J. W. Bess, P. J. Powell, H. J. Issaq, L. J. Schumack, M. K. Grimes, L. E. Henderson and L. O. Arthur, *J. Virol.*, 1992, **66**, 840.
26. B. Maurer and R. M. Flugel, *AIDS Res. Hum. Retroviruses*, 1988, **4**, 467.
27. A. Wlodawer, M. Miller, M. Jaskolski, B. K. Sathyanarayana, E. Baldwin, I. T. Weber, L. M. Selk, L. Clawson, J. Schneider and S. B. Kent, *Science*, 1989, **245**, 616.
28. I. T. Weber, M. Miller, M. Jaskolski, J. Leis, A. M. Skalka and A. Wlodawer, *Science*, 1989, **243**, 928.
29. A. Gustchina and I. T. Weber, *FEBS Lett.*, 1990, **269**, 269.
30. M. A. Navia, P. M. Fitzgerald, B. M. McKeever, C. T. Leu, J. C. Heimbach, W. K. Herber, I. S. Sigal, P. L. Darke and J. P. Springer, *Nature*, 1989, **337**, 615.
31. S. W. Rick, J. W. Erickson and S. K. Burt, *Proteins*, 1998, **32**, 7.
32. L. A. Kohlstaedt, J. Wang, J. M. Friedman, P. A. Rice and T. A. Steitz, *Science*, 1992, **256**, 1783.
33. D. M. Lowe, A. Aitken, C. Bradley, G. K. Darby, B. A. Larder, K. L. Powell, D. J. Purifoy, M. Tisdale and D. K. Stammers, *Biochemistry*, 1988, **27**, 8884.
34. A. Jacobo-Molina, J. Ding, R. G. Nanni, A. D. Clark, X. Lu, C. Tantillo, R. L. Williams, G. Kamer, A. L. Ferris and P. Clark, *Proc. Natl. Acad. Sci. USA*, 1992, **90**, 6320.
35. C. J. Burke, G. Sanyal, M. W. Bruner, J. A. Ryan, R. L. LaFemina, H. L. Robbins, A. S. Zeft, C. R. Middaugh and M. G. Cordingley, *J. Biol. Chem.*, 1992, **267**, 9639.
36. D. J. Hazuda, A. L. Wolfe, J. C. Hastings, H. L. Robbins, P. L. Graham, R. L. LaFemina and E. A. Emini, *J. Biol. Chem.*, 1994, **269**, 3999.
37. B. G. Turner and M. F. Summers, *J. Mol. Biol.*, 1992, **285**, 1.
38. H. K. Deng, D. Unutmaz, V. N. KewalRamani and D. R. Littman, *Nature*, 1997, **388**, 296.
39. G. Alkhatib, F. Liao, E. A. Berger, J. M. Farber and K. W. Peden, *Nature*, 1997, **388**, 238.
40. T. Dragic, V. Litwin, G. P. Allaway, S. R. Martin, Y. Huang, K. A. Nagashima, C. Cayanan, P. J. Maddon, R. A. Koup, J. P. Moore and W. A. Paxton, *Nature*, 1996, **381**, 667.
41. J. P. Moore, *Science*, 1997, **276**, 51.
42. B. J. Doranz, J. Rucker, Y. Yi, R. J. Smyth, M. Samson, S. C. Peiper, M. Parmentier, R. G. Collman and R. W. Doms, *Cell*, 1996, **85**, 1149.
43. M. Samson, F. Libert, B. J. Doranz, J. Rucker, C. Liesnard, C. M. Farber, S. Saragosti, C. Lapoumeroulie, J. Cognaux, C. Forceille, G. Muyldermans, C. Verhofstede, G. Burtonboy, M. Georges, T. Imai, S. Rana, Y. Yi, R. J. Smyth, R. G. Collman, R. W. Doms, G. Vassart and M. Parmentier, *Nature*, 1996, **382**, 722.
44. P. R. Clapham and R. A. Weiss, *Nature*, 1997, **388**, 230.
45. L. M. Mansky and H. M. Temin, *J. Virol.*, 1995, **69**, 5087.
46. A. R. Schroder, P. Shinn, H. Chen, C. Berry, J. R. Ecker and F. Bushman, *Cell*, 2002, **110**, 521.

47. M. Bray, S. Prasad, J. W. Dubay, E. Hunter, K. T. Jeang, D. Rekosh and M. L. Hammarskjold, *Proc. Natl. Acad. Sci. USA*, 1994, **91**, 1256.
48. V. W. Pollard and M. H. Malim, *Annu. Rev. Microbiol.*, 1998, **52**, 491.
49. L. S. Ehrlich, T. Liu, S. Scarlata, B. Chu and C. A. Carter, *Biophys. J.*, 2001, **81**, 586.
50. I. Gross, H. Hohenberg, T. Wilk, K. Wiegers, M. Grattinger, B. Muller, S. Fuller and H. G. Krausslich, *EMBO J.*, 2000, **19**, 103.
51. M. T. Burniston, A. Cimarelli, J. Colgan, S. P. Curtis and J. Luban, *J. Virol.*, 1999, **73**, 8527.
52. S. Scarlata, L. S. Ehrlich and C. A. Carter, *J. Mol. Biol.*, 1998, **277**, 161.
53. T. Murakami and E. O. Freed, *J. Virol.*, 2000, **74**, 3548.
54. Z. Liao, L. M. Cimakasky, R. Hampton, D. H. Nguyen and J. E. Hildreth, *AIDS Res. Hum. Retroviruses*, 2001, **17**, 1009.
55. D. H. Nguyen and J. E. Hildreth, *J. Virol.*, 2000, **74**, 3264.
56. UNAIDS and WHO, *AIDS Epidemic Update: Special Report on HIV/AIDS: December 2006*, UNAIDS, Geneva, 2006.
57. V. T. DeVita, S. Broder, A. S. Fauci, J. A. Kovacs and B. A. Chabner, *Ann. Intern. Med.*, 1987, **106**, 568.
58. J. M. Kilby, G. Sfakianos, N. Gizzi, P. Siemon-Hryczyk, E. Ehrensing, C. Oo, N. Buss and M. S. Saag, *Antimicrob. Agents Chemother.*, 2000, **44**, 2672.
59. A. I. Veldkamp, R. P. van Heeswijk, J. W. Mulder, P. L. Meenhorst, G. Schreij, S. van der Geest, J. M. Lange, J. H. Beijnen and R. M. Hoetelmans, *J. Acquir. Immune Defic. Syndr.*, 2001, **27**, 344.
60. M. Kurowski, B. Kaeser, A. Sawyer, M. Popescu and A. Mrozikiewicz, *Clin. Pharmacol. Ther.*, 2002, **72**, 123.
61. C. Le Tiec, A. Barrail, C. Goujard and A. M. Taburet, *Clin. Pharmacokinet.*, 2005, **44**, 1035.
62. D. J. Hazuda, P. Felock, M. Witmer, A. Wolfe, K. Stillmock, J. A. Grobler, A. Espeseth, L. Gabryelski, W. Schleif, C. Blau and M. D. Miller, *Science*, 2000, **287**, 646.
63. Merck & Co., *Isentress (raltegravir)*, 2007 [package insert].
64. B. Grinsztejn, B. Y. Nguyen, C. Katlama, J. M. Gatell, A. Lazzarin, D. Vittecoq, C. J. Gonzalez, J. Chen, C. M. Harvey and R. D. Isaacs, *Lancet*, 2007, **369**, 1261.
65. C. Hicks and R. M. Gulick, *Clin. Infect. Dis.*, 2009, **48**, 931.
66. R. T. Steigbigel, D. A. Cooper, P. N. Kumar, J. E. Eron, M. Schechter, M. Markowitz, M. R. Loutfy, J. L. Lennox, J. M. Gatell, J. K. Rockstroh, C. Katlama, P. Yeni, A. Lazzarin, B. Clotet, J. Zhao, J. Chen, D. M. Ryan, R. R. Rhodes, J. A. Killar, L. R. Gilde, K. M. Strohmaier, A. R. Meibohm, M. D. Miller, D. J. Hazuda, M. L. Nessly, M. J. DiNubile, R. D. Isaacs, B. Y. Nguyen and H. Teppler, *N. Engl. J. Med.*, 2008, **359**, 339.
67. A. S. Veiga, N. C. Santos and M. A. Castanho, *Recent Pat. Antiinfect. Drug Discov.*, 2006, **1**, 67.
68. F. J. Piacenti, *Pharmacotherapy*, 2006, **26**, 1111.

69. M. Westby, C. Smith-Burchnell, J. Mori, M. Lewis, M. Mosley, M. Stockdale, P. Dorr, G. Ciaramella and M. Perros, *J. Virol.*, 2007, **81**, 2359.

70. L. Xu, A. Pozniak, A. Wildfire, S. A. Stanfield-Oakley, S. M. Mosier, D. Ratcliffe, J. Workman, A. Joall, R. Myers, E. Smit, P. A. Cane, M. L. Greenberg and D. Pillay, *Antimicrob. Agents Chemother.*, 2005, **49**, 1113.

71. C. C. Carpenter, M. A. Fischl, S. M. Hammer, M. S. Hirsch, D. M. Jacobsen, D. A. Katzenstein, J. S. Montaner, D. D. Richman, M. S. Saag, R. T. Schooley, M. A. Thompson, S. Vella, P. G. Yeni and P. A. Volberding, *JAMA*, 1997, **277**, 1962.

72. A. C. Collier, *Adv. Exp. Med. Biol.*, 1996, **394**, 355.

73. C. Schieferstein and T. Buhk, in *HIV Medicine*, ed. C. Hoffmann and B. S. Kamps, Flying Publisher, Wuppertal, 2005.

74. J. W. Erickson and S. K. Burt, *Annu. Rev. Pharmacol. Toxicol.*, 1996, **36**, 545.

75. DHHS Panel of Antiretroviral Guidelines for Adults and Adolescents, *Guidelines for the Use of Antiretroviral Agents in HIV-1-Infected Adults and Adolescents*, DHHS, Washington, DC, 2009.

76. M. A. Wainberg, W. C. Drosopoulos, H. Salomon, M. Hsu, G. Borkow, M. Parniak, Z. Gu, Q. Song, J. Manne, S. Islam, G. Castriota and V. R. Prasad, *Science*, 1996, **271**, 1282.

77. F. Simon, P. Mauclere, P. Roques, I. Loussert-Ajaka, M. C. Muller-Trutwin, S. Saragosti, M. C. Georges-Courbot, F. Barre-Sinoussi and F. Brun-Vezinet, *Nat. Med.*, 1998, **4**, 1032.

78. S. K. Brodine, J. R. Mascola, P. J. Weiss, S. I. Ito, K. R. Porter, A. W. Artenstein, F. C. Garland, F. E. McCutchan and D. S. Burke, *Lancet*, 1995, **346**, 1198.

79. H. Fleury, P. Recordon-Pinson, A. Caumont, M. Faure, P. Roques, J. C. Plantier, E. Couturier, D. Dormont, B. Masquelier and F. Simon, *AIDS Res. Hum. Retroviruses*, 2003, **19**, 41.

80. R. Kantor and D. Katzenstein, *J. Clin. Virol.*, 2004, **29**, 152.

81. A. Holguin, A. Alvarez and V. Soriano, *AIDS*, 2002, **16**, 1163.

82. M. Peeters and P. M. Sharp, *AIDS*, 2000, **14** (Suppl 3), S129.

83. L. Romano, G. Venturi, R. Ferruzzi, M. L. Riccio, P. Corsi, F. Leoncini, A. Vinattieri, L. Incandela, P. E. Valensin and M. Zazzi, *AIDS*, 2000, **14**, 2204.

84. J. Mascola, J. Louwagie, F. E. McCutchan, C. L. Fischer, P. A. Hegerich, K. F. Wagner, A. K. Fowler, J. G. McNeil and D. S. Burke, *J. Infect. Dis.*, 1993, **169**, 48.

85. B. Gaschen, J. Taylor, K. Yusim, B. Foley, F. Gao, D. Lang, V. Novitsky, B. Haynes, B. H. Hahn, T. Bhattacharya and B. Korber, *Science*, 2002, **296**, 2354.

86. M. Peeters, *Transfus. Clin. Biol.*, 2001, **8**, 222.

87. J. Hemelaar, E. Gouws, P. D. Ghys and S. Osmanov, *AIDS*, 2006, **20**, W13.

88. S. Osmanov, C. Pattou, N. Walker, B. Schwardländer and J. Esparza, *J. Acquir. Immune Defic. Syndr.*, 2002, **29**, 184.

89. B. S. Taylor, M. E. Sobieszczyk, F. E. McCutchan and S. M. Hammer, *N. Engl. J. Med.*, 2008, **358**, 1590.

90. D. L. Robertson, J. P. Anderson, J. A. Bradac, J. K. Carr, B. Foley, R. K. Funkhouser, F. Gao, B. H. Hahn, M. L. Kalish, C. Kuiken, G. H. Learn, T. Leitner, F. McCutchan, S. Osmanov, M. Peeters, D. Pieniazek, M. Salminen, P. M. Sharp, S. Wolinsky and B. Korber, *Science*, 2002, **288**, 55.

91. B. Korber, C. Kuiken, B. Foley, B. Hahn, F. McCutchan, J. W. Mellors and J. Sodroski (eds), *Theoretical Biology and Biophysics Group*, Los Alamos National Laboratory, Los Alamos, NM, 1998.

92. M. S. Hirsch, F. Brun-Vezinet, R. T. D'Aquila, S. M. Hammer, V. A. Johnson, D. R. Kuritzkes, C. Loveday, J. W. Mellors, B. Clotet, B. Conway, L. M. Demeter, S. Vella, D. M. Jacobsen and D. D. Richman, *JAMA*, 2000, **283**, 2417.

93. D. J. Hu, T. J. Dondero, M. A. Rayfield, J. R. George, G. Schochetman, H. W. Jaffe, C. C. Luo, M. L. Kalish, B. G. Weniger, C. P. Pau, C. A. Schable and J. W. Curran, *JAMA*, 1996, **275**, 210.

94. R. Kantor and D. Katzenstein, *AIDS Rev.*, 2003, **5**, 25.

95. R. W. Shafer, D. R. Jung, B. J. Betts, Y. Xi and M. J. Gonzales, *Nucleic Acids Res.*, 2000, **28**, 346.

96. S. Spira, M. A. Wainberg, H. Loemba, D. Turner and B. G. Brenner, *J. Antimicrob. Chemother.*, 2003, **51**, 229.

97. D. Paraskevis, E. Magiorkinis, G. Magiorkinis, V. Sypsa, V. Paparizos, M. Lazanas, P. Gargalianos, A. Antoniadou, G. Panos, G. Chrysos, H. Sambatakou, A. Karafoulidou, A. Skoutelis, T. Kordossis, G. Koratzanis, M. Theodoridou, G. L. Daikos, G. Nikolopoulos, O. G. Pybus and A. Hatzakis, *J. Infect. Dis.*, 2007, **196**, 1167.

98. A. C. Palma, F. Araújo, V. Duque, F. Borges, M. T. Paixão and R. Camacho, *Infect. Genet. Evol.*, 2007, **7**, 391.

99. F. Baldanti, S. Paolucci, G. Ravasi, A. Maccabruni, A. Moriggia, G. Barbarini and R. Maserati, *J. Med. Virol.*, 2008, **80**, 947.

100. S. Krauchenco, N. H. Martins, M. Sanches and I. Polikarpov, *J. Enzyme Inhib. Med. Chem.*, 2008, **29**, 1.

101. A. Velazquez-Campoy, M. J. Todd, S. Vega and E. Freire, *Proc. Natl. Acad. Sci. USA*, 2001, **98**, 6062.

102. A. T. Dumans, M. A. Soares, E. S. Machado, S. Hué, R. M. Brindeiro, D. Pillay and A. Tanuri, *J. Infect. Dis.*, 2004, **189**, 1232.

103. Z. Grossman, E. E. Paxinos, D. Averbuch, S. Maayan, N. T. Parkin, D. Engelhard, M. Lorber, V. Istomin, Y. Shaked, E. Mendelson, D. Ram, C. J. Petropoulos and J. M. Schapiro, *Antimicrob. Agents Chemother.*, 2004, **48**, 2159.

104. M. S. Hirsch, F. Brun-Vézinet, R. T. D'Aquila, S. M. Hammer, V. A. Johnson, D. R. Kuritzkes, C. Loveday, J. W. Mellors, B. Clotet, B. Conway, L. M. Demeter, S. Vella, D. M. Jacobsen and D. D. Richman, *JAMA*, 2000, **283**, 2417.

105. L. Vergne, M. Peeters, E. Mpoudi-Ngole, A. Bourgeois, F. Liegeois, C. Toure-Kane, S. Mboup, C. Mulanga-Kabeya, E. Saman, J. Jourdan, J. Reynes and E. Delaporte, *J. Clin. Microbiol.*, 2000, **38**, 3919.

106. R. E. Rose, Y. F. Gong, J. A. Greytok, C. M. Bechtold, B. J. Terry, B. S. Robinson, M. Alam, R. J. Colonno and P. F. Lin, *Proc. Natl. Acad. Sci. USA*, 1996, **93**, 1648.

107. A. Velazquez-Campoy, S. Vega, E. Fleming, U. Bacha, Y. Sayed, H. W. Dirr and E. Freire, *AIDS Rev.*, 2003, **5**, 165.

108. J. M. Baeten, B. Chohan, L. Lavreys, V. Chohan, R. S. McClelland, L. Certain, K. Mandaliya, W. Jaoko and J. D. Overbaugh, *J. Infect. Dis.*, 2007, **195**, 1177.

109. P. J. Kanki, D. J. Hamel, J. L. Sankale, C. Hsieh, I. Thior, F. Barin, S. A. Woodcock, A. Gueye-Ndiaye, E. ZhanG, M. Montano, T. Siby, R. Marlink, I. N. Doye, M. E. Essex and S. M. Boup, *J. Infect. Dis.*, 1999, **179**, 68.

110. L. Bhoopat, L. Eiangleng, S. Rugpao, S. S. Frankel, D. Weissman, S. Lekawanvijit, S. Petchjom, P. Thorner and T. Bhoopat, *Mod. Pathol.*, 2001, **14**, 1263.

111. M. Pope, S. S. Frankel, J. R. Mascola, A. Trkola, F. Isdell, D. L. Birx, D. S. Burke, D. D. Ho and J. P. Moore, *J. Virol.*, 1997, **71**, 8001.

112. C. Yang, M. Li, R. D. Newman, Y. P. Shi, J. Ayisi, A. M. van Eijk, J. Otieno, A. O. Misore, R. W. Steketee, B. L. Nahlen and R. B. Lal, *AIDS*, 2003, **17**, 1667.

113. J. T. Blackard, B. Renjifo, W. Fawzi, E. Hertzmark, G. Msamanga, D. Mwakagile, D. Hunter, D. Spiegelman, N. Sharghi, C. Kagoma and M. Essex, *Virology*, 2001, **287**, 261.

114. B. Renjifo, P. Gilbert, B. Chaplin, G. Msamanga, D. Mwakagile, W. Fawzi and M. Essex, *AIDS*, 2004, **18**, 1629.

115. M. C. Murray, J. E. Embree, S. G. Ramdahin, A. O. Anzala, S. Njenga and F. A. Plummer, *J. Infect. Dis.*, 2000, **181**, 746.

116. N. Tapia, S. Franco, F. Puig-Basagoiti, C. Menendez, P. L. Alonso, H. Mshinda, B. Clotet, J. C. Saiz and M. A. Martinez, *J. Gen. Virol.*, 2003, **84**, 607.

117. L. Monno, L. Scudeller, G. Brindicci, A. Saracino, G. Punzi, A. Chirianni, A. Lagioia, N. Ladisa, S. Lo Caputo and G. Angarano, *Antiviral Res.*, 2009, **83**, 118.

118. V. Pillay, C. Pillay, R. Kantor, F. Venter, L. Levin and L. Morris, *AIDS Res. Hum. Retroviruses*, 2008, **24**, 1449.

119. H. Loemba, B. Brenner, M. A. Parniak, S. Ma'ayan, B. Spira, D. Moisi, M. Oliveira, M. Detorio and M. A. Wainberg, *Antimicrob. Agents Chemother.*, 2002, **46**, 2087.

120. D. Pillay, K. Sinka, P. Rice, B. Peters, J. Clarke and J. Workman, *Antiviral Ther.*, 2000, **5**, 128.

121. A. F. Santos, A. B. Abecasis, A. M. Vandamme, R. J. Camacho and M. A. Soares, *J. Antimicrob. Chemother.*, 2009, **63**, 593.

122. A. C. Palma, A. B. Abecasis, J. Vercauteren, A. P. Carvalho, J. Cabanas, A. M. Vandamme and R. J. Camacho, *Infect. Genet. Evol.*, 2010, **10**, 373.

123. A. T. Dumans, C. C. Barreto, A. F. Santos, M. Arruda, T. M. Sousa, E. S. Machado, E. C. Sabino, R. M. Brindeiro, A. Tanuri, A. J. Duarte and M. A. Soares, *Infect. Genet. Evol.*, 2009, **9**, 62.

124. R. Kantor, D. A. Katzenstein, B. Efron, A. P. Carvalho, B. Wynhoven, P. Cane, J. Clarke, S. Sirivichayakul, M. A. Soares, J. Snoeck, C. Pillay, H. Rudich, R. Rodrigues, A. Holguin, K. Ariyoshi, M. B. Bouzas, P. Cahn, W. Sugiura, V. Soriano, L. F. Brigido, Z. Grossman, L. Morris, A. M. Vandamme, A. Tanuri, P. Phanuphak, J. N. Webe, D. Pillay, P. R. Harrigan, R. Camacho, J. M. Schapiro and R. W. Shafer, *PLoS Med.*, 2005, **2**, e112.

125. A. B. Abecasis, K. Deforche, L. T. Bacheler, P. McKenna, A. P. Carvalho, P. Gomes, A. M. Vandamme and R. Camacho, *J. Antiviral Ther.*, 2006, **11**, 581.

126. L. M. Gonzalez, R. M. Brindeiro, M. Tarin, A. Calazans, M. A. Soares, S. Cassol and A. Tanuri, *Antimicrob. Agents Chemother.*, 2003, **47**, 2817.

127. L. M. Gonzalez, A. F. Santos, A. B. Abecasis, K. Van Laethem, E. A. Soares, K. Deforche, A. Tanuri, R. Camacho, A. M. Vandamme and M. A. Soares, *J. Antimicrob. Chemother.*, 2008, **61**, 1201.

128. J. L. Martinez-Cajas, N. Pant-Pai, M. B. Klein and M. A. Wainberg, *AIDS Rev.*, 2008, **10**, 212.

129. W. P. Bannister, L. Ruiz, C. Loveday, S. Vella, K. Zilmer, J. Kjaer, B. Knysz and A. N. Phillips, A. Mocroft and EuroSIDA Study Group, *Antiviral Ther.*, 2006, **11**, 707.

130. K. Champenois, L. Bocket, S. Deuffic-Burban, L. Cotte, P. André, P. Choisy and Y. Yazdanpanah, *AIDS*, 2008, **22**, 1087.

131. E. R. de Arellano, J. M. Benito, V. Soriano, M. López and A. Holguín, *AIDS Res. Hum. Retroviruses*, 2007, **23**, 891.

132. C. Garrido, N. Zahonero, D. Fernándes, D. Serrano, A. R. Silva, N. Ferraria, F. Antúnes, J. González-Lahoz, V. Soriano and C. de Mendoza, *J. Antimicrob. Chemother.*, 2008, **61**, 694.

133. A. M. Geretti, L. Harrison, H. Green, C. Sabin, T. Hill, E. Fearnhill, D. Pillay, D. Dunn and UK Collaborative Group on HIV Drug Resistance, *Clin. Infect. Dis.*, 2009, **48**, 1296.

134. H. R. Lacerda, L. B. Medeiros, A. M. Cavalcanti, R. A. Ximenes and F. Albuquerque, *Mem. Inst. Oswaldo Cruz*, 2007, **102**, 693.

135. R. M. Coman, A. H. Robbins, M. A. Fernandez, C. T. Gilliland, A. A. Sochet, M. M. Goodenow, R. McKenna and B. M. Dunn, *Biochemistry*, 2008, **47**, 731.

136. A. Alaeus, K. Lidman, A. Sönnerborg and J. Albert, *AIDS*, 1997, **11**, 859.

137. Z. Debyser, E. Van Wijngaerden, K. Van Laethem, K. Beuselinck, M. Reynders, E. De Clercq, J. Desmyter and A. M. Vandamme, *AIDS Res. Hum. Retroviruses*, 1998, **14**, 453.

138. A. Holguín, M. López, M. Molinero and V. Soriano, *J. Clin. Microbiol.*, 2008, **46**, 2918.

139. K. Steegen, S. Luchters, N. De Cabooter, J. Reynaerts, K. Mandaliya, J. Plum, W. Jaoko, C. Verhofstede and M. Temmerman, *J. Virol. Methods.*, 2007, **146**, 178.

140. N. Taylor, I. Schmid, A. Egle, R. Greil, W. Patsch and H. Oberkofler, *AntiviralTher.*, 2009, **14**, 1189.

141. P. J. Kanki, K. U. Travers, S. Mboup, C. C. Hsieh, R. G. Marlink, A. Gueye-Ndiaye, T. Siby, I. Thior, M. Hernandez-Avila and J. L. Sankalé *et al.*, *Lancet*, 1994, **343**, 943.

142. R. Marlink, P. Kanki, I. Thior, K. Travers, G. Eisen, T. Siby, I. Traore, C. C. Hsieh, M. C. Dia and E. H. Gueye et al., *Science*, 1994, **265**, 1587.

143. P. J. Kanki, F. Barin, S. M'Boup, J. S. Allan, J. L. Romet-Lemonne, R. Marlink, M. F. McLane, T. H. Lee, B. Arbeille and F. Denis *et al.*, *Science*, 1986, **232**, 238.

144. J. Ren, L. E. Bird, P. P. Chamberlain, G. B. Stewart-Jones, D. I. Stuart and D. K. StaSmmers, *Proc. Natl. Acad. Sci. USA*, 2002, **99**, 14410.

145. M. Witvrouw, C. Pannecouque, W. M. Switzer, T. M. Folks, E. De Clercq and W. Heneine, *Antiviral Ther.*, 2004, **9**, 57.

146. C. A. Adjé-Touré, R. Cheingsong, J. G. Garcìa-Lerma, S. Eholié, M. Y. Borget, J. M. Bouchez, R. A. Otten, C. Maurice, M. Sassan-Morokro, R. E. Ekpini, M. Nolan, T. Chorba, W. Heneine and J. N. Nkengasong, *AIDS*, 2003, **17** (Suppl. 3), 49.

147. D. Desbois, B. Roquebert, G. Peytavin, F. Damond, G. Collin, A. Bénard, P. Campa, S. Matheron, G. Chêne, F. Brun-Vézinet and D. Descamps, *Antimicrob. Agents. Chemother.*, 2008, **52**, 1545.

148. M. Ntemgwa, B. G. Brenner, M. Oliveira, D. Moisi and M. A. Wainberg, *Antimicrob. Agents Chemother.*, 2007, **51**, 604.

149. B. Rodés, J. Sheldon, C. Toro, V. Jiménez, M. A. Alvarez and V. Soriano, *J. Antimicrob. Chemother.*, 2006, **57**, 709.

150. P. A. Chan, S. E. Wakeman, T. Flanigan, S. Cu-Uvin, E. Kojic and R. Kantor, *AIDS Res. Ther.*, 2008, **14**, 5.

151. E. Y. Takeb, S. Kusagawa and K. Motomura, *Pediatr. Int.*, 2004, **46**, 236.

CHAPTER 16

Design of Capsid-binding Antiviral Agents Against Human Rhinoviruses

CHUAN XIAO,*,a MARK A. MCKINLAYb AND MICHAEL G. ROSSMANNa

a Department of Biological Sciences, Purdue University, 915 W. State Street, West Lafayette, IN 47907, USA; b TetraLogic Pharmaceuticals, 343 Phoenixville Pike, Malvern, PA 19355, USA

1 Introduction

A major goal in the study of pathogenic viruses is the discovery of effective treatments such as developing vaccines or chemotherapeutics. However, it might be impractical or impossible to develop vaccines for viruses that can escape the host's immune system either by rapidly mutating their surface epitopes or by attacking the host's immune system itself. For such viruses, it might be possible to develop small chemotherapeutic compounds that would become effective drugs. The majority of current commercially available antiviral drugs have been developed from compounds initially discovered by screening approaches, not by *ab initio* design. Unlike bacteria that have a distinct prokaryotic biochemistry and can be selectively inhibited by antibiotics, viruses utilize the host cell machinery to replicate themselves, making it more difficult to find a drug inhibiting the virus without affecting the host cell. However, nearly all known viral capsid structures are unique to the virus[1] with no ortholog having been found in cellular components. In this sense, viral capsids

---

*Present address: Department of Chemistry, University of Texas, El Paso, TX 79968, USA.

RSC Biomolecular Sciences No. 21
Structural Virology
Edited by Mavis Agbandje-McKenna and Robert McKenna
© Royal Society of Chemistry 2011
Published by the Royal Society of Chemistry, www.rsc.org

are completely different from viral non-structural proteins such as proteases, helicases, *etc.*, which have cellular counterparts. Hence antiviral compounds targeted to viral capsid proteins are less likely to have undesirable toxic off-target side-effects.

In the past three decades, there have been significant advances in the determination of three-dimensional atomic structures of viruses using X-ray crystallography.[2–4] This structural information has facilitated the systematic modification of promising capsid binding compounds to increase their potency and metabolic stability. An early example of structure-based antiviral drug development was anti human rhinovirus (HRV) capsid binding compounds[5,6] made possible by the structural determination of human rhinovirus 14 (HRV14) (the first animal virus to be determined at near atomic resolution[7]) and other rhinovirus serotypes,[8–11] polioviruses,[12–14] Coxsackieviruses (CV)[15–17] and other enteroviruses.[18] In this chapter, we discuss exclusively the development of pleconaril, the prime example of one of these inhibitors specifically developed for the treatment of HRV infections.

2 The Common Cold

Common cold is a mild upper respiratory illness[19] with symptoms such as sore throat, rhinorrhea, sometimes fever and disturbed sleep.[20,21] Although the symptoms of a common cold are generally mild and are usually resolved within one week in adults, in infants and young children the symptoms may be more severe and can be lethal in patients with comorbidities.[19] The average number of colds caused by viral or bacterial infections is 6–10 annually for preschool children and 2–3 annually for adults in the USA,[22] accounting for 27 million visits to physicians and 161 million days of restricted activities.[23–26] Furthermore, as the common cold is the most frequently encountered infectious disease experienced by humans, the cumulative social cost is considerable with regard to the requirement of over-the-counter drugs and the loss of work hours and school days.[19,27,28] Due to the low fidelity of RNA-dependent RNA polymerase in picornaviruses, the genomic sequence of the virus is constantly changing. The resultant fast mutation rate and the existence of more than 100 different HRV serotypes have made vaccine development impossible. Consequently, substantial effort has been dedicated towards the development of anti-HRV drugs.[5,6]

3 Picornavirus Structure

Picornaviruses, a group of small animal RNA viruses,[29–32] are responsible for more than 50% of all human common colds.[21,22,33–35] HRVs are members of the *Picornaviridae* family and are composed of a protein shell encapsulating a single-stranded positive-sense RNA genome. The icosahedral protein shell or capsid contains 60 copies of each of the four viral proteins, VP1, VP2, VP3 and VP4 (Figure 1). The three VP1, VP2 and VP3 subunits each have a similar

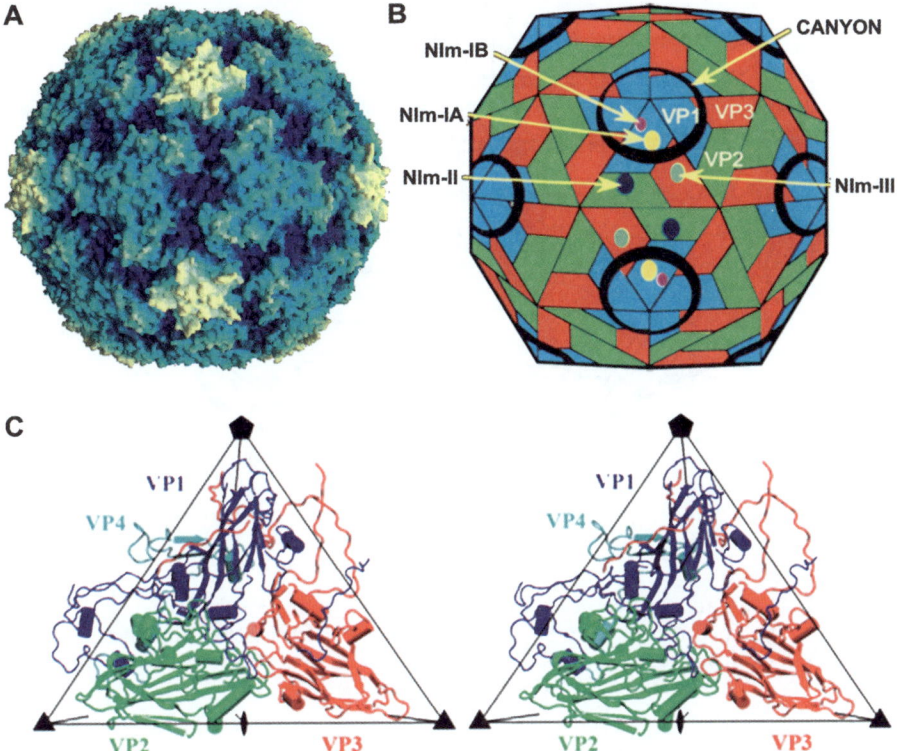

Figure 1 Picornavirus structure. (A) Surface shaded representation of HRV14 showing the canyon and the star-shaped fivefold plateau around the fivefold vertices, composed of five VP1s. (B) Diagrammatic view of HRV showing the icosahedral arrangement of VP1, VP2 and VP3. The canyon running around each fivefold vertex is shaded black. The NIm I, II and III sites are neutralizing immunogenic sites associated primarily with VP1, VP2 and VP3, respectively. (C) Stereo ribbon diagram of one CVA21 icosahedral asymmetric unit. VP1, VP2, VP3 and VP4 are colored blue, green, red and cyan, respectively.

β-barrel, wedge-shaped structure referred to as a 'jelly-roll' fold (Figure 2). The β-strands along each VP peptide are named B–H. Neighboring β-strands are anti-parallel and form two opposing β-sheets on opposite sides of a barrel-like structure. One of these β-sheets is composed of the B–I–D–G sequence of strands and the opposing β-sheet is composed of the C–H–E–F sequence of strands (Figure 2). The VP1, VP2 and VP3 subunits form the external surface of the capsid, with VP1 situated around the icosahedral fivefold axis and VP2 and VP3 forming a pseudo-hexameric structure around the icosahedral three-fold axes (Figure 1B). There is a marked depression 'canyon' surrounding each of the 12 pentameric vertices (Figure 1A and B). The canyon is situated roughly between VP1 forming the 'north rim' and VP2 and VP3 forming the 'south rim'. VP4 is a smaller protein with an extended structure lying at the interface

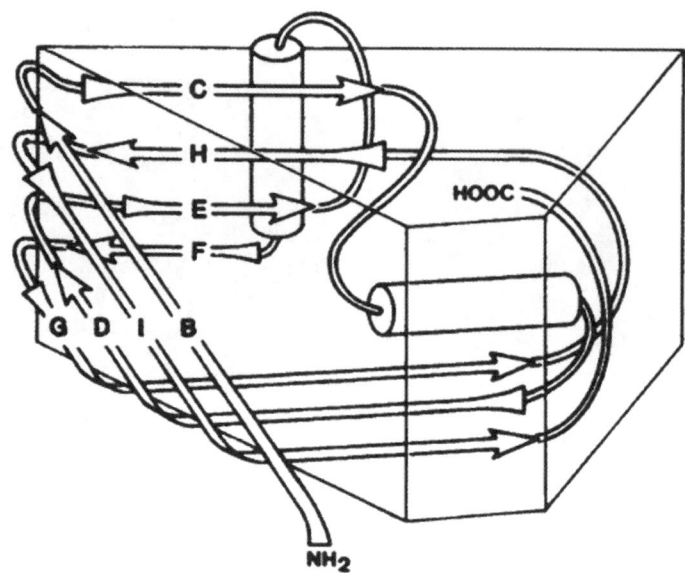

Figure 2 Diagrammatic representation of the wedge-shaped 'jelly-roll' structure using polypeptide fold of one subunit. The same motif is also found in many other viral capsid subunit structures. The β-strands are labeled B to I along the peptide chain. Adapted from Hogle *et al.*[12]

between the capsid and the RNA genome.[3] A post-translationally added myristate moiety at the N-terminus of VP4 may be important for membrane association during infection. The N-termini of VP1 and VP4 are probably extruded from the virus capsid during uncoating and may form a channel in the host cell membrane for genome delivery during the infection.[36,37]

4   Human Rhinovirus Receptors

Like many animal viruses, HRV infection is initialized by binding to cell surface receptor molecules.[38] Amino acid residues on or near the rim of the canyon were shown to be subject to higher rates of mutational change and were found to produce escape mutations to neutralizing antibodies (Figure 1B).[7,39,40] Sequence analysis of major group HRVs has shown conservation of residues on the floor of the canyon. Therefore, it was hypothesized that the canyon is the site of receptor binding,[7,41,42] which was later supported by cryo-electron microscopy (cryo-EM) reconstructions of picornavirus complexes with their receptors[43-46] (Figure 3).

HRVs have been classified into two groups based on their preferred receptor tropism.[47] A major group of HRVs, representing about 90% of all serotypes, use intermolecular adhesion molecule 1 (ICAM-1) as their receptor.[48-51] HRVs belonging to the minor receptor group recognize members of the low-density

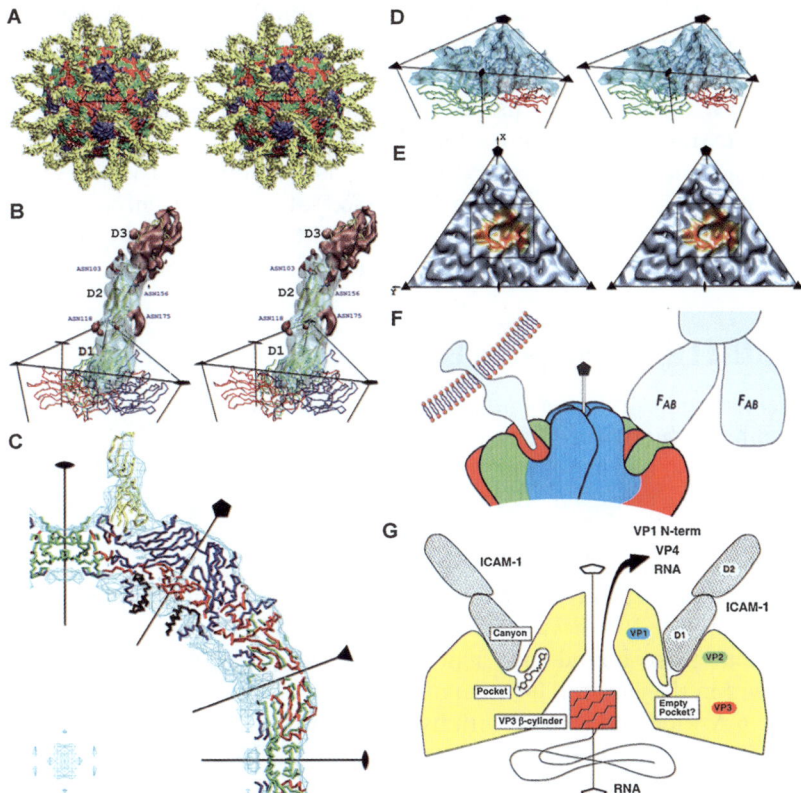

Figure 3 HRV16 complexed with ICAM-1. (A) Stereo view of the surface shaded cryo-EM map of HRV16 complexed with ICAM-1. VP1, VP2, VP3 and ICAM-1 are colored blue, green, red and yellow, respectively. The same color codes are used in the following panels. The black triangle shows the limit of one icosahedral asymmetric unit. (B) Stereo view of one ICAM-1 molecule bound to HRV16. One icosahedral asymmetric unit is outlined in black, showing one copy of the difference density of the ICAM-1 molecule in transparent cyan. ICAM-1 is shown as a ribbon drawing. The four Asn residues that are N-linked to carbohydrates are drawn and labeled in blue. The densities that have not been fitted with the atomic models for the ICAM-1 domain D1 and D2 are shown in solid red, which include the densities of carbohydrates moieties belonging to domain D2, as well as the density of domain D3. (C) The central slab (from -10 to $+10$ Å along the z-axis) of the cryo-EM density (cyan) fitted with the appropriate backbone structures of ICAM-1, VP1, VP2, VP3 and VP4. (D) Stereo view showing the fit of the HRV16 capsid structure into the cryo-EM map surface. VP1, VP2 and VP3 are represented by a trace of their C_α atoms. The density corresponding to ICAM-1 has been removed. (E) Stereo diagram of one icosahedral asymmetric unit of the virus surface of HRV16 (cyan) with the ICAM-1 contact area colored in yellow to red according to the separation distance between the receptor and the virus. (F) Schematic diagram showing the original rationale of the canyon hypothesis. The canyon is accessible to long and narrow receptor molecules but not to bulky antibodies. (G) Schematic diagram showing the competition between the receptor (ICAM-1) and the 'pocket factor'. (A)–(E) adapted from Xiao et al.,[46] (F) from Rossmann[42] and (G) from Kolatkar et al.[44]

lipoprotein family.[52,53] In an immune response to inflammation, ICAM-1 binds to the integrin LFA-1 (lymphocyte-function associated molecule 1) to facilitate adhesion between lymphocytes and endothelial cells, initiating the migration of lymphocytes to the infected tissues.[54] ICAM-1 is a type I membrane protein with five extracellular immunoglobulin-like domains, a single-span transmembrane domain and a short cytoplasmic tail, producing a 190 Å long rod-shaped molecule.[55–58] ICAM-1 is expressed at low concentrations in the membranes of many cells but can be stimulated to have a high expression level by cytokines such as interleukin-1 and tumor necrosis factor.[59]

5   Anti-HRV Compounds

Drug screening has resulted in the discovery of a diversity of small compounds able to inhibit rhinovirus-induced cytopathic effect *in vitro*.[6] These compounds are hydrophobic and have a molecular weight of about 350 Da. Among these compounds is a series of compounds derived from arildone, which were developed at the Sterling Winthrop Research Institute (arm of the Sterling Drug company part of Sanofi Aventis) and identified by 'WIN' numbers.[60] For instance, arildone is WIN 38020 (Figure 4). Early crystallographic studies of HRV14 complexed with various capsid-binding WIN compounds (Figure 5) showed that these bound into a pocket immediately below the floor of the canyon within the jelly-roll β-barrel of VP1.[61] The pocket is approximately 20 Å long and 6 Å wide and is lined with hydrophobic amino acids (Figure 5C). The length of the pocket varies considerably among the different HRV serotypes and other picornaviruses. This pocket is compressed and empty in HRV14[61,62] and HRV3,[10] but is filled with electron density representing a fatty acid-like 'pocket factor' in polioviruses[13,14] and in other picornaviruses such as HRV16,[9] HRV2,[11] CVB3,[15] CVA9[16] and CVA21[17] (Figure 6). The pocket factor might be cellular in origin and might regulate virus assembly and uncoating.[13,61,63,64] Zhang *et al.*[65] showed that the replacement of the pocket factor by an antiviral compound can be more effective during viral assembly rather than in the mature virus.

6   Mechanism of Inhibition by Capsid-binding Compounds: Inhibition of Attachment

Crystallographic studies show that binding of WIN compounds into the pocket of the previously empty HRV14 causes conformational changes in the canyon floor (the receptor attachment site) immediately above the pocket (Figures 3G and 5C). These changes have been correlated with the inhibition of attachment in the presence of the antiviral compounds.[66–68] Thus, when ICAM-1 binds into the canyon, the floor is depressed downwards, which is only possible when the pocket does not contain a well-bound anti-viral compound (Figure 3G). Conversely, when there is a compound in the pocket, the canyon floor is raised upwards inhibiting attachment of ICAM-1 to the virus[64] (Figure 3G).

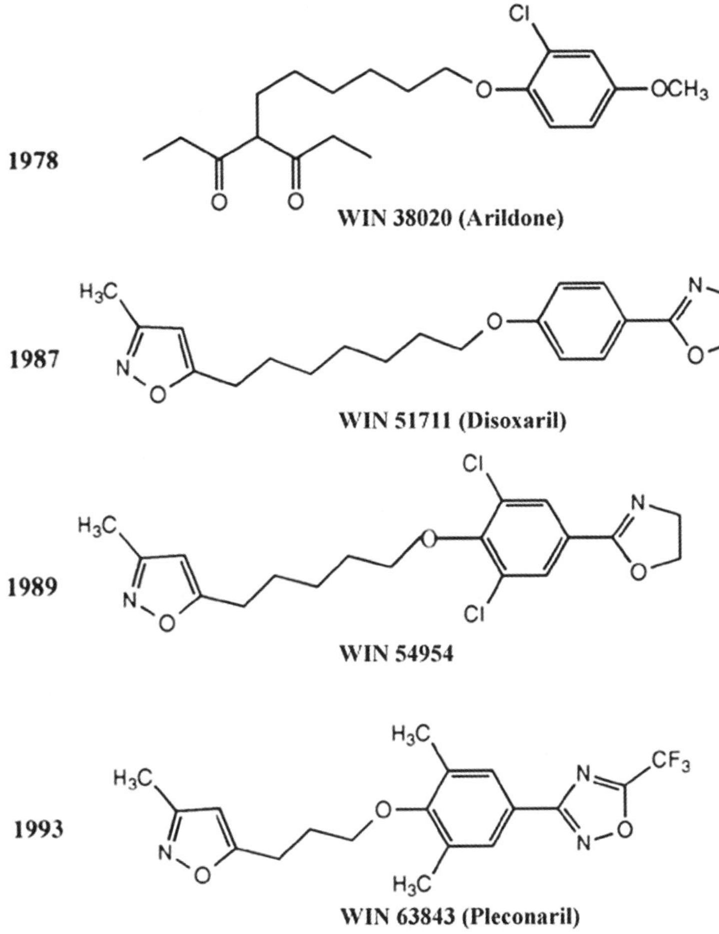

Figure 4 Progression in the development of pleconaril. Arildone was the original lead compound. Disoxaril (WIN 51711) was too long and therefore its activity was limited to a subgroup of serotypes. It failed in Phase 1 clinical tests because it tended to form crystals in the urethra of volunteers. WIN 54954 had low bioavailability and had limited stability, resulting in toxic metabolites. Pleconaril (WIN 63843) had much better and longer lasting bioavailability, and also high efficacy against most serotypes.

7 Mechanism of Inhibition by Capsid-binding Compounds: Inhibition of Uncoating

When the hydrophobic pocket in VP1 is filled with a capsid-binding compound, there is an increase in the thermal stability of the virus,[64,69,70] presumably as a consequence of replacing a hydrophobic molecule in an internal hydrophobic cavity.[71,72] The increase in stability of the virus when complexed with a capsid-binding compound has also been demonstrated by the reduction of exposure of hidden, internal tryptic cleavage sites when the virus is complexed with a capsid-binding compound.[73] The pocket factor may be required to stabilize the virus in transit from one cell to another. However, the delivery of the infectious RNA into the cytoplasm must require a destabilizing step that might be affected by expulsion of the pocket factor during receptor molecule-mediated virus uncoating. Presumably, the destabilization of the virus on cell attachment is made possible by the displacement of a sufficient number of pocket factors when the receptor competes with the overlapping binding sites on the viral surface. Progressive recruitment of receptors is sufficient to trigger the release of VP4s to initiate the entry process of the virus.[64]

8   Drug-resistant Mutants

Drug-resistant (compensation) mutants can be selected by growing HRVs in the presence of antiviral capsid-binding compounds. Two classes of drug-resistant mutants were observed: a high-resistant (HR) class with a frequency of once in about every 2.5×10^4 virions and a low-resistant (LR) class with a 10–30-fold higher frequency.[66] These mutant viruses were shown to have mostly single amino acid substitutions.[68,69] The HR mutations were mapped to the wall of the drug-binding pocket (Figures 5C and 7). The side-chains of these residues point directly into the pocket and were invariably replaced by bulkier groups, thereby probably hindering the entry or seating of the drug within the binding pocket. In contrast, all the LR mutations were mapped near the canyon floor and increased the affinity of ICAM-1 to the virus. Therefore, it was concluded that ICAM-1 binds better to the mutant viruses than the capsid-binding compounds[66,74,75] (Figure 7). Some HRV mutants were found to be drug dependent, which required the drug to be present in the cell culture media during viral replication.[6]

9   Computational Analyses

The atomic structure of the virus–drug complex determined by X-ray crystallography provides a static snapshot of a dynamic macromolecular assembly.

Figure 5   Structure of a capsid-binding compound bound into the VP1 hydrophobic pocket. (A) Ribbon diagram of one asymmetric unit of HRV14, with VP1, VP2 and VP3 colored blue, green and red, respectively. The capsid-binding compound density is shown in yellow within VP1. (B) Schematic representation of the VP1 fold of HRV14. The binding site of an antiviral WIN compound is shown in green within the hydrophobic interior of VP1. (C) A WIN compound is shown in contact with many hydrophobic residues that surround the pocket. The positions of compensation mutations are also shown. Some of the mutations lie within the pocket (pink) whereas others lie on the canyon floor (blue). The solvent-accessible entrance to the binding pocket or 'pore' is marked on the left-hand side of the figure. The other end of the pocket, the 'toe', is enclosed and is more hydrophobic. (A) and (B) adapted from Rossmann[64] and (C) from Hadfield *et al.*[75]

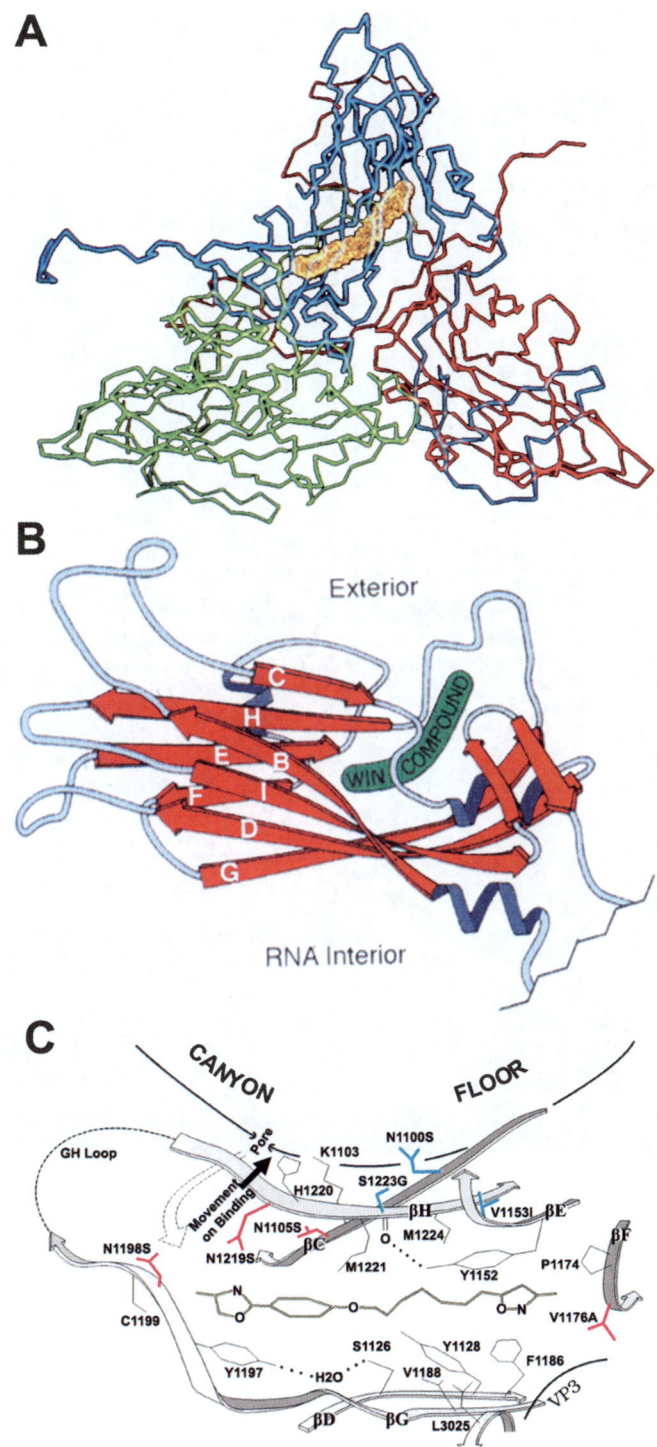

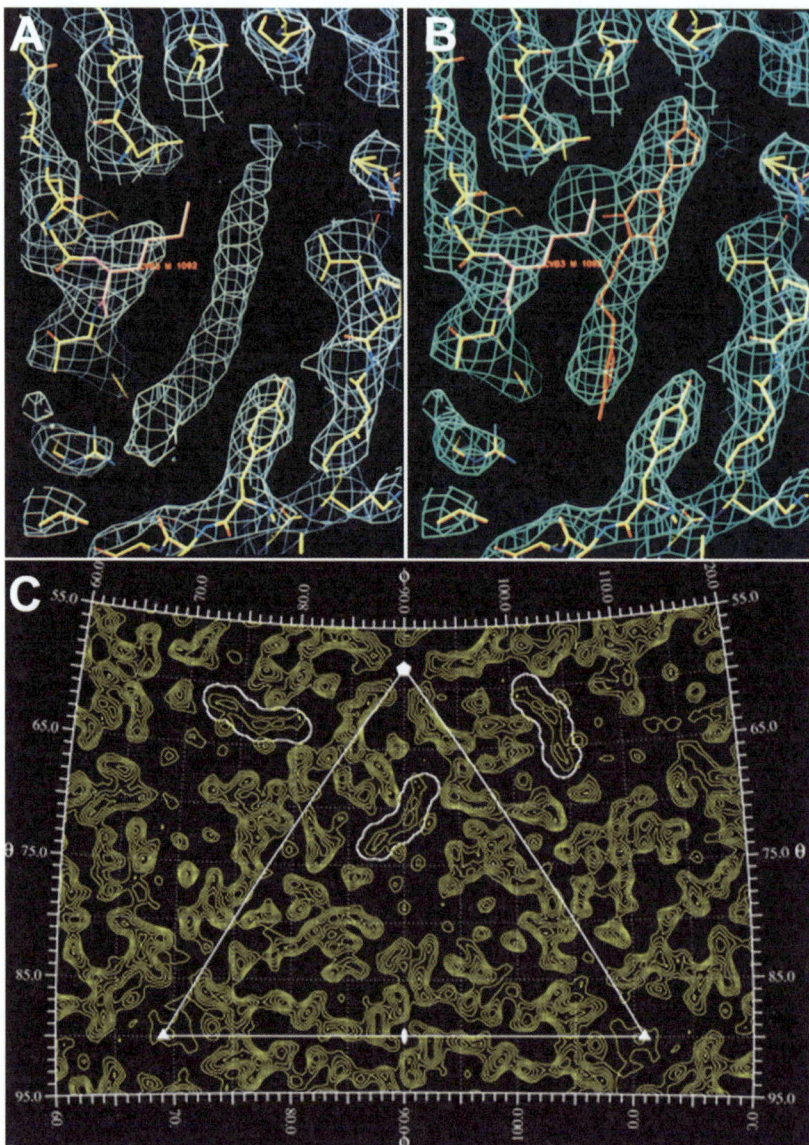

Figure 6 Electron density in the vicinity of the hydrophobic pocket within VP1. (A) Coxsackie virus B3 (CVB3) electron density superimposed with the structure of VP1 surrounding the pocket factor. (B) CVB3 complexed with the iodinated antiviral compound WIN 66393. The view is the same as shown in (A). (C) The location of the 'pocket factor' in a 3.2 Å resolution electron density map of CVA21 crystals shown as a stereographic projection of the density from a spherical section at a fixed radius of 129.6 Å. The density of the pocket factor is outlined in white. (A) and (B) adapted from Muckelbauer *et al.*[15] and (C) from Xiao and Rossmann.[91]

Molecular dynamic calculations based on the structure have investigated the energy constraints of the conformational changes and particle stability on drug binding.[76,77] However, such calculations are limited by the magnitude and complexity of the virus structure.

Computer-assisted structure-based drug design is a developed and specialized science.[78–80] However, the successful development of pleconaril was based largely on graphical interpretation of numerous picornavirus–drug complex structures together with biological and chemical knowledge[5] of what determines the potency and spectrum of picornavirus sensitivity and also metabolic stability (to avoid rapid clearance from the body) and solubility (to increase oral bioavailability).

10 Development of an Effective Anti-HRV Drug

The original arildone compound (WIN 38020) (Figure 4) was examined for its effects on poliovirus in 1978 and was shown to inhibit the uncoating of the virus.[81] The compound was later shown to prevent poliovirus-induced paralysis and death in mice.[82] Although fairly active against poliovirus and a few other enteroviruses, arildone lacked activity against most rhinoviruses. In a screen for rhinovirus-specific agents at Sterling Winthrop, an analog of arildone (WIN 41137) was discovered with modest activity against rhinoviruses. Modifications were made in an effort to improve the potency and spectrum of rhinovirus activity, and the result was WIN 51711 (disoxaril). This compound completed preclinical development and entered Phase 1 clinical trails in 1987. In 1986, crystallographic studies showed that the compound acted by binding to the virus capsid.[61] Disoxaril, however, was rapidly and extensively metabolized, with the metabolites crystallizing in the urethra of volunteers. Further modifications produced WIN 54954, which was also rapidly metabolized, producing a rash in some subjects during a Phase 2 clinical trial in 1989. Additional modifications were made to improve the metabolic stability and resulted in WIN 63843 (pleconaril) in 1993. The addition of a trifluoromethyl group on the oxadiazole ring (Figure 4) protected the compound from rapid metabolic degradation. Furthermore, the length of the aliphatic chain in the compound was shortened to increase the potency against a broader spectrum of viruses[6] that included viruses with short pockets such as HRV1A[8,83] and long pockets such as HRV14.[61,62]

Pleconaril was found to have good oral bioavailability and metabolic stability and was efficacious in mice infected with Coxsackievirus A9, A21 and B3.[84] Pleconaril was evaluated in two large Phase 3 clinical trials and was the first antiviral compound shown to be effective in shortening the duration and severity of the common cold and also the period of viral shedding[85] (Table 1). The alleviation of cold illness in 65% of 1363 patients enrolled with a picornavirus infection was reduced by 14% (from 7.3 to 6.3 days in the two studies combined). The magnitude of the clinical benefit was far greater in the patients infected with a more sensitive strain of rhinovirus.[86] Patients infected with a highly susceptible virus (50% effective concentration $\leq 0.38\,\mu g\,mL^{-1}$) had the duration of their illness shortened by 1.9–3.9 days, whereas those infected with a virus of lower sensitivity obtained no benefit[84] (Table 2). Despite the fact that the primary and secondary endpoints were met in the two large Phase 3 trials, pleconaril failed to obtain FDA approval because of a CYP 3A drug interaction that was reported in women taking oral contraceptives after 2 weeks of dosing in a 6 week prophylaxis study that was conducted while the new drug application was under review. The concern was that even though no interaction was detected during the 5 days of dosing intended for treatment indication, the magnitude of clinical benefit and the potential for patients to use the drug beyond the prescribed labeling made the risk benefit insufficient.

An intranasal form or pleconaril is currently under development by Schering-Plough. The low dose used in an intranasal dosage form would not be expected to cause the CYP 3A drug interaction. In addition, since 2001, at least 475 patients have been treated with pleconaril under a compassionate use protocol, many of whom have noted anecdotal evidence of a treatment benefit. The symptoms of these patients had varied from chronic meningoencephalitis, encephalitis/meningoencephalitis, myocarditis to neonatal enteroviral disease.[87–90] Although pleconaril was not approved, the clinical studies showed that an antiviral drug can shorten the duration and severity of a generally mild and self-limiting disease in healthy persons and reduce viral shedding. Studies

---

Figure 7   Schematic representation of the competitive binding among receptor, pocket factor and WIN compound. Crystallographically and electron microscopically determined structures are colored in yellow, red and green, respectively. Hypothetical structures are shaded in gray. (A) In wild-type HRV14, the pocket factor binds weakly and is not observed in crystallographic studies. When WIN compounds bind into the pocket, they deform the roof of the pocket, which is also the floor of the canyon. This inhibits the attachment of the virus to the ICAM-1 receptor, hence presumably the binding affinity of WIN compound is greater than that of ICAM-1. When ICAM-1 recognizes the canyon floor, the putative pocket factor must be displaced by ICAM-1 and, hence, the binding affinity of ICAM-1 is greater than that of pocket factor. (B) High-frequency low-resistant compensation mutants (LR mutants) of HRV14 cluster around the canyon walls and floor (crosses) and increase the affinity of ICAM-1 for the virus. Although WIN compounds can bind to the virus, they do not inhibit infectivity. Thus, the binding affinity of the mutant virus to ICAM-1 is greater than that of WIN compound. (C) Low-frequency high-resistant compensation mutants (HR mutants) of HRV14 line the hydrophobic binding pocket that reduces the affinity of the WIN compounds for the virus. Since the affinity of the receptor of ICAM-1 for the virus is unchanged, the equilibrium moves in favor of receptor binding rather than WIN compound binding. (D) Wild-type HRV16 contains a pocket factor. This can be replaced by WIN compounds, which inhibit attachment. Hence, in this case, the affinity of HRV16 for WIN compounds is greater than that of ICAM-1, which is greater than that of pocket factor. Adapted from Rossmann[64] and Hadfield *et al.*[75]

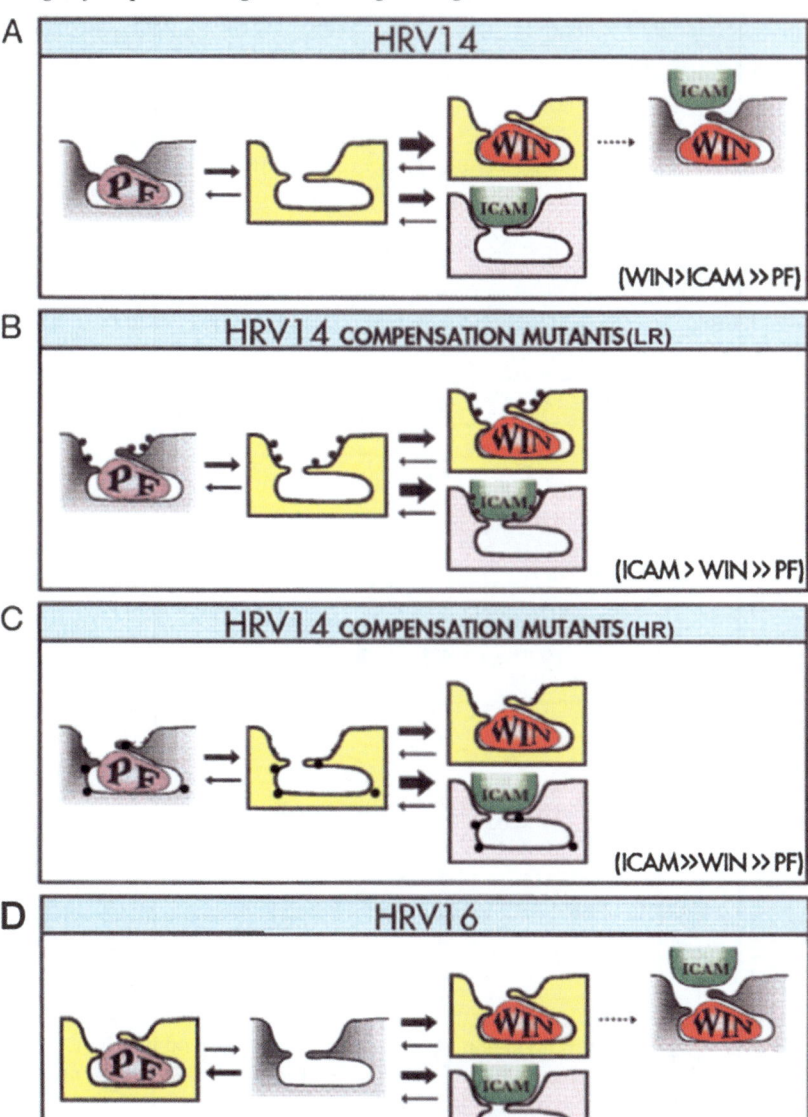

Table 1 Effect of pleconaril on duration of the common cold in picornavirus-infected patients[a].[85]

Subject group variable	Study 843-043			Study 843-044			Studies combined		
	Placebo group	Pleconaril group	P	Placebo group	Pleconaril group	P	Placebo group	Pleconaril Group	P
Positive RT-PCR results (ITT-I population)									
No. of patents	326	337		356	344		682	681	
Time to event (days)									
25th percentile	4.1	3.3		4.3	3.3		4.3	3.3	
Median	7.2	6.6	0.037	7.7	6.2	0.001	7.3	6.3	<0.001
75th percentile	11.7	10.8		12.3	10.4		12.0	10.8	
Negative RT-PCR results									
No. of patients	200	189		168	176		368	365	
Time to event (days)									
25th percentile	3.3	3.3		3.7	3.7		3.3	3.4	
Median	5.9	6.1	0.639	5.9	6.0	0.776	5.9	6.0	0.591
75th percentile	10.1	10.9		13.8	11.3		11.4	11.3	
All randomized subjects (ITT population)									
No. of patients	526	526		524	520		1050	1046	
Time to event (days)									
25th percentile	3.8	3.3		4.1	3.4		3.9	3.3	
Median	6.9	6.4	0.201	7.1	6.2	0.015	7.0	6.3	0.009
75th percentile	11.2	10.8		12.3	10.9		11.9	10.8	

[a]Alleviation of illness is defined as the absence of rhinorrhea and presence of no or mild other cold symptoms for ≥ 48 h without use of cold symptom relief medication. ITT, intention-to-treat; ITT-I, intention-to-treat-infected.

Table 2 Degree of clinical benefit is related to the sensitivity of the infecting virus.[84]

Quintile	EC$_{50}$ range ($\mu g\,mL^{-1}$)	Placebo		Pleconaril		
		No. of samples	Median No. of days (95% CI[b])	No. of samples	Median No. of days (95% CI[b])	P[a]
1	0.0008–0.005	68	6.7 (5.6–7.8)	80	4.8 (3.8–5.8)	<0.05
2	>0.005–0.025	78	8.8 (7.5–10.1)	72	4.9 (3.9–5.9)	<0.05
3	>0.025–0.096	78	9.0 (7.8–10.2)	69	6.4 (5.3–7.5)	<0.05
4	>0.096–0.38	81	7.9 (6.8–9.0)	67	5.9 (4.5–7.3)	0.14
5	>0.38– ≥ 3.8	75	5.7 (4.6–6.8)	76	9.4 (8.3–10.5)	<0.05

[a]Mann–Whitney two-tailed Student's *t*-test.
[b]CI, confidence interval.

in the future should focus on patients with underlying illnesses such as asthma, where the consequences of a rhinovirus cold can be fairly severe.

11 Acknowledgments

Many individuals were involved in the work that eventually led to the development of pleconaril. Foremost among these is Guy Diana, a chemist of exceptional skill and insight. Others include Frank Dutko, James Groarke, Michael Otto and Dan Pevear at Sterling Winthrop, Beverly Heinz, Roland Rueckert, Debborah Shepard at the University of Wisconsin and Eddy Arnold, John Badger, Jodi Bibler-Muckelbauer, Michael Chapman, Andrea Hadfield, Sangsoo Kim, Marcos Olivera, Alan Simpson (deceased), Thomas Smith, Gerd Vriend, Ying Zhang and Rui Zhao at Purdue University. The work was supported mostly by NIH grants to M.G.R. (AI 11219) and to Roland Rueckert (AI 31960 and AI 24939), but also a grant from Sterling Winthrop, NSF instrumentation grants to M.G.R., a Purdue Showalter grant to M.G.R. and a Purdue reinvestment grant and a Lucille P. Markey award for structural biology at Purdue University to M.G.R. We thank Carol Greski for help with the preparation of the manuscript.

References

1. M. Krupovič and D. H. Bamford, *Nat. Rev. Microbiol.*, 2008, **6**, 941.
2. S. C. Harrison, *Adv. Virus Res.*, 1983, **28**, 175.
3. M. G. Rossmann and J. E. Johnson, *Annu. Rev. Biochem.*, 1989, **58**, 533.
4. D. H. Bamford, J. M. Grimes and D. I. Stuart, *Curr. Opin. Struct. Biol.*, 2005, **15**, 655.
5. M. A. McKinlay and M. G. Rossmann, *Annu. Rev. Pharmacol. Toxicol.*, 1989, **29**, 111.
6. M. A. McKinlay, D. C. Pevear and M. G. Rossmann, *Annu. Rev. Microbiol.*, 1992, **46**, 635.
7. M. G. Rossmann *et al.*, *Nature*, 1985, **317**, 145.
8. S. S. Kim *et al.*, *J. Mol. Biol.*, 1989, **210**, 91.
9. M. A. Oliveira *et al.*, *Structure*, 1993, **1**, 51.
10. R. Zhao *et al.*, *Structure*, 1996, **4**, 1205.
11. N. Verdaguer, D. Blaas and I. Fita, *J. Mol. Biol.*, 2000, **300**, 1179.
12. J. M. Hogle, M. Chow and D. J. Filman, *Science*, 1985, **229**, 1358.
13. D. J. Filman *et al.*, *EMBO J.*, 1989, **8**, 1567.
14. K. N. Lentz *et al.*, *Structure*, 1997, **5**, 961.
15. J. K. Muckelbauer *et al.*, *Structure*, 1995, **3**, 653.
16. E. Hendry *et al.*, *Struct. Fold. Des.*, 1999, **7**, 1527.
17. C. Xiao *et al.*, *Structure*, 2005, **13**, 1019.
18. M. Smyth *et al.*, *Nat. Struct. Biol.*, 1995, **2**, 224.
19. I. M. Mackay, *J. Clin. Virol.*, 2008, **42**, 297.

20. J. M. Gwaltney Jr, J. O. Hendley, G. Simon and W. S. Jordan Jr, *JAMA*, 1967, **202**, 494.
21. E. Arruda, A. Pitkaranta, T. J. Witek Jr, C. A. Doyle and F. G. Hayden, *J. Clin. Microbiol.*, 1997, **35**, 2864.
22. J. M. Gwaltney Jr, in *Principles and Practices of Infectious Diseases*, ed. G. L. Mandell, R. G. Douglas and J. E. Bennett, Wiley, New York, 1985, p. 351.
23. S. J. Sperber and F. G. Hayden, *Antimicrob. Agents Chemother.*, 1988, **32**, 409.
24. W. J. McIsaac, N. Levine and V. Goel, *J. Fam. Pract.*, 1998, **47**, 366.
25. N. Rosenstein *et al.*, *Pediatrics*, 1998, **101**, 181.
26. R. Gonzales *et al.*, *Ann. Intern. Med.*, 2001, **134**, 479.
27. A. G. Mainous III, W. J. Hueston and J. R. Clark, *J. Fam. Pract.*, 1996, **42**, 357.
28. J. S. Bertino, *Am. J. Med.*, 2002, **112**, Suppl 6A, 42S.
29. R. R. Rueckert, in *Fields Virology*, ed. B. N. Fields, D. M. Knipe and P. M. Howley, Lippincott-Raven, Philadelphia, PA, 1996, p. 609.
30. A. M. Q. King *et al.*, in *Virus Taxonomy: Classification and Nomenclature of Viruses: Seventh Report of the International Committee on Taxonomy of Viruses*, Academic Press, San Diego, CA, 2000, p. 657.
31. V. R. Racaniello, in *Fields Virology*, ed. D. M. Knipe and P. M. Howley, Vol. 1, Lippincott Williams and Wilkins, Philadelphia, PA, 2001, p. 685.
32. B. L. Semler and E. Wimmer, *Molecular Biology of Picornaviruses*, ASM Press, Washington, DC, 2002, **Vol. 19**, p. 502.
33. R. J. Phillpotts and D. A. Tyrrell, *Br. Med. Bull.*, 1985, **41**, 386.
34. A. S. Monto, E. R. Bryan and S. Ohmit, *J. Infect. Dis.*, 1987, **156**, 43.
35. M. J. Makela *et al.*, *J. Clin. Microbiol.*, 1998, **36**, 539.
36. M. Chow *et al.*, *Nature*, 1987, **327**, 482.
37. D. M. Belnap *et al.*, *J. Virol.*, 2000, **74**, 1342.
38. E. Rieder and E. Wimmer, in *Molecular Biology of Picornaviruses*, ed. B. L. Semler and E. Wimmer, ASM Press, Washington, DC, 2002, p. 61.
39. B. Sherry and R. Rueckert, *J. Virol.*, 1985, **53**, 137.
40. B. Sherry, A. G. Mosser, R. J. Colonno and R. R. Rueckert, *J. Virol.*, 1986, **57**, 246.
41. M. G. Rossmann, *Viral Immunol.*, 1989, **2**, 143.
42. M. G. Rossmann, *J. Biol. Chem.*, 1989, **264**, 14587.
43. N. H. Olson *et al.*, *Proc. Natl. Acad. Sci. USA*, 1993, **90**, 507.
44. P. R. Kolatkar *et al.*, *EMBO J.*, 1999, **18**, 6249.
45. M. G. Rossmann, Y. He and R. J. Kuhn, *Trends Microbiol.*, 2002, **10**, 324.
46. C. Xiao *et al.*, *J. Virol.*, 2004, **78**, 10034.
47. C. R. Uncapher, C. M. DeWitt and R. J. Colonno, *Virology*, 1991, **180**, 814.
48. J. M. Greve *et al.*, *Cell*, 1989, **56**, 839.
49. D. E. Staunton *et al.*, *Cell*, 1989, **56**, 849.
50. J. E. Tomassini *et al.*, *Proc. Natl. Acad. Sci. USA.*, 1989, **86**, 4907.
51. R. J. Kuhn and M. G. Rossmann, in *Molecular Biology of Picornaviruses*, ed. B. L. Semler and E. Wimmer, ASM Press, Washington, DC, 2002, p.85.

52. D. Schober, P. Kronenberger, E. Prchla, D. Blaas and R. Fuchs, *J. Virol.*, 1998, **72**, 1354.
53. D. Blaas, in *Molecular Biology of Picornaviruses*, ed. B. L. Semler and E. Wimmer, ASM Press, Washington, DC, 2002, p. 93.
54. A. van de Stolpe and P. T. van der Saag, *J. Mol. Med.*, 1996, **74**, 13.
55. D. E. Staunton, M. L. Dustin, H. P. Erickson and T. A. Springer, *Cell*, 1990, **61**, 243.
56. J. Bella, P. R. Kolatkar, C. W. Marlor, J. M. Greve and M. G. Rossmann, *Proc. Natl. Acad. Sci. USA*, 1998, **95**, 4140.
57. J. M. Casasnovas, T. Stehle, J. H. Liu, J. H. Wang and T. A. Springer, *Proc. Natl. Acad. Sci. USA*, 1998, **95**, 4134.
58. Y. Yang et al., *Mol. Cell*, 2004, **14**, 269.
59. K. A. Roebuck and A. Finnegan, *J. Leukoc. Biol.*, 1999, **66**, 876.
60. K. S. Kim, V. J. Sapienza and R. I. Carp, *Antimicrob. Agents Chemother.*, 1980, **18**, 276.
61. T. J. Smith *et al.*, *Science*, 1986, **233**, 1286.
62. J. Badger *et al.*, *Proc. Natl. Acad. Sci. USA*, 1988, **85**, 3304.
63. O. Flore, C. E. Fricks, D. J. Filman and J. M. Hogle, *Semin. Virol.*, 1990, **1**, 429.
64. M. G. Rossmann, *Protein Sci.*, 1994, **3**, 1712.
65. Y. Zhang *et al.*, *J. Virol.*, 2004, **78**, 11061.
66. B. A. Heinz *et al.*, *J. Virol.*, 1989, **63**, 2476.
67. D. C. Pevear *et al.*, *J. Virol.*, 1989, **63**, 2002.
68. D. A. Shepard, B. A. Heinz and R. R. Rueckert, *J. Virol.*, 1993, **67**, 2245.
69. B. A. Heinz, D. A. Shepard and R. R. Rueckert, in *Use of X-ray Crystallography in the Design of Antiviral Agents*, ed. W. G. Laver and G. Air, Academic Press, San Diego, CA, 1990, p. 173.
70. J. K. Bibler-Muckelbauer *et al.*, *Virology*, 1994, **202**, 360.
71. A. E. Eriksson, W. A. Baase, J. A. Wozniak and B. W. Matthews, *Nature*, 1992, **355**, 371.
72. A. E. Eriksson *et al.*, *Science*, 1992, **255**, 178.
73. J. K. Lewis, B. Bothner, T. J. Smith and G. Siuzdak, *Proc. Natl. Acad. Sci. USA*, 1998, **95**, 6774.
74. J. Badger *et al.*, *J. Mol. Biol.*, 1989, **207**, 163.
75. A. T. Hadfield *et al.*, *J. Mol. Biol.*, 1995, **253**, 61.
76. D. K. Phelps and C. B. Post, *J. Mol. Biol.*, 1995, **254**, 544.
77. Y. Li, Z. Zhou and C. B. Post, *Proc. Natl. Acad. Sci. USA*, 2005, **102**, 7529.
78. G. Scapin, *Curr. Pharm. Des.*, 2006, **12**, 2087.
79. N. Huang and M. P. Jacobson, *Curr. Opin. Drug Discov. Dev.*, 2007, **10**, 325.
80. R. L. Mancera, *Curr. Opin. Drug Discov. Dev.*, 2007, **10**, 275.
81. J. J. McSharry, L. A. Caliguiri and H. J. Eggers, *Virology*, 1979, **97**, 307.
82. M. A. McKinlay, J. V. Miralles, C. J. Brisson and F. Pancic, *Antimicrob. Agents Chemother.*, 1982, **22**, 1022.
83. K. H. Kim *et al.*, *J. Mol. Biol.*, 1993, **230**, 206.

84. D. C. Pevear, T. M. Tull, M. E. Seipel and J. M. Groarke, *Antimicrob. Agents Chemother.*, 1999, **43**, 2109.
85. F. G. Hayden *et al.*, *Clin. Infect. Dis.*, 2003, **36**, 1523.
86. D. C. Pevear *et al.*, *Antimicrob. Agents Chemother.*, 2005, **49**, 4492.
87. E. Aradottir, E. M. Alonso and S. T. Shulman, *Pediatr. Infect. Dis. J.*, 2001, **20**, 457.
88. A. D. Webster, *J. Clin. Virol.*, 2005, **32**, 1.
89. J. Callen and B. A. Paes, *Adv. Neonatal Care*, 2007, **7**, 238.
90. K. Katsibardi *et al.*, *Eur. J. Pediatr.*, 2008, **167**, 97.
91. C. Xiao and M. G. Rossmann, *J. Struct. Biol.*, 2007, **158**, 182.

CHAPTER 17

Viral Vectors for Gene Delivery

DAVID J. DISMUKE,[a] STEVEN J. GRAY,[a] MATTHEW L. HIRSCH,[a] RICHARD SAMULSKI[a] AND NICHOLAS MUZYCZKA[b]

[a] UNC Gene Therapy Center, University of North Carolina, Chapel Hill, NC 27599, USA; [b] UF Powell Gene Therapy Center, University of Florida College of Medicine, Gainesville, FL 32610, USA

1 Adenoviral Vectors

Adenoviruses have become one of the best characterized viruses since their discovery in the early 1950s.[1-3] A number of properties of adenovirus (Ad) make it an attractive vector for gene therapy. Over 50 human serotypes have been identified, with most disease manifestations being relatively mild. Thus far, the group C virus Ad5 and Ad2 are the most common serotypes and the first used to generate Ad vectors.[4] Ad vectors can be produced in high titers ($\sim 10^5$ mature virus particles per cell) and are capable of efficient and rapid transduction in a wide variety of dividing and non-dividing cells, including liver, lung, skeletal muscle, neuronal, spleen and heart. It is also relatively straightforward to manipulate the double-stranded DNA genome. Recombinant adenovirus (rAd) vectors can be generated by replacing a number of viral early genes. The E1 region, which contains the Ad-transforming genes and is necessary for efficient transcription and DNA replication, is invariably deleted. In addition, one or more of the other Ad early regions, including E2A, E3 and E4, can be deleted using a variety of reverse genetic techniques. Helper dependent (also called 'gutless') Ad vectors have also been developed in which all of the Ad coding regions have been removed.[5] These vectors can package more than 30 kb of non-viral DNA.

RSC Biomolecular Sciences No. 21
Structural Virology
Edited by Mavis Agbandje-McKenna and Robert McKenna
© Royal Society of Chemistry 2011
Published by the Royal Society of Chemistry, www.rsc.org

Despite their advantages, rAd vectors also have major issues which need to be overcome. First, most people have been exposed to Ad2 and Ad5 and have developed humoral and cell-mediated immunity against the virus; the immune response is largely directed towards the viral capsid. Second, the broad tropism of Ad vectors limits their therapeutic potency due to the dissemination of the vector to non-target organs,[6–9] a problem common to all viral vector systems. Third, the genetic material is maintained in the nucleus of target cells in episomal form, so in rapidly dividing cells gene expression is gradually lost. For gene replacement therapy, this necessitates the repeated administration of the vector, which increases the chance of eliciting an immune response. Finally, Ad induces robust innate and adaptive immune responses, including a strong cytotoxic T cell response. This is apparently because most Ad vectors that retain Ad coding regions continue to synthesize Ad proteins once inside the cell, which induces a CTL response. Thus, even in non-dividing cells, Ad mediated gene expression is short lived (2–4 weeks) *in vivo*. The problems with Ad vector safety were tragically observed with the death of a study patient in 1999.[10,11] In an effort to reduce the immune response and also engineer the virus for specific tropism, many groups have altered the Ad capsid proteins. To accomplish these goals, researchers rely heavily on the structural information about the virus.

Capsid Structure and Viral Entry

The structure of the Ad capsid has been determined by both electron microscopy and X-ray crystallography and has considerably improved understanding of the virus and the ability to manipulate the vectors.[12] Adenoviruses are non-enveloped viruses that have a pseudo-icosahedral symmetry of $T = 25$.[13] The central capsid is approximately 125 Å in diameter, with long fibers (120–330 Å) that extend from the 12 vertices at each fivefold axis. Adenovirus is composed of at least 11 different structural proteins. The three major coat proteins include the 240 hexon trimers that form the face of the icosahedron, the penton base protein that forms a pentameric complex at the 12 vertices of the capsid and the fiber protein that forms a trimeric fiber that extends approximately 330 Å from the penton base in Ad2 and Ad5.[14] A number of minor coat proteins also play important roles in the virus lifecycle and help stabilize the capsid; these include proteins IIIa, VI, VIII and IX.[15] An additional group of proteins are located inside the capsid and interact with the double-stranded DNA genome to form a compact core structure. These include proteins V, VII and X. Finally, the Ad terminal protein is covalently attached to each 5′-OH end of the Ad genome. This protein serves as the primer during Ad DNA replication.

The Ad fiber protein is composed of three domains, an N terminal domain that interacts with the penton base, a fiber shaft of variable length and a knob domain at the end of the fiber. The virus mediates attachment to target cells through the interaction of the fiber knob domain with the Coxsackievirus and adenovirus receptor (CAR).[16] At least in cell culture, this appears to be the primary mechanism for cell surface attachment. Subsequent internalization occurs through clathrin-coated pit endocytosis that is induced by binding of a penton base RGD motif to integrin receptors.[17,18] Other cell surface binding proteins have been identified as well for some serotypes. These include heparan sulfate glycosoaminoglycan,[19] which binds to a KKTK sequence present in the fiber shaft of Ad5, and also blood coagulation factors VII, FIX and FX and complement component C4-binding protein (C4BP), presumably through interactions with the fiber proteins.[12] The interaction of adenovirus with CAR and also the $\alpha V\beta 5$ receptor has been evaluated structurally.[20,21] Interaction of the integrins with pentons, in addition to the acidic environment of the endosome, produces a change in capsid structure that results in the release of fiber, penton and some internal structural proteins, protein IIIa and VI. Protein VI has been shown to have an amphipathic alpha helical domain that is capable of lysing the endosome and releasing the partially disaggregated Ad capsids into cytoplasm.[22] The capsid is then transported to the nuclear pore via microtubules, where further disassembly occurs and the DNA enters the nucleus.

Using Structural Information to Modify Ad Tropism

Since the fiber proteins have been shown to mediate attachment and are also hot spots for neutralizing antibodies, a substantial amount of research has focused on altering this region of the viral capsid. It has also been documented that the hexon protein is immunogenic and that modifications of the capsid surface can decrease immune recognition. Several approaches have been taken. Genetic modification of the fiber knob HI loop and C-terminus, the penton RGD-containing loop, the hypervariable region 5 (HVR5) loop of hexon and the C-terminus of protein IX have been performed to introduce targeting peptides.[23-34] Another strategy is the use of bispecific molecular adaptors to bridge attachment to target cells.[35,36] Insertion of novel binding domains into the fiber knob, for example, the Fc-binding domain of protein A, which allows the attachment of monoclonal targeting antibodies, has also been effective.[37,38] Tropism can also be altered, and some immune response to adenovirus may be evaded, by utilizing vectors containing chimeric capsids that contain swapped fibers or hexons.[12,39] A challenging but promising approach is the replacement of the fiber knob with exogenous trimerization domains containing novel binding ligands.[40] Of the minor proteins, IX has recently been shown to be dispensable and can tolerate modifications that can alter virus tropism.[34,41-45] Protein IX is notable in that it can tolerate large ligand insertions (including GFP, HSV-TK and luciferase) without affecting viral stability or infectivity. Other methods for increasing immune evasion are the use of chemical agents, such as PEG, to coat and shield the vector, or the use of non-human serotypes, which are not quickly recognized by the adaptive arm of the immune system.[46-49]

In principle, once the natural tropism of type C adenoviruses, which is mediated by CAR and integrin binding, is abolished by mutation, a new targeting ligand should create a virus that infects only a specific cell type.

In practice, it has been relatively easy to show significant changes in tropism in cell culture experiments but not *in vivo*. Although the CAR receptor is clearly important for viral entry in cell culture, ablation of CAR binding by mutation has not shown significant changes in biodistribution in animal experiments when the mutant viruses were compared with wild-type Ad.[50,51] Similar results were seen with penton base RGD mutations which ablated integrin binding.[52] This suggests that infection in the host occurs by alternative receptors or is modified by mechanical sequestration. To develop consistent targeting, it will be necessary to understand the mechanism of entry *in vivo*. Furthermore, the addition of a novel binding epitope to the capsid is not itself sufficient to insure targeting. The infection process also involves an interaction with an endosomal receptor that leads to cell entry and to modifications of the capsid that insure correct trafficking. Nevertheless, much of what we know about Ad capsid structure and function has emerged from gene therapy attempts to engineer targeting.

2   Retroviral Vectors

While there is obvious irony in manipulating viruses from the same family as HIV to treat disease, lentiviral vectors or lentivectors (LVs) have quickly become an important and promising tool in gene therapy.[53–58] LVs are an appealing option due to the ability of the vector to transduce quiescent cells and elicit little immune response.[59] This is in contrast to murine retroviruses, which were developed earlier for gene therapy,[60–65] but were found to require dividing cells for transduction. The packaging capacity of LVs is relatively large; it can package approximately 8 kb of a foreign gene provided that certain *cis* elements are incorporated, including the long terminal repeats (LTRs), the primer binding site (pbs), the 3′ polypurine tract and the packaging signal.[56] The central polypurine tract (cPPT) and the rev responsive element (RRE) have also been found to improve transduction or vector yield.[66] The transgene is transferred into the nucleus of the target cell and is stably integrated into the genome, thereby insuring long-term expression even in dividing cells. Self-inactivating vectors (SIN) have also been developed in which the integrated genome is incapable of transcriptional activation from the LTR promoter and relies on a heterologous promoter. This reduces the chance of accidental activation of downstream genes in the chromosome. Initially, LVs were derived from HIV,[53] but other lentiviral systems have been developed by using feline immunodeficiency virus (FIV),[67] simian immunodeficiency virus (SIV),[68] bovine immunodeficiency virus (BIV), equine infectious anemia virus (EIAV)[69] and caprine arthritis encephalitis virus (CAEV).[70] These vectors share many common structural characteristics[70,71]

Capsid Structure

LVs are enveloped vectors that contain surface glycoproteins embedded in the viral membrane, which mediate attachment and entry of the particle.

The particles have a diameter of approximately 100 nm and have a character-istic electron-dense, cone-shaped core made of a shell of approximately 1500 capsid (CA) proteins.[72,73] Within the core, the two copies of the single-stranded, positive-polarity RNA genome are complexed with the viral proteins matrix (MA), nucleocapsid (NC), protease (PR), integrase (IN) and reverse transcriptase (RT). Our understanding of viral function has been greatly facilitated by determination of the protein structures for CA, MA, NC, PR, IN and RT.[74–80]

LVs can be produced by two main methods: a split-genome system in which multiple plasmids (three or more) are transfected into producer cells or by utilizing packaging cell lines which express the packaging and structural genes.[58] In either case, it is necessary to provide the transgene cassette sepa-rately from the genes for Gag, Pol and the viral envelope (Env). This eliminates the possibility of producing a replication competent recombinant virus. Deletion of accessory genes such as *tat*, *nef*, *vpu*, *vpr* and *vif* has also been used to reduce the risk of replication competent virus. Gag is produced as a poly-protein which is cleaved during virus maturation to form the structural proteins MA, CA and NC, while a less abundant Gag-Pol precursor yields the RT, IN and PR enzymes. One of the difficulties of using lentivirus vectors for gene therapy is that, in contrast to Ad and AAV, the current production systems for lentivectors produce low viral yields (approximately 10^3 per cell) and are not scalable.

One problem that has been realized with LVs (and also other retroviral vectors) is that normal transduction involves integration of the LTR-containing transgene into the host cell chromosome. While this ability of the vector causes long-term gene expression and retention of the gene as the cell divides, inte-gration occurs mostly near active genes,[81–83] which poses a potential safety concern since it can be oncogenic.[84,85] Unlike murine retroviruses, lentivirus vectors have not been associated with oncogenesis either during natural HIV or animal virus infections or as the result of gene therapy animal experiments.[86,87] However, both murine retrovirus and lentivirus vectors have been reported to undergo gene silencing, presumably due to epigenetic modification of the integrated genome.[88–90] Some groups[91–95] have investigated the use of non-integrating lentiviral vectors (NILVs). Although mutation of IN appears to enable the vector to mediate efficient gene expression from the episomal gen-ome, it is unclear how these LVs compare with other non-integrating vectors. Another possible option is the development of site-specific integration using different systems such as zinc finger nucleases, and also the Sleeping Beauty and phiC31 transposons.[96–99]

Using Structural Information to Target Lentiviruses

Lentiviruses, like other enveloped viruses, insert their genetic material into cells by fusing their membranes with the cell membrane. This typically involves a two-step Type 1 fusion mechanism. A viral envelope protein or surface unit

(SU) is used for cell surface attachment and this causes a change in the meta-stable complex of SU with a viral envelope transmembrane protein (TM) that fuses the viral and cell membranes. Type 1 fusion generally can occur at neutral pH. In contrast, other enveloped viruses use a pH-dependent process (Type 2 fusion), in which a lower pH is required to convert the surface proteins to a conformation capable of membrane fusion. While the lentiviral env proteins, gp41 and gp120, can be used for vector production, the tropism conferred by the endogenous glycoproteins is largely limited to lymphocytes and other cells of hematopoietic origin to the surface receptors that targeted by HIV. This changed when several groups demonstrated that the fusogenic G protein from vesicular stomatitis virus (VSV-G) could be substituted for the env proteins of HIV.[55,100,101] Therefore, it is now customary to pseudotype vectors with the envelope proteins from other viruses, such as VSV, Ross River, Ebola, influenza, Sindbis, measles, ecotropic and amphotropic retroviruses and others, which dramatically increases the types of cells LVs can target.[71,102] In general, it has been found that a wide variety of fusion proteins, including the baculovirus GP64 protein,[103,104] can be incorporated into lentivirus envelopes, provided that the C-terminal portion of the protein is short and does not interact with the viral core. GP64 has the advantage that it is non-toxic when expressed in cells, thus simplifying lentivirus production. It is also less likely to promote infection of antigen positive cells and interaction with complement.[104]

Although pseudotyping lentiviruses has been relatively easy and has expanded the types of tissues that lentivectors can infect, targeting lentivectors to specific cell types has proven to be more problematic. Envelope proteins must perform the dual function of cell attachment and membrane fusion. This involves the formation of a metastable state in the viral env complex during virus assembly. Insertion of foreign epitopes into envelope proteins designed to retarget the virus have often reduced their ability to carry out membrane fusion or reduced titers during virus production.[71] Nevertheless, several groups have made progress with both pH-independent and pH-dependent fusion proteins. These groups have successfully targeted certain cell types by engineering novel proteins into the pseudotyped glycoprotein complex.[103-110] Notably, in some cases large peptides, such as single-chain monoclonal antibodies, have been successfully engineered into HIV envelopes while retaining acceptable viral titers and were found to target the appropriate cell type[110] *in vitro*. In addition, high-throughput screening methods for targeting ligands have been established.[111]

3 AAV Vectors

AAV vectors were first shown to be capable of achieving persistent, long-term gene expression by Hermonat and Muzyczka.[112] They demonstrated that the capsid gene could be replaced with an exogenous transgene driven by a heterologous promoter and the recombinant could be packaged by a plasmid cotransfection system in which the missing capsid gene was supplied *in trans*. Subsequent work demonstrated that all of the AAV coding regions could be removed and supplied *in trans* and that only the 145 bp terminal repeat sequences were essential *in cis* for vector production.[113,114] The virus codes for only two open reading frames (rep and cap) whose gene products are provided *in trans* during vector production. However, the virus also requires the expression of helper virus genes from a heterologous virus, either herpesvirus or adenovirus.[115] These can be supplied either by coinfection with the helper virus or by cotransfection of the helper virus genes on an expression plasmid. Interest in AAV vectors was minimal until Xiao *et al.*[116] and Byrne and colleagues[117] demonstrated that injection of AAV vectors into muscle tissue could produce efficient gene transfer and long-term expression *in vivo*. Others demonstrated long-term expression in the eye, brain and liver.[118–123] The longest uninterrupted expression engineered with AAV vectors thus far has been in the eye; Hauswirth and colleagues[124] have demonstrated expression of the rpe65 gene in the deficient dog model for more than 11 years with no decrease in the level of expression.

Like Ad vectors, AAV recombinant genomes do not integrate into host chromosomes, but persist as episomes in most tissues. The frequency of integration in liver is approximately 1% and in muscle, eye and brain is undetectable.[125–127] The reason for the persistent expression of AAV vectors, compared with Ad vectors, is not completely clear, but is likely due to at least three things. First, AAV vectors are completely deleted for viral genes, hence there is frequently no detectable CTL response to AAV capsids, only a circulating antibody response. Second, AAV produces a mild and frequently undetectable innate immune response. Third, because AAV vectors do not integrate, epigenetic shut-off of gene expression due to chromosome integration, which is common with retroviruses, has rarely been seen with AAV vectors. AAV vectors have also benefited from the development of rapid and efficient production methods,[128–130] including recently developed scalable methods that make commercial gene therapy applications feasible.[131,132] Recently, human trials for rpe65 deficiency using AAV vectors have demonstrated that this vector system might be useful for treating human disease.[133–136]

Capsid Structure

AAV has a T = 1 non-enveloped icosahedral capsid about 260 Å in diameter, containing 60 capsid proteins.[115] It is, therefore, the simplest capsid of the three vector systems and in principle may be the easiest to manipulate. A single open reading frame coding for the capsid gene uses alternative splicing and alternative start codons to express three capsid genes (VP1, VP2 and VP3) with overlapping amino acid sequences in which the two larger proteins (VP1 and VP2) have additional N-terminal sequences. The unique N-terminal portion of VP1, the largest capsid protein, contains a phospholipase A2 (PLA2) motif that has enzymatic activity and is required for viral infection.[137] The N-terminal portions of both VP1 and VP2 contain nuclear localization signals that appear to be essential for directing the capsid to the nucleus.[138,139] The three capsid proteins have an approximate ratio of 1:1:10 (VP1:VP2:VP3). Over 100 different serotypes in humans, primates and other mammals have been identified[140] and pseudotyping of AAV vectors using the original AAV2 genome vector backbone is readily accomplished by substituting an alternative serotype capsid gene during production.[141] The atomic structures of several AAV serotypes have been determined[142–147] (as discussed in Chapters 1 and 6).

Viral Entry

Some of the details of viral entry are known. Like Ad, AAV uses at least two cell surface receptors, one for binding to the cell surface, usually a glycan, and a second receptor that promotes clathrin-mediated endocytosis. To date, most of the information about AAV viral entry has been gathered with studies of AAV2. In 1996, several cell lines were reported to be non-permissive to AAV and transduction of permissive cell lines could be abrogated by prior trypsinization, suggesting that a protein receptor was also necessary.[148,149] *In vitro* assays demonstrated AAV2 binding to a 150 kDa unknown protein that was only detected in cells permissive for transduction.[149] Additional characterization suggested that N-linked glycans are required for particle binding, which led to the first proposal that a 150 kDa proteoglycan serves as the receptor for AAV2. Summerford and Samulski[150] identified the first AAV receptor, heparan sulfate proteoglycan, for AAV2. Further biochemical and genetic evidence demonstrated that the heparan sulfate (HS) moiety, specifically, was necessary for AAV2 infection.[150] HS is a highly sulfonated linear polysaccharide nearly ubiquitously expressed on cell surfaces. Soluble heparan sulfate inhibits AAV2 infection in cell culture[151] and heparin chromatography is routinely used to affinity purify the virus.[152] Together this established heparan sulfate as an essential part of AAV2 entry into cells. However, it is not known whether HS proteoglycans can mediate both particle attachment and internalization as has been seen for some other viruses.

Subsequently, Srivastava and colleagues[153] demonstrated an independent role for the human fibroblast growth factor receptor 1 (FGFR1) as a coreceptor in conjunction with HSPG. High-efficiency binding of the fibroblast growth factor (FGF) to FGFR1 first requires low-affinity HSPG binding. Therefore, it is possible that AAV2 binding to the cell surface follows a route that involves an interaction with HSPG followed by binding to FGFR1 and subsequent internalization. In these experiments, it was shown that the FGFR1 and HSPG ligand, FGF, could inhibit the binding of AAV2 particles to the cell surface and consequently the degree of particle-mediated transduction. In addition, Summerford et al.[154] used viral overlay and immunoprecipitation experiments to demonstrate a direct interaction between AAV2 particles and αVβ5 integrin. Investigations using cell lines deficient or enhanced for αVβ5 integrin expression suggested that the integrin played a role in particle internalization. Sanlioglu et al.[155] confirmed αVβ5 integrin-mediated endocytosis of AAV2 and showed that it occurs through a Rac1 and PI3 kinase activation cascade.

They suggested that this directed viral movement along the cytoskeletal network to the nucleus. More recently, other groups have identified additional coreceptors for AAV2, hepatocyte growth factor receptor (c-Met),[156] integrin α5β1[157] and laminin.[158] In the case of c-Met, labeled virus tracking demonstrated that c-Met overexpression did not alter the ability of AAV2 particles to bind the cell surface, but did substantially increase both particle internalization and transgene synthesis.[156] This is consistent with the concept of two receptors, one required for attachment and the other for internalization, along with the notion that AAV2 particles may use different coreceptors depending on the cell type. The putative α5β1 integrin–capsid interaction appears mediated by a highly conserved asparagine–glycine–arginine (NGR) motif that upon mutagenesis, decreased the transduction efficiency of AAV2. Because α5β1 was able to internalize AAV2 in the absence of αVβ1, a model was proposed in which AAV2 binding to cell surface HS results in a reversible capsid conformation that primes the NGR motif for α5β1 integrin (and possibly others) binding.[157] Consistent with this, a cryo-EM study[159] showed that binding of heparan sulfate to AAV2 capsids produced secondary changes at the fivefold pore and internally within the capsid.

Prior to the publication of AAV2's crystal structure, a comprehensive mutagenic capsid analysis implicated two amino acid clusters (509–522 and 561–591) important for HS binding in an *in vitro* assay.[160] Of particular interest were R to A substitutions at amino acids 585 and 588 that were suspected to contribute directly to reduced HS binding. The elucidation of AAV2's crystal structure localized this region, 585–RGNR–588, in an external surface loop region that demonstrates variability in other AAV serotypes known to not bind heparin.[145] For example, a swap of AAV2's amino acid residues 585–590 into a structurally equivalent location in the AAV5 capsid conferred the new property of HS binding;[161] AAV5 normally relies on interactions with sialic acid, not HS.[162] Additional AAV2 capsid mutagenesis further defined additional residues involved in HS binding including positions 484, 487 and 532 in addition to 585 and 588.[161,163] When considered structurally, these five amino acids create a basic patch on the surface spike of the threefold axis of symmetry and demonstrate low overall sequence conservation among other serotypes. This is consistent with the observations that positively charged regions are commonly used in heparin binding due to interactions with the negatively charged sugar sulfates.[164]

Receptors for other AAV serotypes have also been identified. AAV1 and AAV3 bind to α-2,3 and α-2,6 N-linked sialic acids.[165] AAV5 binds to α-2,3 N-linked sialic acid, whereas AAV4 binds 2,3 O-linked sialic acid.[162] In addition, platelet-derived growth factor receptor (PDGFR) has been identified as a coreceptor for AAV5[166] and the 37/67 kDa laminin receptor (LamR) has been suggested as a receptor for AAV2, -3, -8 and -9.[158]

As mentioned earlier, viral entry occurs through clathrin-mediated endocytosis. Once the virus has entered the early endosome, structural changes occur that expose the N-terminal sequences within VP1 and VP2,[139] exposing PLA2 activity and nuclear localization signals. However, the role of PLA2 and the exact mechanism of trafficking to the nucleus is not clear. Evidence exists for immediate release from the early endosome, and also trafficking to recycling endosomes, lysosomes and for some serotypes, the Golgi compartment, before nuclear entry.[115,167] There is also no definitive agreement on whether intact virus particles enter the nucleus prior to uncoating of the nucleic acid.[139,168] Curiously, AAV vectors can persist as intact particles for extended periods of time in infected cells.[169] Additionally, once the nucleic acid material has been released, the single-stranded DNA must synthesize the complementary strand before gene expression can occur. These events presumably account for the slow onset of gene expression that is seen *in vivo*, taking up to 3 months to reach the steady state.[170]

Using Structural Information to Engineer Improved AAV Vectors

Although the different AAV serotypes have shown varied tissue tropism *in vivo*, all of the serotypes infect a wide variety of cell types. As with the other vectors, the development of simple methods to engineer cell type-specific infection would be a major step forward in gene therapy applications. Given the relative simplicity of the AAV capsid, many groups have tried to target AAV2 capsids to specific cell types. These include simple amino acid changes, introduction of peptide motifs, cross-dressing the particle and shuffling of the capsid proteins. Another complementary approach is the use of inducible or tissue-specific promoters; this topic has been addressed in several review articles.[171–173]

The simplest approach to alter tropism of AAV is to make single amino acid substitutions in the VPs. Most of the substitutions that have been reported have resulted in vectors with decreased functionality.[160,174,175] However, several studies have shown more beneficial changes through mutating individual amino acids in Vp3. A lysine-to-glutamate mutation at position 531 in AAV1 imparts the ability to bind heparan sulfate and increases liver tropism of this sero-type.[176] Structure–function comparisons of sequenced capsid regions from numerous isolates have also identified key amino acids that when altered can improve the vector.[176,177] Recently, an evaluation of tyrosine-to-phenylalanine mutation of surface-exposed residues in Vp3 suggested that these mutations drastically improved the transduction ability of the mutant vectors.[178] These examples provide evidence that there are likely other single amino acid changes that would increase vector production and transduction to specific tissues. Also, retargeting AAV vectors will likely require mutations that decrease binding to primary receptors in combination with other modifications that mediate novel interactions with target calls.

Targeting Peptide Insertions

In an attempt to direct transduction of specific cell types, numerous groups have inserted peptide sequences into the AAV capsid. Since the capsid structure of AAV was not solved until recently,[145] early work relied on mutation at random sites or information from canine parvovirus (CPV), feline panleuko-penia virus and B19, which are related parvoviruses. From these studies, it was shown that insertions were tolerated in VP1,[137,160,179,180] at the N-terminal end of VP2,[160,174,180] at the N-terminal end of VP3,[181] at various proposed loops within VP3[160,174,180,182–184] and at the C-terminal end of VP3.[48] Viruses with insertions at these positions produced wild-type yields of viral particles, which retained wild-type infectivity. Based on these observations and the crystal structure determination of AAV2, numerous reports have demonstrated that rational design of peptide insertions could successfully target diverse cell types and tissues.[160,174,183–191] One site in the capsid that has been repeatedly shown to be amenable to modification lies in VP3 at position 587.[182] Since it has been demonstrated that this location is critical for heparin binding, insertions here have the possibility to cause simultaneous loss of normal tropism and targeting to specific receptors.

Peptide Libraries

A recent, powerful adaptation of the peptide insertion method is the con-struction and screening of peptide libraries.[192–195] Thus far, these screens have utilized vectors with a seven amino acid peptide library constructed into the 587 position in Vp3. This approach has identified several unique peptides which mediate increased transduction to several cell types, including acute myeloid leukemia, B-cell chronic lymphocytic leukemia, coronary endothelial, lung carcinoma, prostate carcinoma, rat cardiomyoblasts and primary human venous endothelial cells.

Although these approaches are very promising, they are limited in the length of the peptide that can be introduced; the size limit that is structurally allowed is normally around 15 amino acids. To expand the ability of AAV vectors to target specific cell types it may be necessary to incorporate much larger proteins into the vector surface. Promising work from several labora-tories has shown that this may be feasible. These groups have reported the ability to engineer particles that contain large insertions of up to 32 kDa into the N-terminus of VP2,[168,196–198] if VP1 and VP3 are offered *in trans*. The fused proteins in these constructs are usually externalized through the fivefold pores and are able to extend considerably from the capsid surface. Although these studies did not utilize proteins that directly mediate viral targeting, this approach may provide a novel strategy to present relatively large ligands on the capsid.

In general, the rational peptide design approach suffers from the fact that the virus often retains the binding sites for its natural receptors. Thus, with some exceptions,[199] peptide insertions have often expanded viral tropism to new cell types without significantly affecting the normal tropism. Moreover, the new ligand might promote cell binding but not necessarily promote entry, resulting in lower infectivity.

Capsid Libraries

As mentioned earlier, over 100 variants of AAV have been isolated.[140] These variants have individual tropisms that might potentially be exploited for therapeutic purposes. Although very useful, each variant evolved for the purpose of virus persistence and propagation and not necessarily for gene therapy. For example, a gene therapy vector for the treatment of muscular dystrophy, with the primary targets being skeletal and smooth muscle, would be very useful. However, AAV vectors, even those with high muscle tropism, have a dominant tropism for the liver.[200] If the component of the capsid responsible for liver tropism was ablated while retaining or enhancing muscle tropism, the resulting vector would be suitable for systemic administration at a lower dose than wild-type AAV serotypes.

A second goal of capsid modification is to evade neutralizing antibodies from the host immune system. An estimated 25–30% of the human population have neutralizing antibodies (NABs) to AAV2, the most common serotype used in AAV-based human clinical trials.[201–203] Antibodies to other serotypes have also been found in the human population. The presence of circulating NABs would preclude a significant portion of the population from being eligible to receive an AAV-based therapeutic agent. Moreover, once a person has been injected with an AAV vector, they will be immunized against that serotype, preventing re-administration of the same vector. For these reasons, engineering the capsids for immune evasion is a high priority for vector development.

However, our understanding of the virus–cell interactions that determine tropism are far from complete. Viral tropism is likely to be the result of both cell entry and post-entry processing steps specific to the target tissue. In order to advance the creation of novel therapeutic vectors, innovative strategies have been developed to bypass rational design and use random capsid libraries to select for characteristics that would make AAV a safer and more effective gene therapy vector.

An early approach used the natural propensity of AAV vectors to recombine with similar serotypes. Heparin-negative mutants of AAV2 were coinfected with AAV3, which also binds HSP, to rescue AAV2/3 recombinants that recovered heparin binding and infectivity on HeLa cells.[204] Another approach was to mix capsids of different serotypes to combine tropisms. In this approach, chimeric capsids are recovered containing a mixture of the two capsid proteins.[141,205] These approaches demonstrated what was becoming clear from sequence comparisons and X-ray crystal structure studies of AAV serotypes, namely that there was a high degree of conservation of the capsid sequences in the regions of the capsid required for capsid assembly and structural integrity. As a consequence, it was possible to use capsid shuffling techniques to create libraries of new capsids from the existing serotypes.

The generation of random AAV capsid libraries, termed 'directed evolution', was pioneered by Schaffer and co-workers,[206,207] borrowing from DNA shuffling techniques described by Stemmer.[208,209] Multiple variations of the AAV-directed evolution process have since been utilized, but the overall strategy is similar. First the capsid genes are randomly cleaved or mutagenized and mixed to form a library of pooled capsid variants in the context of a replication-competent backbone. Next, this library is subjected to multiple rounds of selective pressure. At the end, the recovered library clone(s) should be enhanced for whatever characteristic was selected for, above that of the parent serotype(s). Described methods for producing the library include random mutagenesis of the capsid gene of a single serotype by error-prone PCR, randomly mixing capsids from multiple serotypes by DNA shuffling or a combination of the two methods.[208,210]

Directed evolution of AAV capsids has been successfully employed to create AAV variants resistant to circulating NABs.[194,206,207,211,212] In each case, a library was produced using a combination of error-prone PCR and/or DNA shuffling with a diversity of up to 2.5×10^7 pooled clones. Then the library of infectious AAV variant was incubated with neutralizing serum and applied to cultured cells, such that those AAVs that had lower affinity for the NABs but still retained tropism for the cells would have a competitive advantage. Using this strategy, it was possible to map the antigenic epitopes and identify amino acid mutations that would alter those epitopes without affecting tropism. For example, Perabo *et al.*[194] screened a randomly mutated AAV2 library with three rounds of selection against human serum and infectivity of HeLa cells. Mutations of amino acids R459 and N551 were identified in a majority of recovered clones. In addition to mapping NAB epitopes, a variant containing both mutations was 5.5-fold more resistant to pre-immunized human sera. An improved strategy was to add an element of 'evolution' beyond simply screening random mutations. Maheshri *et al.*[206] employed a step of staggered extension PCR between rounds of selection, so that enhancing genetic elements could be combined in an additive fashion. After two rounds of selection, a clone with mutations of E12A, K258N, T567S, N587I and T716A was recovered that showed nearly 100-fold resistance to NABs. Four of the five mutations lie within NAB epitopes or regions previously characterized as playing a role in immune evasion.

Shuffled libraries composed of multiple serotypes were also shown to be resistant to pooled human sera, as expected from the chimeric makeup of the recovered clones and the lower seroprevalence of the non-human AAVs in the human population.[211,212] Interestingly, even when shuffled libraries were not selected for NAB evasion, the recovered clones still had a unique serological profile.[212,213] For example, Li *et al.*[213] recovered a single dominant clone after directed evolution on CS-1 cells which was a chimera composed of AAV1, -2, -8 and -9 (dubbed '1829'). Antisera from AAV1-, AAV8- and AAV9-immunized mice did not cross-react with chimera 1829 and antisera from AAV2-immunized mice was cross-reactive at a 25-fold reduced NAB titer. In addition to having novel and/or enhanced tropism, chimeric AAV variants might be new serotypes by formal definition, with antigenic epitopes not seen in nature. This would give them a distinct advantage over traditional AAV gene therapy vectors using capsids found in nature.

A major rationale for directed evolution of AAV capsids is to modify or enhance the tropism of the virus. Maheshri *et al.*[206] demonstrated that a negative selection for heparin binding of an AAV2-based library could select for novel amino acid mutations that disrupted HSP binding but did not affect virus production or viability. The 1829 clone that Li *et al.*[213] recovered from *in vitro* selection on CS-1 hamster melanoma cells had an enhanced tropism for CS-1 cells greatly exceeding any of the parent serotypes. Moreover, the 1829 clone had a unique profile of transduction *in vivo* after tail vein injection or direct brain injection, in some cases dramatically different from any of the parents. Structural modeling and rational mutagenesis of the chimeric 1829 capsid suggested that AAV2 contributed to the ability of the mutant to bind heparin and AAV9 was necessary (but not sufficient) for the tropism of melanoma cells. Grimm *et al.*[211] used a different AAV shuffled library to select for human hepatocyte tropism and NAB evasion simultaneously. The dominant recovered chimera, termed 'AAV-DJ', was a mixture of AAV2, -8 and -9. Interestingly, AAV-DJ showed an *in vitro* tropism superior to that of any parent serotype for a variety of transformed cells derived from human liver, kidney, cervix, retina, skin and lung. *In vivo*, AAV-DJ had strong tropism for liver and reduced tropism for lung, brain, pancreas, gut and muscle, compared with AAV8 and -9.

Directed evolution of AAV capsids provides a relatively high-throughput screening process to identify and/or create structural motifs to confer unique tropisms to AAV capsids. The capsid library approach to generating new and/ or enhanced tropism and NAB evasion can be faster than traditional modification of the AAV capsid by rational design and it also promises to expand greatly the diversity of potential therapeutic AAV vectors. A major strength of the library approach is the stringent selection that generates AAV capsids with targeted tropism for a cell or tissue of interest. To date, most selections have been done in cell culture and efficient *in vivo* methods of selection are still in their infancy. Because tropism in cell culture has often been a poor predictor of *in vivo* tropism, animal models for selection will be essential. In addition, the strength of the directed evolution approach is also a potential theoretical limitation. If a vector is found with high specificity for a mouse tissue or transformed human cell, that tropism does not necessarily translate to the same target cell in a human or non-human primate. The pioneering studies in directed evolution of the AAV capsid over a span of 5 years offer great promise, but they still require optimization and validation before transition to viable clinical vectors. That said, structure–function analysis of the chimeric capsids can accelerate the general understanding of the role of specific structural domains on the capsid surface, leading to the generation of better vectors by rational design. In closing, the combination of lower immunogenicity with higher specific tropism is the promise of the next-generation gene therapy vectors.

References

1. W. P. Rowe, R. J. Huebner, L. K. Gilmore, R. H. Parrott and T. G. Ward, *Proc. Soc. Exp. Biol. Med.*, 1953, **84**, 570.

2. W. C. Russell, *J. Gen. Virol.*, 2000, **81**, 2573.
3. L. Jager and A. Ehrhardt, *Curr. Gene Ther.*, 2007, **7**, 272.
4. K. Van Doren, D. Hanahan and Y. Gluzman, *J. Virol.*, 1984, **50**, 606.
5. C. Volpers and S. Kochanek, *J. Gene Med.*, 2004, **6**(Suppl 1), S164.
6. S. L. Brody, M. Metzger, C. Danel, M. A. Rosenfeld and R. G. Crystal, *Hum. Gene Ther.*, 1994, **5**, 821.
7. R. D. McCoy *et al.*, *Hum. Gene Ther.*, 1995, **6**, 1553.
8. J. N. Lozier, M. E. Metzger, R. E. Donahue and R. A. Morgan, *Blood*, 1999, **94**, 3968.
9. M. Christ *et al.*, *Hum. Gene Ther.*, 2000, **11**, 415.
10. D. W. Knorr, *Hum. Gene. Ther.*, 2000, **11**, 1591.
11. E. Marshall, *Science*, 1999, **286**, 2244.
12. S. K. Campos and M. A. Barry, *Curr. Gene Ther.*, 2007, **7**, 189.
13. P. L. Stewart, R. M. Burnett, M. Cyrklaff and S. D. Fuller, *Cell*, 1991, **67**, 145.
14. J. van Oostrum and R. M. Burnett, *J. Virol.*, 1985, **56**, 439.
15. J. Vellinga, S. Van der Heijdt and R. C. Hoeben, *J. Gen. Virol.*, 2005, **86**, 1581.
16. J. M. Bergelson *et al.*, *Science*, 1997, **275**, 1320.
17. T. J. Wickham, P. Mathias, D. A. Cheresh and G. R. Nemerow, *Cell*, 1993, **73**, 309.
18. B. Salone *et al.*, *J. Virol.*, 2003, **77**, 13448.
19. M. C. Dechecchi *et al.*, *J. Virol.*, 2001, **75**, 8772.
20. M. C. Bewley, K. Springer, Y. B. Zhang, P. Freimuth and J. M. Flanagan, *Science*, 1999, **286**, 1579.
21. C. Y. Chiu, P. Mathias, G. R. Nemerow and P. L. Stewart, *J. Virol.*, 1999, **73**, 6759.
22. C. M. Wiethoff, H. Wodrich, L. Gerace and G. R. Nemerow, *J. Virol.*, 2005, **79**, 1992.
23. J. Crompton, C. I. Toogood, N. Wallis and R. T. Hay, *J. Gen. Virol.*, 1994, **75**(Pt 1), 133.
24. M. A. Barry, W. J. Dower and S. A. Johnston, *Nat. Med.*, 1996, **2**, 299.
25. R. Pasqualini and E. Ruoslahti, *Nature*, 1996, **380**, 364.
26. T. J. Wickham, P. W. Roelvink, D. E. Brough and I. Kovesdi, *Nat. Biotechnol.*, 1996, **14**, 1570.
27. T. J. Wickham *et al.*, *J. Virol.*, 1996, **70**, 6831.
28. I. Dmitriev *et al.*, *J. Virol.*, 1998, **72**, 9706.
29. V. Krasnykh *et al.*, *J. Virol.*, 1998, **72**, 1844.
30. L. M. Work, P. N. Reynolds and A. H. Baker, *Genet. Vaccines Ther.*, 2004, **2**, 14.
31. S. A. Nicklin, S. J. White, C. G. Nicol, D. J. Von Seggern and A. H. Baker, *J. Gene Med.*, 2004, **6**, 300.
32. H. Wu *et al.*, *J. Virol.*, 2005, **79**, 3382.
33. L. Mailly *et al.*, *Mol. Ther.*, 2006, **14**, 293.

34. I. P. Dmitriev, E. A. Kashentseva and D. T. Curiel, *J. Virol.*, 2002, **76**, 6893.
35. I. Dmitriev, E. Kashentseva, B. E. Rogers, V. Krasnykh and D. T. Curiel, *J. Virol.*, 2000, **74**, 6875.
36. Z. B. Zhu *et al.*, *Virology*, 2004, **325**, 116.
37. N. Korokhov *et al.*, *J. Virol.*, 2003, **77**, 12931.
38. C. Volpers *et al.*, *J. Virol.*, 2003, **77**, 2093.
39. D. M. Roberts *et al.*, *Nature*, 2006, **441**, 239.
40. V. W. van Beusechem *et al.*, *Gene Ther*, 2000, **7**, 1940.
41. R. J. Parks, *Mol. Ther.*, 2005, **11**, 19.
42. L. P. Le et al., *Mol. Imaging*, 2004, **3**, 105.
43. R. A. Meulenbroek, K. L. Sargent, J. Lunde, B. J. Jasmin and R. J. Parks, *Mol. Ther.*, 2004, **9**, 617.
44. J. Li, L. Le, D. A. Sibley, J. M. Mathis and D. T. Curiel, *Virology*, 2005, **338**, 247.
45. L. P. Le *et al.*, *J. Natl. Cancer Inst.*, 2006, **98**, 203.
46. C. R. O'Riordan *et al.*, *Hum. Gene Ther.*, 1999, **10**, 1349.
47. M. A. Croyle, N. Chirmule, Y. Zhang and J. M. Wilson, *J. Virol.*, 2001, **75**, 4792.
48. M. A. Croyle, N. Chirmule, Y. Zhang and J. M. Wilson, *Hum. Gene Ther.*, 2002, **13**, 1887.
49. F. Paillard, *Hum. Gene Ther.*, 1997, **8**, 2007.
50. R. Alemany and D. T. Curiel, *Gene Ther.*, 2001, **8**, 1347.
51. T. Smith *et al.*, *Mol. Ther.*, 2002, **5**, 770.
52. H. Mizuguchi *et al.*, *Gene Ther.*, 2002, **9**, 769.
53. K. A. Page, N. R. Landau and D. R. Littman, *J. Virol.*, 1990, **64**, 5270.
54. G. L. Buchschacher Jr and A. T. Panganiban, *J. Virol.*, 1992, **66**, 2731.
55. L. Naldini *et al.*, *Science*, 1996, **272**, 263.
56. G. L. Buchschacher Jr and F. Wong-Staal, *Blood*, 2000, **95**, 2499.
57. N. Loewen and E. M. Poeschla, *Adv. Biochem. Eng. Biotechnol.*, 2005, **9**, 169.
58. A. S. Cockrell and T. Kafri, *Mol. Biotechnol.*, 2007, **36**, 184.
59. A. Follenzi, G. Sabatino, A. Lombardo, C. Boccaccio and L. Naldini, *Hum. Gene Ther.*, 2002, **13**, 243.
60. P. K. Bandyopadhyay and H. M. Temin, *Mol. Cell. Biol.*, 1984, **4**, 749.
61. D. J. Donoghue, C. Anderson, T. Hunter and P. L. Kaplan, *Nature*, 1984, **308**, 748.
62. R. C. Willis *et al.*, *J. Biol. Chem.*, 1984, **259**, 7842.
63. C. L. Cepko, B. E. Roberts and R. C. Mulligan, *Cell*, 1984, **37**, 1053.
64. J. Doehmer *et al.*, *Proc. Natl. Acad. Sci. USA*, 1982, **79**, 2268.
65. C. J. Tabin, J. W. Hoffmann, S. P. Goff and R. A. Weinberg, *Mol. Cell. Biol.*, 1982, **2**, 426.
66. S. Wurtzer *et al.*, *J. Virol.*, 2006, **80**, 3679.
67. E. M. Poeschla, F. Wong-Staal and D. J. Looney, *Nat. Med.*, 1998, **4**, 354.
68. D. Negre *et al.*, *Gene Ther.*, 2000, **7**, 1613.

69. K. Mitrophanous *et al.*, *Gene Ther.*, 1999, **6**, 1808.
70. J. C. Olsen, *Somat. Cell Mol. Genet.*, 2001, **26**, 131.
71. B. Bartosch and F. L. Cosset, *Curr. Gene Ther.*, 2004, **4**, 427.
72. V. M. Vogt and M. N. Simon, *J. Virol.*, 1999, **73**, 7050.
73. J. A. Briggs, T. Wilk, R. Welker, H. G. Krausslich and S. D. Fuller, *EMBO J.*, 2003, **22**, 1707.
74. T. R. Gamble *et al.*, *Science*, 1997, **278**, 849.
75. C. P. Hill, D. Worthylake, D. P. Bancroft, A. M. Christensen and W. I. Sundquist, *Proc. Natl. Acad. Sci. USA*, 1996, **93**, 3099.
76. J. C. Chen *et al.*, *Proc. Natl. Acad. Sci. USA*, 2000, **97**, 8233.
77. N. Morellet *et al.*, *J. Mol. Biol.*, 1994, **235**, 287.
78. B. K. Ganser-Pornillos, A. Cheng and M. Yeager, *Cell*, 2007, **131**, 70.
79. J. Ding *et al.*, *Structure*, 1995, **3**, 365.
80. N. Thanki *et al.*, *Protein Sci.*, 1992, **1**, 1061.
81. R. S. Mitchell *et al.*, *PLoS Biol.*, 2004, **2**, E234.
82. A. R. Schroder *et al.*, *Cell*, 2002, **110**, 521.
83. R. Daniel and J. A. Smith, *Hum. Gene Ther.*, 2008, **19**, 557.
84. S. Hacein-Bey-Abina *et al.*, *J. Clin. Invest.*, 2008, **118**, 3132.
85. U. P. Dave, N. A. Jenkins and N. G. Copeland, *Science*, 2004, **303**, 333.
86. M. Themis *et al.*, *Mol. Ther.*, 2005, **12**, 763.
87. E. Montini *et al.*, *Nat. Biotechnol.*, 2006, **24**, 687.
88. A. Hotta and J. Ellis, *J. Cell. Biochem.*, 2008, **105**, 940.
89. J. He, Q. Yang and L. J. Chang, *J. Virol.*, 2005, **79**, 13497.
90. A. Hofmann *et al.*, *Mol. Ther.*, 2006, **13**, 59.
91. J. Vargas Jr, G. L. Gusella, V. Najfeld, M. E. Klotman and A. Cara, *Hum. Gene Ther.*, 2004, **15**, 361.
92. S. J. Nightingale *et al.*, *Mol. Ther.*, 2006, **13**, 1121.
93. R. J. Yanez-Munoz *et al.*, *Nat. Med.*, 2006, **12**, 348.
94. N. J. Philpott and A. J. Thrasher, *Hum. Gene Ther.*, 2007, **18**, 483.
95. M. Bayer *et al.*, *Mol. Ther.*, 2008, **16**, 1968.
96. Z. Ivics and Z. Izsvak, *Curr. Gene Ther.*, 2006, **6**, 593.
97. M. P. Calos, *Curr. Gene Ther.*, 2006, **6**, 633.
98. E. A. Moehle *et al.*, *Proc. Natl. Acad. Sci. USA*, 2007, **104**, 3055.
99. A. Lombardo *et al.*, *Nat. Biotechnol.*, 2007, **25**, 1298.
100. R. K. Akkina *et al.*, *J. Virol.*, 1996, **70**, 2581.
101. J. Reiser *et al.*, *Proc. Natl. Acad. Sci. USA*, 1996, **93**, 15266.
102. J. Cronin, X. Y. Zhang and J. Reiser, *Curr. Gene Ther.*, 2005, **5**, 387.
103. M. Kumar, B. P. Bradow and J. Zimmerberg, *Hum. Gene Ther.*, 2009, **14**, 67.
104. C. A. Schauber, M. J. Tuerk, C. D. Pacheco, P. A. Escarpe and G. Veres, *Gene Ther.*, 2004, **11**, 266.
105. G. H. Guibinga and T. Friedmann, *Mol. Ther.*, 2005, **11**, 645.
106. D. M. Markusic, A. Kanitz, R. P. Oude-Elferink and J. Seppen, *Hum. Gene Ther.*, 2007, **18**, 673.
107. N. Pariente *et al.*, *Mol. Ther.*, 2007, **15**, 1973.
108. N. Pariente, S. H. Mao, K. Morizono and I. S. Chen, *J. Gene Med.*, 2008, **10**, 242.

109. K. I. Joo and P. Wang, *Gene Ther.*, 2008, **15**, 1384.
110. S. Funke *et al.*, *Mol. Ther.*, 2008, **16**, 1427.
111. C. J. Buchholz, L. J. Duerner, S. Funke and I. C. Schneider, *Comb. Chem. High Throughput Screen.*, 2008, **11**, 99.
112. P. L. Hermonat and N. Muzyczka, *Proc. Natl. Acad. Sci. USA*, 1984, **1**, 6466.
113. S. K. McLaughlin, P. Collis, P. L. Hermonat and N. Muzyczka, *J. Virol.*, 1988, **62**, 1963.
114. R. J. Samulski, L. S. Chang and T. Shenk, *J. Virol.*, 1989, **63**, 3822.
115. K. I. Berns and C. R. Parrish, in *Fields Virology*, **Vol. 2**, ed. D. M. Knipe and P. M. Howley, Wolters Kluwer Lippincott Williams and Wilkins, New York, 2007, p. 2437.
116. X. Xiao, J. Li and R. J. Samulski, *J. Virol.*, 1996, **70**, 8098.
117. P. D. Kessler *et al.*, *Proc. Natl. Acad. Sci. USA*, 1996, **93**, 14082.
118. M. G. Kaplitt *et al.*, *Nat. Genet.*, 1994, **8**, 148.
119. X. Xiao, J. Li, T. J. McCown and R. J. Samulski, *Exp. Neurol.*, 1997, **144**, 113.
120. R. J. Mandel, S. K. Spratt, R. O. Snyder and S. E. Leff, *Proc. Natl. Acad. Sci. USA*, 1997, **94**, 14083.
121. J. G. Flannery *et al.*, *Proc. Natl. Acad. Sci. USA*, 1997, **94**, 6916.
122. R. O. Snyder *et al.*, *Nat. Genet.*, 1997, **16**, 270.
123. S. Ponnazhagan *et al.*, *Gene*, 1997, **190**, 203.
124. G. M. Acland *et al.*, *Mol. Ther.*, 2005, **12**, 1072.
125. H. Nakai *et al.*, *J. Virol.*, 2001, **75**, 6969.
126. B. C. Schnepp, K. R. Clark, D. L. Klemanski, C. A. Pacak and P. R. Johnson, *J. Virol.*, 2003, **77**, 3495.
127. B. C. Schnepp, R. L. Jensen, K. R. Clark and P. R. Johnson, *J. Virol.*, 2009, **83**, 1456.
128. X. Xiao, J. Li and R. J. Samulski, *J. Virol.*, 1998, **72**, 2224.
129. S. Zolotukhin *et al.*, *Methods*, 2002, **28**, 158.
130. D. Grimm, M. A. Kay and J. A. Kleinschmidt, *Mol. Ther.*, 2003, **7**, 839.
131. M. Urabe, C. Ding and R. M. Kotin, *Hum. Gene Ther.*, 2002, **13**, 1935.
132. W. Kang *et al*, *Gene Ther.*, 2009, **16**, 229.
133. A. V. Cideciyan *et al.*, *Proc. Natl. Acad. Sci. USA*, 2008, **105**, 15112.
134. A. M. Maguire *et al.*, *N. Engl. J. Med.*, 2008, **358**, 2240.
135. W. W. Hauswirth *et al.*, *Hum. Gene Ther.*, 2008, **19**, 979.
136. J. W. Bainbridge *et al.*, *N. Engl. J. Med.*, 2008, **358**, 2231.
137. A. Girod *et al.*, *J. Gen. Virol.*, 2002, **83**, 973.
138. J. C. Grieger, S. Snowdy and R. J. Samulski, *J. Virol.*, 2006, **80**, 5199.
139. F. Sonntag, S. Bleker, B. Leuchs, R. Fischer and J. A. Kleinschmidt, *J. Virol.*, 2006, **80**, 11040.
140. G. Gao, L. H. Vandenberghe and J. M. Wilson, *Curr. Gene Ther.*, 2005, **5**, 285.
141. J. E. Rabinowitz *et al.*, *J. Virol.*, 2004, **78**, 4421.
142. M. DiMattia *et al.*, *Acta Crystallogr., Sect. F*, 2005, **61**, 917.
143. H. J. Nam *et al.*, *J. Virol.*, 2007, **81**, 12260.

144. E. Padron *et al.*, *J. Virol.*, 2005, **79**, 5047.
145. Q. Xie *et al.*, *Proc. Natl. Acad. Sci. USA*, 2002, **99**, 10405.
146. E. B. Miller *et al.*, *Acta Crystallogr., Sect. F*, 2006, **62**, 1271.
147. O. Quesada *et al.*, *Acta Crystallogr., Sect. F*, 2007, **63**, 1073.
148. S. Ponnazhagan et al., *J. Gen. Virol.*, 1996, **77**, 1111.
149. H. Mizukami, N. S. Young and K. E. Brown, *Virology*, 1996, **217**, 124.
150. C. Summerford and R. J. Samulski, *J. Virol.*, 1998, **72**, 1438.
151. A. Handa, S. Muramatsu, J. Qiu, H. Mizukami and K. E. Brown, *J. Gen. Virol.*, 2000, **81**, 2077.
152. S. Zolotukhin *et al.*, *Gene Ther.*, 1999, **6**, 973.
153. K. Qing *et al.*, *Nat. Med.*, 1999, **5**, 71.
154. C. Summerford, J. S. Bartlett and R. J. Samulski, *Nat. Med.*, 1999, **5**, 78.
155. S. Sanlioglu *et al.*, *J. Virol.*, 2000, **74**, 9184.
156. Y. Kashiwakura *et al.*, *J. Virol.*, 2005, **79**, 609.
157. A. Asokan, J. B. Hamra, L. Govindasamy, M. Agbandje-McKenna and R. J. Samulski, *J. Virol.*, 2006, **80**, 8961.
158. B. Akache *et al.*, *J. Virol.*, 2006, **80**, 9831.
159. H. C. Levy *et al.*, *J. Struct. Biol.*, 2009, **165**, 146.
160. P. Wu *et al.*, *J. Virol.*, 2000, **74**, 8635.
161. S. R. Opie, K. H. Warrington Jr, M. Agbandje-McKenna, S. Zolotukhin and N. Muzyczka, *J. Virol.*, 2003, **77**, 6995.
162. N. Kaludov, K. E. Brown, R. W. Walters, J. Zabner and J. A. Chiorini, *J. Virol.*, 2001, **75**, 6884.
163. A. Kern et al., *J. Virol.*, 2003, **77**, 11072.
164. B. Mulloy and R. J. Linhardt, *Curr. Opin. Struct. Biol.*, 2001, **11**, 623.
165. Z. Wu, E. Miller, M. Agbandje-McKenna and R. J. Samulski, *J. Virol.*, 2006, **80**, 9093.
166. G. Di Pasquale *et al.*, *Nat. Med.*, 2003, **9**, 1306.
167. W. Ding, L. Zhang, Z. Yan and J. F. Engelhardt, *Gene Ther.*, 2005, **12**, 873.
168. K. Lux *et al.*, *J. Virol.*, 2005, **79**, 11776.
169. C. E. Thomas, T. A. Storm, Z. Huang and M. A. Kay, *J. Virol.*, 2004, **78**, 3110.
170. S. Song *et al.*, *Proc. Natl. Acad. Sci. USA*, 1998, **95**, 14384.
171. N. Vilaboa and R. Voellmy, *Curr. Gene Ther.*, 2006, **6**, 421.
172. E. D. Papadakis, S. A. Nicklin, A. H. Baker and S. J. White, *Curr. Gene Ther.*, 2004, **4**, 89.
173. D. M. Nettelbeck, *J. Mol. Med.*, 2008, **86**, 363.
174. W. Shi, G. S. Arnold and J. S. Bartlett, *Hum. Gene Ther.*, 2001, **12**, 1697.
175. N. DiPrimio, A. Asokan, L. Govindasamy, M. Agbandje-McKenna and R. J. Samulski, *J. Virol.*, 2008, **82**, 5178.
176. M. Limberis *et al.*, *Mol. Ther.*, 2006, **13**, S264.
177. L. H. Vandenberghe, M. H. Yang, J. Johnston, G. Gao and J. M. Wilson, *Mol. Ther.*, 2005, **11**, S195.
178. L. Zhong *et al.*, *Proc. Natl. Acad. Sci. USA*, 2008, **105**, 7827.

179. P. L. Hermonat, M. A. Labow, R. Wright, K. I. Berns and N. Muzyczka, *J. Virol.*, 1984, **51**, 329.
180. J. E. Rabinowitz, W. Xiao and R. J. Samulski, *Virology*, 1999, **265**, 274.
181. M. Hoque *et al.*, *Biochem. Biophys. Res. Commun.*, 1999, **266**, 371.
182. A. Girod *et al.*, *Nat. Med.*, 1999, **5**, 1438.
183. M. Grifman *et al.*, *Mol. Ther.*, 2001, **3**, 964.
184. S. A. Nicklin *et al.*, *Mol. Ther.*, 2001, **4**, 174.
185. S. A. Loiler *et al.*, *Gene Ther.*, 2003, **10**, 1551.
186. W. Shi and J. S. Bartlett, *Mol. Ther.*, 2003, **7**, 515.
187. S. J. White *et al.*, *Circulation*, 2004, **109**, 513.
188. L. M. Work *et al.*, *Mol. Ther.*, 2004, **9**, 198.
189. L. Gigout *et al.*, *Mol. Ther.*, 2005, **11**, 856.
190. W. Shi, A. Hemminki and J. S. Bartlett, *Gynecol. Oncol.*, 2006, **103**, 1054.
191. X. Shi, G. Fang, W. Shi and J. S. Bartlett, *Hum. Gene Ther.*, 2006, **17**, 353.
192. N. A. Huttner *et al.*, *Gene Ther.*, 2003, **10**, 2139.
193. O. J. Muller *et al.*, *Nat. Biotechnol.*, 2003, **21**, 1040.
194. L. Perabo *et al.*, *J. Gene Med.*, 2006, **8**, 155.
195. D. A. Waterkamp, O. J. Muller, Y. Ying, M. Trepel and J. A. Kleinschmidt, *J. Gene Med.*, 2006, **8**, 1307.
196. Q. Yang *et al.*, *Hum. Gene Ther.*, 1998, **9**, 1929.
197. K. H. Warrington Jr *et al.*, *J. Virol.*, 2004, **78**, 6595.
198. J. C. Grieger, J. S. Johnson, B. Gurda-Whitaker, M. Agbandje-McKenna and R. J. Samulski, *J. Virol.*, 2007, **81**, 7833.
199. O. J. Muller *et al.*, *Cardiovasc. Res.*, 2006, **70**, 70.
200. C. Zincarelli, S. Soltys, G. Rengo and J. E. Rabinowitz, *Mol. Ther.*, 2008, **16**, 1073.
201. L. H. Vandenberghe and J. M. Wilson, *Curr. Gene Ther.*, 2007, **7**, 325.
202. C. L. Halbert *et al.*, *Hum. Gene Ther.*, 2006, **17**, 440.
203. W. Xiao *et al.*, *J. Virol.*, 1999, **73**, 3994.
204. D. E. Bowles, J. E. Rabinowitz and R. J. Samulski, *J. Virol.*, 2003, **77**, 423.
205. B. Hauck, L. Chen and W. Xiao, *Mol. Ther.*, 2003, **7**, 419.
206. N. Maheshri, J. T. Koerber, B. K. Kaspar and D. V. Schaffer, *Nat. Biotechnol.*, 2006, **24**, 198.
207. D. V. Schaffer and N. Maheshri, *Conf. Proc. IEEE Eng. Med. Biol. Soc.*, 2004, **5**, 3520.
208. W. P. Stemmer, *Nature*, 1994, **370**, 389.
209. W. P. Stemmer, *Proc. Natl. Acad. Sci. USA*, 1994, **91**, 10747.
210. J. T. Koerber, N. Maheshri, B. K. Kaspar and D. V. Schaffer, *Nat. Protoc.*, 2006, **1**, 701.
211. D. Grimm *et al.*, *J. Virol.*, 2008, **82**, 5887.
212. J. T. Koerber, J. H. Jang and D. V. Schaffer, *Mol. Ther.*, 2008, **16**, 1703.
213. W. Li *et al.*, *Mol. Ther.*, 2008, **16**, 1252.

Subject Index